Monographs on
Theoretical and Applied Genetics 7

Edited by
R. Frankel (Coordinating Editor), Bet-Dagan
G. A. E. Gall, Davis · M. Grossman, Urbana
H. F. Linskens, Nijmegen · R. Riley, London

W. Gottschalk   G. Wolff

Induced Mutations in Plant Breeding

With 42 Figures and 45 Tables

Springer-Verlag
Berlin Heidelberg New York Tokyo 1983

Professor Dr. WERNER GOTTSCHALK
Dr. GISELA WOLFF
Institut für Genetik der Universität Bonn
Kirschallee 1
5300 Bonn/FRG

ISBN 3-540-12184-6 Springer-Verlag Berlin Heidelberg New York Tokyo
ISBN 0-387-12184-6 Springer-Verlag New York Heidelberg Berlin Tokyo

Library of Congress Cataloging in Publication Data. Gottschalk, Werner. Induced mutations in plant breeding. (Monographs on theoretical and applied genetics; 7) Bibliography: p. Includes index. 1. Plant mutation breeding. I. Wolff, G. (Gisela), 1943–. II. Title. III. Series. SB123.G67 1983 631.5'3 83-357

This work is subject to copyright. All rights are reserved, whether the whole or part of the material is concerned specifically those of translation, reprinting, re-use of illustrations, broadcasting, reproduction by photocopying machine or similar means, and storage in data banks. Under § 54 of the German Copyright Law where copies are made for other than private use, a fee is payable to 'Verwertungsgesellschaft Wort', Munich.

© by Springer-Verlag Berlin Heidelberg 1983

Printed in Germany.

The use of registered names, trademarks, etc. in this publication does not imply, even in the absence of a specific statement, that such names are exempt from the relevant protective laws and regulations and therefore free for general use.

Offsetprinting and bookbinding: Brühlsche Universitätsdruckerei, Giessen
2131/3130-543210

Preface

Mutation breeding has been introduced into modern plant breeding in the early 1940's. In spite of pessimistic predictions, the application of experimental mutagenesis has led to encouraging results demonstrating that mutation breeding is a well-functioning method in many crops. So far, more than 500 varieties, developed by means of induced mutations, have been officially released; others have been approved for registration. Many mutants with characters of agronomic interest cannot be utilized directly because of their unsatisfying yielding capacities, or of other negative traits which are partly due to the pleiotropic action of the mutant genes. Sometimes their negative selection value can be overcome by transferring them into the genomes of other varieties. According to experience available, the efficiency of mutant genes can considerably vary depending on the genotypic background in which they become effective. The interactions between mutant genes and genotypic background cannot be predicted. Therefore, mutants with valuable traits should be crossed with many varieties and strains in order to discern positive and negative interactions. In this way, genotypes can be selected in which the mutant gene is able to express its action without showing negative by-effects. This procedure has been used for about 10 years by combining the methods of mutation and crossbreeding.

Mutation breeding is predominantly used in annual diploid and allopolyploid self-fertilizing crops, while it causes much more difficulties in cross-pollinating species. Especially rapid results have been obtained in ornamentals, because not only generative but also somatic mutations can be utilized in this material. Of special interest is applied mutagenesis in that small group of crops which are unsuitable for crossbreeding because of full sterility. In these cases, induction of mutations is the only way to increase the genetic variability within a short period.

A particularly intensively studied field of applied mutagenesis refers to the alteration of seed proteins, to some extent also of other storage substances. In spite of immense expenditure of time and work, the practical success with regard to the selection of "protein mutants" and related genotypes is small. The initiation of international research programs, however, has led to an immense broadening of our knowledge on the relations between gene action and metabolic processes. This holds true not only for this branch of mutation breeding. On the contrary, the application of experimental mutagenesis to crops in general has provided a huge amount of basic information about the behavior of mutant genes, which is of interest for plant genetics in general.

Since 1940, thousands of papers have been published dealing with the various problems of applied mutagenesis. In the frame of the present book, it was not possible to discuss all the findings available in detail. We have considered the literature since 1965, but references on review papers are given containing older results. Moreover, references to plenty of additional information are given in the tables of the book.

Bonn, April 1983   Werner Gottschalk
                                                                       Gisela Wolff

Contents

1	Introduction.	1
2	Methods for Inducing Mutations.	10
2.1	Mutagenic Agents and Related Problems.	10
2.2	The Chimerical Structure of the M_1 Plants.	14
3	The Selection Value of Mutant Genes	18
4	The Seed Production of Mutants and the Alteration of Quantitative Characters	23
4.1	The Alteration of Quantitative Characters	23
4.2	Mutants with Increased Seed Yield	26
4.3	Released or Approved Mutant Varieties	30
5	The Utilization of Mutants in Crossbreeding.	32
5.1	The Incorporation of Mutant Genes into the Genomes of Varieties or Strains	32
5.2	The Joint Action of Mutant Genes	34
5.2.1	Negative Interactions.	35
5.2.2	Positive Interactions.	40
6	The Alteration of the Shoot System by Means of Mutations.	43
6.1	Mutants with Reduced Plant Height: Erectoides Types, Semidwarfs, Dwarfs.	43
6.1.1	Barley.	44
6.1.2	Rice	46
6.1.3	Bread and Durum Wheat; Other Gramineae	48
6.1.4	Dicotyledonous Crops	51
6.2	Mutants with Increased Plant Height	52
6.3	Mutants with Altered Stem Structure	53
6.3.1	Branching, Tillering	54
6.3.2	Stem Bifurcation.	54
6.3.2.1	Bifurcated Mutants	55
6.3.2.2	Bifurcated Recombinants	57

6.3.3	Stem Fasciation	58
6.3.3.1	Fasciated Mutants	58
6.3.3.2	Fasciated Recombinants	62
6.3.4	Mutations in Fiber Plants	64
7	**Alterations of Flower Shape, Color and Function**	**65**
7.1	Flower Shapes and Flower Colors in Ornamentals	66
7.2	Inflorescences	67
7.3	Genetic Male Sterility	68
8	**Leaf Mutants of Agronomic Interest**	**71**
9	**Mutations Affecting the Root System**	**74**
10	**The Alteration of Flowering and Ripening Times**	**75**
10.1	Earliness	75
10.2	Lateness	82
10.3	Changes of the Photoperiodic Reaction	83
11	**Mutations in Vegetatively Propagated Crops and Ornamentals**	**85**
12	**Heterosis**	**88**
13	**Disease Resistance**	**91**
13.1	Resistance Against Fungi, Bacteria, and Viruses	91
13.1.1	Barley	91
13.1.2	Rice	94
13.1.3	Bread and Durum Wheat	96
13.1.4	Oats	97
13.1.5	Maize	97
13.1.6	Pearl Millet	97
13.1.7	Sugarcane	97
13.1.8	Dicotyledonous Crops	100
13.2	Resistance Against Animal Pathogens	102
13.3	Herbicide Tolerance	102
14	**Drought Resistance, Heat Tolerance, Winterhardiness**	**103**
15	**Shattering and Shedding Resistance**	**104**
16	**The Pleiotropic Gene Action as a Negative Factor in Mutation Breeding**	**105**

| 16.1 | The Alteration of Pleiotropic Patterns Under the Influence of Changed Genotypic Background or Environment... | 105 |
| 16.2 | Mutations of Closely Linked Genes............. | 108 |

| 17 | **The Penetrance Behavior of Mutant Genes as a Negative Factor**................................ | 112 |

18	**The Adaptability of Mutants to Altered Environmental Conditions**.............................	116
18.1	The Reaction of Mutants to Different Natural Environments................................	116
18.2	The Reaction of Mutants Under Controlled Phytotron Conditions.................................	123

19	**The Alteration of Morphological and Physiological Seed Characters**...........................	126
19.1	Seed Size...............................	126
19.2	Seed Shape..............................	127
19.3	Seed Color..............................	128
19.4	Physiological Seed Characters	128

20	**The Alteration of Seed Storage Substances**.........	130
20.1	Seed Proteins	130
20.1.1	The Characterization of Seed Proteins...........	131
20.1.2	Factors Influencing Protein Content and Composition .	132
20.1.2.1	Environmental Factors.....................	132
20.1.2.2	Endogenous Factors.......................	135
20.1.3	Seed Protein Content of Different Varieties of the Same Species	140
20.1.4	Alteration of Seed Proteins Through Mutant Genes ...	142
20.1.4.1	Protein Mutants in Cereals...................	142
20.1.4.2	Protein Mutants in Legumes	158
20.2	Seed Carbohydrates.......................	165
20.2.1	Maize	166
20.2.2	Barley and Other Cereals....................	168
20.2.3	Peas	169
20.3	Seed Lipids	170

| 21 | **Other Plant Substances**...................... | 172 |

22	**The Nutritional Value of Mutants**...............	174
22.1	Maize Mutants	174
22.2	Barley Mutants..........................	176

22.3	Sorghum Genotypes	176
22.4	Pea Mutants	177
23	**General Aspects of Mutation Breeding with Regard to the Improvement of Seed Storage Substances**	178
	References	182
	Subject Index	231

1 Introduction

The permanently increasing number of released or approved varieties, developed by means of mutant genes, demonstrates the effectiveness of the methods of applied mutagenesis in a convincing way (Fig. 1, Table 1). According to the mode of development, the agronomically utilized mutant material can be subdivided into three groups as follows:

1. Selected mutants with favorable properties are directly developed into varieties. This has preferably been done during the early period of mutation breeding; but even nowadays many mutant varieties derive directly from prospective mutants. This holds particularly true for those crops the breeding of which has not yet reached a high level. In such cases, the broadening of the genetically conditioned variability can be achieved relatively easily within a short period. Some of the new genotypes obtained may be of direct economic value. In general, however, this method proved to be not as successful as originally expected.

2. Prospective mutants are incorporated into crossbreeding programs in which the mutant genes are utilized indirectly. For the majority of the sexually propagated crops suited for mutation breeding findings are available, demonstrating thereby that the indirect use of mutant genes may be more propitious for reaching distinct aims of plant breeding than their direct use. This method is widely used, preferably in cereals. Impressive examples for the potentialities of this method are the development of a relatively large number of commercial *barley* varieties from a very small number of mutants in Sweden and Czechoslovakia. Many mutant genes of different crops are just being transferred into the genomes of varieties or strains, and the genotypes selected are tested with regard to their agronomic usefulness. There is no doubt that many of them will lead to new cultivars in near future. Findings are available in *bread* and *durum wheat, rice, peanuts, peas,* and *tomatoes* among others.

3. Different favorable mutants are crossed with each other in order to combine specific mutant genes in recombinant lines. This method is only used to a small extent at present. It will preferably be used in those crops in which collections with hundreds or even thousands of mutants are already available, i.e., in some cereals and pulses.

Abbreviations: EMS, ethyl methane sulfonate; MMS, methyl methane sulfonate; DES, diethyl sulfate; DMS, dimethyl sulfate; EI, ethylene imine; EO, ethylene oxide; HA, hydroxylamine; MH. maleic hydrazide; ENH, NEU, ethyl nitroso urea; MNU, NMU, N-methyl-N-nitroso urethane or urea, respectively

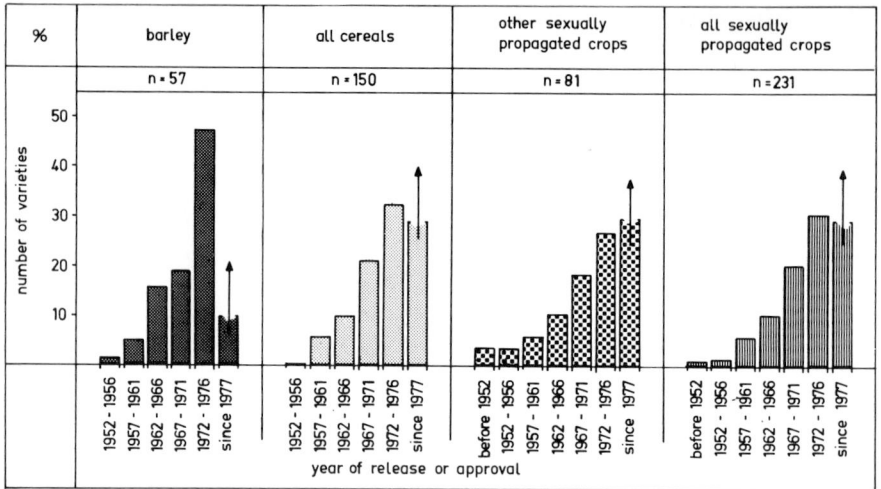

Fig. 1. Increase of the number of mutant varieties during the past decades.
Twenty-one released mutant varieties of sexually propagated crops are not considered in this graph, details of which have just been published in Mutation Breeding Newsletter 19:14–19, 1982. They belong to the following species:

– *Triticum aestivum:*	6 varieties	– *Glycine max:*	1 variety
– *Hordeum vulgare:*	4 varieties	– *Lupinus luteus:*	1 variety
– *Oryza sativa:*	2 varieties	– *Lycopersicon esculentum:*	1 variety
– *Zea mays:*	1 variety	– *Gossypium* sp.:	1 variety
– *Pisum sativum:*	2 varieties	– *Nicotiana tabacum:*	1 variety
– *Cicer arietinum:*	1 variety		

In spite of the positive results available, many geneticists and plant breeders are still very sceptical with regard to the effectiveness of the application of mutagenesis in plant breeding. It is true that only little success was had during a relatively long initial period. This is due to the fact that plenty of positive and negative experiences were necessary for learning to handle these methods in a promising way. Moreover, geneticists, plant breeders, phytopathologists, and biochemists had to be convinced of the usefulness of these methods; this needs time and patience. At present, about 500 officially released varieties, derived from experimentally induced mutants, are available. In many cases, the expenditure in time was considerably less than the time necessary for developing a variety by using conventional methods. The effectiveness of these methods becomes especially clear in some ornamentals from which whole groups of mutant varieties have been developed within a few years. All these findings and results demonstrate that the methods of applied mutagenesis are well suited for supplementing the conventional breeding procedures. This especially is valid for annual diploid and allopolyploid autogamous crops and for some vegetatively propagated species inclusive many ornamentals, whereas mutation breeding in allogamous species is only in its very beginnings.

Our knowledge on some fundamental aspects of the mutational process has been widened during the past years, but it is not yet clear, whether the findings obtained in

Introduction

Table 1. Mutagens used for developing 485 released or approved varieties of different crops and ornamentals

Mutagens	Sexually propagated crops		Vegetatively propagated crops		All crops	
	Number of varieties	%	Number of varieties	%	Number of varieties	%
X rays	87	36.3	125	51.0	212	43.7
Gamma rays	94	39.2	103	42.1	197	40.6
^{32}P beta rays	2	0.8	–	–	2	0.4
Neutrons	29	12.1	9	3.7	38	7.9
Combination of different rays	2	0.8	5	2.0	7	1.4
Total physical mutagens	214	89.2	242	98.8	456	94.0
Chemical mutagens	23	9.6	2	0.8	25	5.2
Combination of physical and chemical mutagens	3	1.2	1	0.4	4	0.8
Grand total	240	100.0	245	100.0	485	100.0

distinct species treated with distinct mutagens can be generalized. An important theoretical and practical problem refers to the possibility that different mutagens may induce different kinds of effects at the chromosome level. It is the opinion of many geneticists that spontaneous and induced mutations do not differ qualitatively from each other. This hypothesis is supported by the fact that nearly the whole natural genetic variability of *barley* has been reproduced by means of induced mutations and by recombination following crossing (Stubbe 1967). Similar experiences have been made in many other crops. There are numerous examples demonstrating that experimentally obtained mutants proved to be identical with corresponding spontaneously arisen genotypes available in the collections of institutes or breeders. There are, however, certain discrepancies in this respect.

The question whether mutants, showing monohybrid segregations, are homozygous for "true" gene mutations or for minute deficiencies, is still an open problem. According to Ichikawa (1965), the majority of his gamma-ray-induced mutations in hexaploid *wheat* are deficiencies or other chromosomal aberrations rather than gene mutations. This seems to be also valid for X rays, neutrons, and alkylating mutagenetic chemicals such as EMS and EI (Bender 1970: Auerbach and Kilbey 1971). On the other hand, findings obtained by Dumanović et al. (1968) in *wheat* indicate, that low and medium gamma ray doses cause a relatively high proportion of useful mutants with normal yielding properties, whereas high doses lead to mutants with negative effects. Similar experiences with X rays, gamma rays, and EMS were made by Yonezawa and Yamagata (1977) in *rice* and *barley*. It is difficult to understand that plant organisms, homozygous for the loss of a certain amount of genetic material, are fully vital.

Our results in *Pisum* indicate likewise that many of our X-ray- and neutron-induced mutants are obviously not characterized by minute deficiencies. In the frame of our radiation genetic experiments, about 800 "gene mutations" and 120 translocations were isolated. A large number of them was maintained by propagating the translocation-heterozygous plants. In their offspring, plants homozygous for the translocated chromosomes were selected and developed into strains. Twenty-five different strains of this group were studied with regard to vitality and fertility. Only one of them was distinctly inferior to the mother variety. All the other ones were fully fertile and vital (Gottschalk 1978a). This is in contrast to the situation in *Drosophila* where translocation homozygosity leads often to lethality. The high degree of physiological effectiveness observed in our *Pisum* material could hardly be expected if the radiation-induced translocations would have led to the loss of genes.

Another important problem refers to the specificity of the mutation spectra produced by different mutagens. According to Smith (1972) there is little, if any, evidence in higher plants for mutagenetic specificity among physical mutagens. It may exist between specific physical and chemical mutagens. Some clear examples are available in *barley* (Gustafsson 1972; Lundqvist 1976, 1978). It is, however, very difficult to evidence this specificity because a large number of mutants, obtained in parallel treatments with different mutagens, is needed for obtaining the empirical data. The value of some experiences published in this field should therefore not be overestimated. Also the biochemical situation of the treated nuclei seems to be of importance with regard to the effects of distinct mutagens. DES, given at middle G_1 stage of the interphase nuclei of *barley* seeds, was found to give the highest rates of induced mutations. The most promising mutants for breeding purposes, however, were obtained when the mutagen was applied in late S stage (Yamaguchi et al. 1974).

For practical use of applied mutagenesis, it is of interest to consider that different cultivars of the same crop can differ markedly from each other with regard to their susceptibility against distinct mutagens. A specific variety of *Lens culinaris*, for instance, was found to be relatively resistant to gamma ray and NMU treatment, whereas two other cultivars were sensitive giving a wide range of different types of mutants (Sharma and Sharma 1979a). Similar experiences were made in *Cicer arietinum* (Ahmad and Godward 1981) and *Phaseolus vulgaris* (Al-Rubeai and Godward 1981) among others. There is no doubt that genetically conditioned differences with regard to both the radiosensitivity in general and the mutation rates in particular exist between different strains of many crops.

Many efforts are still necessary in order to clarify the problems just mentioned. This holds also true for the fact that often mutants are obtained which resemble the existing varieties, arisen by conventional cross breeding, without being identical with them. These cases are possibly not exclusively due to mutational processes but also to alterations in the pattern of gene arrangement and intragenic recombination (Kulshrestha and Mathur 1978; *Triticum aestivum*).

A considerable expenditure in space, time, and money is necessary if one intends to obtain a specific group of mutations or even a single desired mutation. This may be demonstrated by the following examples:

Introduction

- Three blast-resistant *rice* mutants were selected in Japan out of an M_2 generation of 51,530 plants (Yamasaki and Kawai 1968).
- Screening of 951,000 M_2 *barley* plants in Denmark resulted in selecting 5 M_1 plants giving rise to powdery mildew-resistant M_2 seedlings (Jörgensen 1975).
- Testing more than 2.5 million M_2 plants of *barley* in the German Democratic Republic resulted in selection of 95 mildew-resistant mutants (Hentrich 1977).
- Seven mildew resistant *barley* mutants were selected from 1,200,000 M_2 plants in Japan (Yamaguchi and Yamashita 1979).
- From more than 6 million M_2 plants of *Mentha piperita*, grown in the United States, 7 wilt-resistant strains were developed (Murray 1969, 1971).
- 500,000 stems with dormant buds of *coastcross 1 bermuda grass* hybrids were irradiated, giving rise to one mutant with improved winterhardiness (Burton et al. 1980).

Similar estimates have been given by Yonezawa (1975): For the reliable selection of a mutation for a distinct quantitative character, at least 200,000–250,000 M_2 plants should be grown. Comprehensive M_2 generations of *barley*, obtained after gamma ray and sodium azide treatment, were evaluated in U.S.A. with the aim to select distinct single locus mutations. In these model experiments, the respective mutants, induced by NaN_3, occurred with the following frequencies:

- 2.7 per 10,000 M_2 seedlings for "waxy endosperm;"
- 1.0 per 10,000 M_2 seedlings for "vine (gigas)."

Similar results were obtained with regard to mutations causing distinct physiological anomalies in *barley* and *peas* (Kleinhofs et al. 1978). These experiments are insofar encouraging as they demonstrate that it is in principle possible to get a desired mutation in the frame of mutation treatments. The prospects are even better if we consider that many plant characters are controlled by polymeric systems. The desired character occurs already if one single gene mutates out of the polygenic group of 30, 50, or more genes. Impressive examples for the existence of such polygenic groups are:

- The *eceriferum mutants* of *barley* showing certain alterations of the wax coating on different organs such as stem, leaf blade or sheath, and spike. In Svalöf (Sweden), 1310 mutants of this type have been isolated so far. They were referred to 65 different loci randomly distributed over the seven chromosomes of the genome by diallel analysis. Thousand and sixty-one induced *eceriferum* mutants are recessive, 18 are dominant. Details concerning this giant amount of work have been published by Lundqvist et al. (1968), Lundqvist (1976, 1978), Fester and Søgård (1969).
- The *erectoides mutants* of *barley* showing shortened internodes combined with lodging resistance, dense spikes, certain alterations of the root system and some other characters. Until the end of the 1960's, more than 700 *erectoides* mutants have been genetically analyzed in Svalöf. They were found to refer to 26 different loci, some of them having more than 30 multiple alleles (Gustafsson 1969; Persson and Hagberg 1969).

Some of the *erectoides* mutants are of direct importance for barley breeding. It can be assumed that a similar polygenic situation is realized with regard to other traits of agronomic interest which should be obtained relatively easily in mutation experiments. This holds for instance true for male sterility. Another aid, not yet utilized so far in applied mutagenesis, is the pronounced mutagen specifity of distinct loci of polygenic systems. Results in this field are available in the *eceriferum* and *erectoides* mutants of *barley* (Gustafsson 1972; Lundqvist 1976, 1978).

Theoretically, each gene which is of any agronomic interest can mutate. Therefore, a wide spectrum of mutants can be expected in mutation experiments which have to be tested with regard to their usefulness for specific aims of breeding. This holds true for all yielding characters, furthermore for all kinds of resistance and tolerance, for the flowering and ripening behavior, and for many other traits influencing the breeding value of a crop directly or indirectly. It was already mentioned that desired mutants can be in principle obtained if the total number of the mutants selected is very high. It is, however, not yet possible in higher plants to induce specific mutations in distinct loci; the mutational event is still a matter of chance.

A comparably new aspect in applied mutagenesis is the quantitative and qualitative alteration of seed storage substances, such as proteins and carbohydrates, to some extent also of specific other substances deposited in various plant organs. Special emphasis is directed to seed proteins because a part of diseases due to malnutrition and undernourishment is related to insufficient protein supply. An increase and improvement of these substances could bridge the "protein gap." These considerations especially become valid under the aspect that 70% of consumed proteins derive from plants (IAEA 1979).

The importance of this problem becomes evident from the fact that the International Atomic Energy Agency (IAEA) in cooperation with the Food and Agriculture Organization of the United Nations (FAO) started a voluminous research program on this topic in 1968. The aim was to find ways for increasing the world's protein resources by means of selection and mutation. The findings published on *opaque* and *floury maize mutants* in the early 1960's (Mertz et al. 1964; Nelson et al. 1965) were the first promising results in this field, demonstrating that by gene mutation seed protein improvement is possible in principle. Later on, protein research programs were started in many countries and mutants were selected differing from their initial lines with regard to their seed proteins.

"Protein mutants" can be subdivided into two groups: Those genotypes which differ quantitatively from their control lines in seed protein content, and those in which deviations in the composition of seed protein subfractions are found. The distribution of amino acids in the latter group shows distinct deviations from the normal pattern, which in the first group cannot be detected.

Quantitative traits, such as protein content, are polygenically controlled and highly influenced by environmental factors. The genotype determines a certain potential of ontogenetic development; environment in the broadest sense determines the dimension of this development, resulting in the visible and measurable phenotype. In *wheat*, for example, Dhaliwal (1977) assumes that hundreds of structural genes may be involved in the control of seed protein content. Also the findings in other crops have shown that several genes are directly involved in protein production. Furthermore, various plant characters, such as seed size and number of seeds per plant, which are likewise genetically controlled, can influence the amount of deposited seed proteins. In addition, certain environmental factors, such as temperature and manuring, have a proven effect not only on the amount but even on the composition of the seed proteins. These interactions are the reasons for the difficulties in analyzing the "quantitative protein mutants." The action of genes controlling the amount of seed proteins is influenced by environmental factors in a way that the reliable selection of mutants of this category becomes difficult. It is necessary to evaluate several replications of mutants and initial lines in a comparable way. These replications should be grown in the same year as well as in different years. In this way, variation ranges can be calculated which are a better criterion than single values for characterizing the respective genotypes. This presupposition, however, is often not fulfilled. Therefore, one should be careful in judging the protein situation of those mutants from which data of only one generation are available.

With regard to the "qualitative protein mutants," the results are more clear. In this field, extensive research work has been done during the last years and much basic knowledge about the biochemical interrelations in these mutants has been published. It is remarkable that qualitative deviations in mutants till now are preferably found in *cereals,* while there are hardly any in *pulses.* Furthermore, the comparison of the deviations found in these cereal mutants reveal that the mechanism which causes the qualitative deviations is rather similar in all the genotypes analyzed.

As proteins are direct gene products, mutations are immediately reflected in the respective polypeptides. With regard to other plant substances, the situation is different. They are synthesized under the influence of special enzymes which are likewise genetically controlled. The relations between deviations in these substances and the respective mutated genes are indirect and thus are more complicated.

Some examples may demonstrate that an immense expenditure in material, work, and time is necessary for selecting protein mutants. The following number of genotypes with favorably altered seed proteins was isolated in different cereals:

- 20 *barley* mutants with increased lysine content from 14,776 M_2 plants (Doll et al. 1974).
- 2 *barley* mutants from 10,000 M_2 or M_3 plants (Scholz 1972).
- 8 *barley* mutants from 2,455 M_2 plants (Krausse et al. 1974).
- Some *bread wheat* mutants from 25,000 M_2 plants (Parodi and Nebreda 1979), whereas no mutants of this category were found within 15,000 samples studied by Johnson et al. (1973).
- 2 *rice* mutants from 6,600 M_2 plants (Tanaka and Hiraiwa 1978).
- 1 *sorghum* mutant with increased lysine content from 23,000 M_2 heads (Axtell et al. 1974).
- From the *pearl millet,* 16,770 samples were analyzed, but no protein mutants could be isolated (Rabson et al. 1978).

Mutated genes that influence carbohydrates have been known for a long time. Of practical importance in this connection are genotypes, in which the sugar content is increased; these forms are utilized as vegetables such as *sugary maize.* In these genotypes in general, a block in starch synthesis is supposed being responsible for the higher sugar content. Differences in the proportion of amylose and amylopectin in starch, as compared to normal conditions, are found as being caused by single mutant genes, too. In some of these cases deviations in the enzyme makeup, controlling starch synthesis, could be made responsible for the alterations.

Mutated genes with effect on seed oil content are hardly known and only little information is available in this field. Besides these three important groups of substances, several single mutants in various groups are known, where the respective genes alter the amount of special components. Most of these substances interfere with the nutritional value of the crop. Such substances are for example tannins in *sorghum,* that reduce the availability of the seed proteins. A particularly interesting example is BOAA (β-N-oxalyl-α,β-diamino-propionic acid) in *Lathyrus* seeds which acts as a neurotoxin if not completely destroyed by heat before consumption. Mutants in *Lathyrus sativus* are known in which this unfavorable component is reduced.

In the present survey, only gene mutations are considered. The utilization of chromosome mutations in plant breeding is still in its very beginnings. Some experiences are available in *barley* in Sweden and U.S.A. demonstrating that certain lines, homozygous for translocated chromosomes, are superior to their chromosomally normal mother varieties. The problem has been discussed by Gustafsson et al. (1966, 1971); Hagberg and Hagberg (1971); Hagberg et al. (1972); Künzel and Scholz (1972).

The number of papers, published in the field of the application of mutagenesis in plant breeding, is so large that not all of them can be considered in the present publication. The reasons for these restrictions are not only due to the huge number of data available but also to the difficulties in obtaining certain foreign periodicals. In order to restrict the number of references in a way appropriate to the present book, preferably papers published since 1965 are referred to. Some publications are available, in which the problems, methods, and results of mutation breeding are discussed and reviewed in general. Plenty of information on this field is compiled in publications presented by Gustafsson (1965, 1969), Gaul (1965a, c); Matsuo and Yamaguchi (1967); Khvostova (1967, 1978); Sigurbjörnsson (1968); Scarascia Mugnozza (1969a, b); Swaminathan (1969a, b, 1971); Černý (1970); Brock (1971); Favret (1972); Gaul et al. (1972); Gottschalk (1978f, 1979a, 1980b). This history of mutation breeding along with early results in *barley, rice, wheat,* and *soybean* was described by Sigurbjörnsson and Micke (1969, 1974); Brock (1965); Scossiroli (1965). They discussed the prospects of mutagenesis with regard to the alteration of quantitative characters. Surveys on the application of mutagenesis for reaching specific aims of plant breeding (resistance, straw stiffness and others) are mentioned in the various chapters of the present paper. Furthermore, review papers have been published considering the utilization of induced mutations in distinct crops. Examples are as follows:

— barley (Gaul 1965c; Sigurbjörnsson 1976);
— rice (Gustafsson and Gadd 1966);
— bread wheat (Khvostova et al. 1969; Konzak 1972, 1976; Konzak et al. 1973);
— durum wheat (Bagnara 1971);
— oats (Gustafsson and Gadd 1965e);
— sorghum (Sree Ramulu 1975);
— *Poa pratensis* (Gustafsson and Gadd 1965c);
— sugarcane (Heinz 1973);

- sweet potatoes (Gustafsson and Gadd 1965b);
- legumes in general (Gottschalk 1971b; Blixt and Gottschalk 1975; Gottschalk and Wolff 1977);
- peanuts (Gustafsson and Gadd 1965d; Gregory 1968);
- lupins (Gustafsson and Gadd 1965a);
- soybeans (Koo 1972);

Older review papers with many references were published by Gustafsson (1942, 1947, 1951, 1963); Gustafsson and v. Wettstein (1958); Konzak (1956, 1957); MacKey (1956, 1961); Sparrow and Konzak (1958); Gaul (1958, 1961, 1963, 1964); Smith (1958); Hoffmann (1959); Prakken (1959); Röbbelen (1959); Stubbe (1959); Swaminathan (1963); Nilan (1964), among others. For the reasons mentioned above, it was not possible to discuss details of the methods used in mutation breeding. Quantitative and qualitative alterations of the seed proteins and of other storage substances under the influence of mutant genes are referred in the second part of this book.

Lists of released or approved mutant varieties have been given by Sigurbjörnsson and Micke (1969, 1974). They are supplemented in the issues of the *Mutation Breeding Newsletter* edited since 1972 by the International Atomic Energy Agency in Vienna. Lists of commercialized mutant cultivars of asexually propagated crops were published by Sigurbjörnsson and Micke (1973) and by Broertjes and van Harten (1978).

In the papers just mentioned, plenty of generally valid and detailed information is discussed; moreover, a large number of references on older findings are considered. The methods of experimental mutation research, the situation of the M_1 generation, the different types of mutations, and the plant characters which can be improved by means of induced mutations, are summarized in the *Manual of Mutation Breeding* published by the International Atomic Energy Agency (1977b). Moreover, a number of volumes of symposia organized by this agency is available (1966, 1968a, b, 1969a, b, 1970a, b, 1971, 1972, 1973a, b, 1974a, b, 1975a, b, 1976a–d, 1977a, c, 1978, 1979, 1980, 1981). Further information on problems of mutation breeding is found in the proceedings of some other symposia organized by:

- the Institut für Kulturpflanzenforschung (Gatersleben 1966; Erwin-Baur-Gedächtnisvorlesungen IV),
- the Indian Agricultural Research Institute (New Delhi 1973),
- the Bulgarian Academy of Sciences (Sofia 1978),
- the Institute of Radiation Breeding (Ohmiya, Japan, 1978, 1979a, b).

In the field of the genetics of seed storage substances, general and special aspects have been published by Creech (1968); Munck (1972); Kaul (1973); Axtell et al. (1974); Johnson and Lay (1974); Pollmer et al. (1974); Milner (1975); Müntz (1975); Bhatia and Rabson (1976); Muhammed et al. (1976); Munck (1976); Przybylska (1976); Sichkar (1976); Doll (1977); Ewertson (1977); Frey (1977); Eggum (1978); Bahl et al. (1979); Blixt (1979); Hagberg et al. (1979); Luse and Rachie (1979); Mitra and Bhatia (1979); Mitra et al. (1979); Röbbelen (1979); Thomson and Doll (1979); Nelson (1980); Tallberg (1982). Further surveys on biochemistry, genetics and nutritional value of seed proteins have been edited by Gottschalk and Müller (1983).

2 Methods for Inducing Mutations

Hundreds of papers have been published dealing with a great variety of different methods for the induction of gene mutations. In most cases, the frequency of chlorophyll mutants in the M_2 generations was used as test for judging the effectiveness of the mutagens used. For the problems dealt with in the present book it is not necessary to discuss these methodological questions in detail. Many information in a summarized form is given in the *Manual of Mutation Breeding* (International Atomic Energy Agency 1977b). Further details are found in the chapters "Mutation" in *Progress in Botany* (Gottschalk 1975, 1977b, 1979d, 1981d). Methodological aspects of mutagenesis in vegetatively propagated species have been discussed in a symposium of the International Atomic Energy Agency (1973a), furthermore by Nakajima (1977), and by Broertjes and van Harten (1978). I shall give only a brief survey, adding some recently published findings.

2.1 Mutagenic Agents and Related Problems

As physical mutagens, X rays, gamma rays, fast and thermal neutrons are widely used, whereas only few examples are known that beta rays, deriving from [^{32}P] and [^{35}S] have been successfully utilized in applied mutagenesis. Protons were found to be as effective as fast neutrons and more effective than gamma rays (Tarasenko 1977b; *potatoes*). The mutagenic efficiency of gamma rays can be increased by irradiating the seeds at extremely low temperatures (Nakai and Saito 1979; *rice*). Moreover, many findings exist demonstrating that the mutation frequency obtained by the various rays can be positively influenced by specific kinds of pre- or posttreatment. Recurrent irradiation has been found to be considerably more effective than a single treatment (Loose 1979; *Rhododendron*). The utilization of radiation-induced mutants for successive use of further irradiations proved to be an especially successful method for obtaining a large number of new mutants. In this way, hundreds of *Chrysanthemum* mutants were produced in the Netherlands, some of them being of direct floricultural interest (Broertjes et al. 1980).

During the past decade, the utilization of mutagenic chemicals has been strongly strengthened. Out of the great number of different compounds which proved to be mutagenic, only very few are used in applied mutagenesis. The most important ones are ethyl methane sulfonate (EMS), diethyl sulfate (DES), ethylene imine (EI), propane sultone, N-methyl-N-nitroso urethane (MNU) and some related substances. An unusually powerful mutagen is sodium azide (NaN$_3$; Nilan et al. 1973; Conger and Carabia 1977; Sander and Muehlbauer 1977). With regard to the mutagenic efficiency, these chemicals are comparable to the physical mutagens used, or even better. The efficiency of all the mutagens mentioned can be improved by combining them in specific ways.

In sexually propagated crops, usually seeds are treated. Although the mutagenic chemicals are easily available, rays are preferred by many geneticists. Their main advantage lies in the fact that the seeds are irradiated in a dry condition and that they can be handled for sowing like untreated material. If they have been treated with soluble solutions of chemical compounds, however, they are very susceptible and need much more care than dry seeds. Another disadvantage following application of some of the most potent chemical mutagens is that their negative effects on the vitality and fertility of the M$_1$ plants are considerably stronger than the corresponding effects of mutagenic rays. This is a very important problem in the practical performance of applied mutagenesis, which has not yet been solved.

In Table 1, a survey on the mutagens used for developing 485 released or approved mutant varieties is given. Only 9.6% of the group of sexually and 0.8% of the group of vegetatively propagated crops and ornamentals have been obtained by chemicals. This does not mean, however, that the chemical mutagens are less important in mutation breeding than the physical ones. The development of a commercial variety needs much time, and 10 or 15 years ago more rays than chemicals were used for inducing mutations. Nowadays, chemicals are widely utilized and it can be expected that an essentially higher proportion of commercialized varieties will originate from chemomutagenesis in future, at least in the sexually propagated crops.

From theoretical reasons, haploid cell and suspension cultures of diploid or allopolyploid crops should be an ideal material for the application of mutagens. As each gene is only present once in such a single cell, no heterozygosity for mutant genes occurs. Thus, the mutants are already discernible in the M$_1$ generation, provided that the problem of their regeneration from the isolated cells is solved. Prospective mutants can be selected and their chromosome number has to be doubled by colchicine. Mutants derived from single haploid cells have already been isolated in *Datura innoxia* (Weber and Lark 1979) and *Nicotiana sylvestris* (Malepszy et al. 1977) among others. Mutants from diploid cell cultures were produced in *soybean* (Weber and Lark 1979) and *tobacco* (Chaleff 1980). A review on these problems with many references was given by Thomas et al. (1979). In applied mutagenesis, however, these methods are not yet utilized on a broader scale because they involve obviously too many difficulties until the mutants are available.

The methods of applied mutagenesis are particularly important for triploid crops. Because of their sterility, they cannot be used for crosses. Thus, the induction of mutations is an essential point for increasing their genetic variability. Some *turf* and *forage Bermudagrasses* belong to this group. They represent sterile triploids between 4n *Cynodon dactylon* and 2n *C. nlemfuencis* or 2n *C. transvaalensis*. By treating stems, rhizomes or stolons with gamma rays, prospective mutants were obtained showing an improvement with regard to leaf size, internode length, spreading rate, nematode resistance, herbicide tolerance and winterhardiness. One of them, *Tifway-2 Bermudagrass*, is a released variety (Burton 1976, 1981; Burton et al. 1980). Similar results were obtained in the triploid *St. Augustinegrass* (Powell and Toler 1980).

In single cases, the combination of hybridization and induced mutation may be a method for improving a crop, but so far only a few experiences exist in this field. Preliminary results are available in *Vigna unguiculata* (Virupakshappa et al. 1980).

About 99% of all mutants are due to recessive mutations. Therefore, the mutant genes are not discernible in the M_1 plants. The selection of mutants can only begin in the M_2 generation. Very often, however, the mutant genes are in the M_2 families not yet present in the homozygous but only in the heterozygous condition. In these cases, they become for the first time discernible in the M_3 sibs (details in the next chapter). Occasionally, they appear even later. Examples are known in *wheat* (Borojević 1975); we have made similar experiences in our *pea* material. The causes of this unexpected situation are not yet clear. About 1% of all induced mutations are dominant ones. They have been found in *rice* (long culm; Okuno and Kawai 1978b), *rye* (short straw; Kuckuck and Peters 1979), *peas* (long and short internodes), among others.

Relatively often, several mutations are induced in the same intial cell of a treated embryo. In these cases, the mutant genes can be separated from each other in later generations, and genotypes homozygous for only one of them are obtained. In exceptional cases, up to ten or even more mutational events have been found in the growing points of the treated embryos. In principle different from this situation are those cases in which several very closely linked or even neighboring genes mutate simultaneously, pretending pleiotropic effects (details in Chap. 16). A specific behavior is the ± simultaneous mutation of a large group of identical genes in different embryos. This situation is realized in some fasciated *Pisum* mutants and will be discussed in Sect. 6.3.

Mutation breeding is mainly carried out in annual diploid and allopolyploid autogamous crops. In allogamous crops, it is much more time consuming and remains still in the infancy. It is necessary to make the initial material homozygous for a possibly large number of gene pairs by means of inbreeding. The inbred lines are used for the mutagenic treatment. In some cross-pollinating crops, such as *maize,* this is not an additional difficulty because many inbred lines are available for utilizing the heterosis effect. In other allogamous species, however, these presuppositions may be a serious handicap for beginning such a program. This is particularly valid for self-incompatible species. Single examples on mutants of agronomic interest of cross-fertilizing species are given in the various chapters of the review paper. Some more experiences have been made in the following crops:

– *Secale cereale* (different inbred lines were found to differ from each other with regard to mutation frequency after EMS treatment; Müntzing and Bose 1969). In 1975, a mutant variety was released in Finland obtained by gamma irradiation.
– *Phleum pratense, Phleum nodosum* (very high mutation frequency after gamma ray, neutron, and EMS treatment; Blixt 1976a).
– *Lolium temulentum, Lolium multiflorum* (dwarfs, early flowering mutants, different spike mutants; Malik and Mary 1971, 1974).
– *Sorghum vulgare* (a mutant with 27% more green crop than the control on the average of 4 years; Barabás 1965).
– *Beta vulgaris.* A mutation breeding program in sugar beet has been initiated in the German Democratic Republic; preliminary results are available (Melzer and Sackewitz 1979).

- *Brassica campestris* var. *dichotoma* (early flowering mutants, genotypes with increased number of seeds per siliqua; Kumar 1972).
- *Helianthus annuus* (earliness; Remussi and Gutierrez 1965). In the Soviet Union, a mutant sunflower variety has been released in 1977.
- *Trifolium pratense* (multifoliate leaves; Jaranowski and Broda 1978).
- *Melilotus albus.* The sweet clover could become a valuable forage crop for soils of very poor quality. Some mutants of interest for agronomic purposes, particularly genotypes with strongly reduced coumarine content, are available (Scheibe and Micke 1967; Micke 1969a, b, 1976).

In *maize,* a mutation breeding program has already been initiated in 1955 in Nebraska, U.S.A., with an aim to improve the grain yield. An increased additive variance for prolificacy and grain yield was observed after treatment with thermal neutrons. In M_{10}, the mean values for yield of selected lines of irradiated and control material were still equal but greater future gains for the treated material are expected (Gardner 1969). A voluminous mutation breeding program is running in Hungary; some genotypes of agronomic interest have already been selected (Pásztor 1978, 1979). Four officially released mutant maize varieties are available in the Peoples Republic of China (1974, 1979; gamma rays, fast neutrons), another three in Czechoslovakia (1979; gamma rays).

In special cases, mutants can be used indirectly for studying questions of agronomic interest. This holds for instance true for determining the rates of natural outcrossing of distinct crops. The broad bean *(Vicia faba)* is a facultatively cross-pollinating species, which is often handled as a self-pollinating crop because of the low degree of outcrossing. This question can be studied reliably by means of mutants homozygous for recessive genes, which are grown in different distances from the varieties or strains tested. After self- or sister-pollination of the mutant, homozygosity for the mutant gene will occur in the respective seeds resulting in mutant plants. If the seeds of the mutant, however, arose by cross-pollination, they are heterozygous for the mutant gene, the plants developing from them being phenotypically normal. In this way, the cross-pollination rates of *Vicia faba minor* were determined using a recessive unifoliate mutant as parent for the seed production. These leaf mutants can already be distinguished from the normal plants in very early stages of ontogenetic development; thus, a high number of fertilization events can be evaluated on a small greenhouse area. The cross-fertilization rates of the cultivar tested ranged between 16% and 46% when the two genotypes were grown closely together. This rate declines rapidly with increasing distance between the genotypes being about 5% only in intervals of 8–10 m (Gottschalk 1960b, 1978e). The *broad bean* should therefore be handled as a cross-pollinating crop in genetic and mutation genetic experiments. The *garden pea,* on the other hand, is cleistogamous and is thus a reliable self-pollinator. The natural outcrossing rate of our *Pisum* material was found to range between 0.5% and 1.6% only, using the method just mentioned.

2.2 The Chimerical Structure of the M_1 Plants

The chimerical structure of the M_1 plants has often negative consequences for the performance of mutation experiments, especially in crops with a relatively low propagation coefficient. The number of initial cells, present in the embryonic growing points, is a characteristic feature of each species. They are responsible for producing the shoot system of the M_1 plants following mutagenic treatment of the seeds. Most species contain several initial cells in their growing points. A distinct mutation will only occur in one of them; the others can likewise mutate but not with regard to the same gene. If a crop has a single initial cell and if a mutation has been induced in this cell, the M_1 plant arising from this seed is genetically uniform in all its organs, representing a hybrid heterozygous for the mutant gene. Theoretically, a 3:1 segregation should be expected in its progeny. Very often, however, more unfavorable segregations are obtained which are characterized by a more or less pronounced deficit of mutants probably due to zertation. The same situation can be expected in crops with several initial cells in all those cases in which only one of them survives, whereas the others die as a consequence of the mutagenic treatment. If, however, two initial cells survive, one of them containing a mutant gene, a chimerical M_1 plant arises. This can also occur in those cases in which the young embryo has several lateral growing points besides the apical main growing point. Those parts of the plant which develop from the mutant initial cell are heterozygous for the mutant gene, whereas the other parts are genetically normal. Segregation can only be obtained from the heterozygous parts of the plant which, however, cannot be distinguished from the nonmutated parts because of the recessiveness of the mutated gene. Thus, all the seeds of each M_1 plant have to be harvested, giving rise to a considerable deficit of mutant plants in the respective M_2 families. It is not possible to estimate the segregations expected conclusively, because this deficit depends on some criteria which can often not be determined in detail (Fig. 2):

1. The normal number of initial cells, present in the embryonic growing points, is not reliably known for many crops. Moreover, the number of surviving initial cells cannot be determined directly. It can only be estimated indirectly from the segregations in the M_2 and in later generations.

2. The degree of the diplontic selection during ontogenetic development of the M_1 plants cannot be determined reliably. In many M_1 plants, the heterozygous sector is smaller than expected because of a reduced activity of the heterozygous meristematic cells resulting in an additional deficit of recessives in the M_2 families.

3. Additional difficulties can be caused by zertation, i.e., by a reduced growth speed of the pollen tubes containing the mutant genes in relation to the tubes with the nonmutant dominant alleles. The consequence of this behavior consists in a reduced fertilization chance for the mutant germ cells.

With increasing number of initial cells in the growing points of the crops used, the situation becomes more and more unclear (Fig. 2). The generally valid regularity is as follows:

The Chimerical Structure of the M₁ Plants

	1	2	3	4	5
histogenetic initial situation	only one initial cell present	3 initial cells present, one of them mutated	3 initial cells present, one mutated; competition during ontogenesis	3 initial cells present; only the mutant cell surviving	6 initial cells present; one of them mutated
genetic constitution of the M₁ plant	no chimera	chimera; large mutated sector	chimera; smaller mutated sector	no chimera	chimera; very small mutated sector
M₂ segregation	~ 3:1	~ 8:1	~ 12:1	~ 3:1	~ 30:1

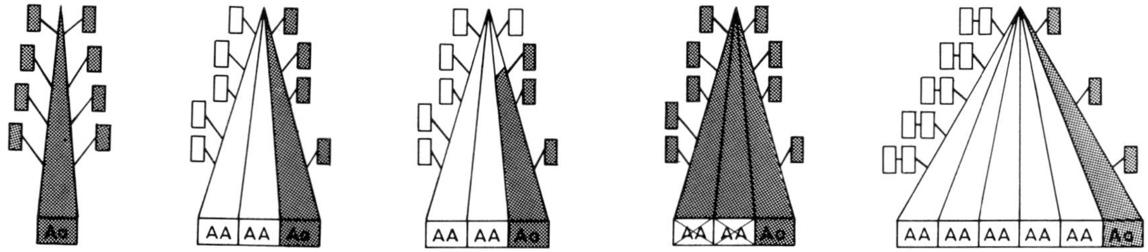

Fig. 2. Dependence of the M₂ segregations of recessive mutants on the genetic constitution of the M₁ plants as illustrated by different kinds of chimerism

- The higher the number of surviving initial cells,
- the smaller the mutant sector of the M_1 plant,
- the lower the proportion of mutant plants in the respective M_2 family.

In mutational experiments with the aim to develop collections with a possibly large number of different mutants, there is a summarizing effect of three negative factors:

1. The mutation rate increases with increasing radiation dosage or concentration of the chemical compound used, but the surviving rate of the treated embryos decreases. Moreover, the vitality and the seed production of the M_1 plants decrease. In small M_2 families, the 3:1 segregation can often not manifest itself.

2. Because of the chimerical structure of the M_1 plants, a deficit of recessives occurs. Even in larger M_2 families, the mutant gene becomes often not discernible.

3. Another deficit of recessives, due to zertation, is a characteristic peculiarity of many mutant genes. It has no connection to the chimerical status of the M_1 plants; on the contrary, it appears in the segregating families of all generations, reducing the proportion of mutants even more. Examples for *Pisum* have been given by Gottschalk (1964).

As a consequence of this situation, mutant genes are present in many M_2 families not yet in the homozygous but only in the heterozygous state, giving rise to segregations of the following type:

— 12 AA : 2 Aa : 0 aa

Thus, the mutant gene cannot be discerned because of its recessiveness. In many mutation experiments, the selection of mutants is carried out exclusively in M_2, and all the families which do not contain any mutants are eliminated. In this way, many mutant genes get definitely lost. A quantitative evaluation of mutation experiments with regard to the true frequency of mutants present can only be done if all the M_2 families — the segregating and the nonsegregating ones — are used for growing a giant M_3 generation. This will often be impossible because of lack of space.

Some of the consequences of the negative factors just discussed are presented in Fig. 3 for our irradiated *Pisum* material which can be regarded as an example valid for many other crops. The species *Pisum sativum* has several initial cells in the embryonic growing points, their exact number being not yet known. Nevertheless, a monohybrid segregation was relatively often found in the M_2 families, demonstrating thereby that only one of the initial cells present had survived. The response of our pea variety to different X-ray doses is presented in the right-hand part of Fig. 3 with regard to different yielding criteria. The seed production decreases rapidly with increasing dose; this hold also true for the proportion of germinating and surviving M_1 plants (left-hand part of the figure). The negative consequences of these additive effects become especially clear, if we do not use the mean values for the number of seeds per M_1 plant after having used different X-ray doses, but for the number of seeds per M_1 plot. In one of our test trials, 300 seeds per dose were irradiated with 5, 7, 9, 11, 13, and 15 kr. From the plot containing the 15-kr plants, only about 3% of that amount of seeds were harvested which was obtained from the 5-kr plot. Because of these difficulties, we have used a dose of 11 kr for our main trials, giving a relatively high mutation rate,

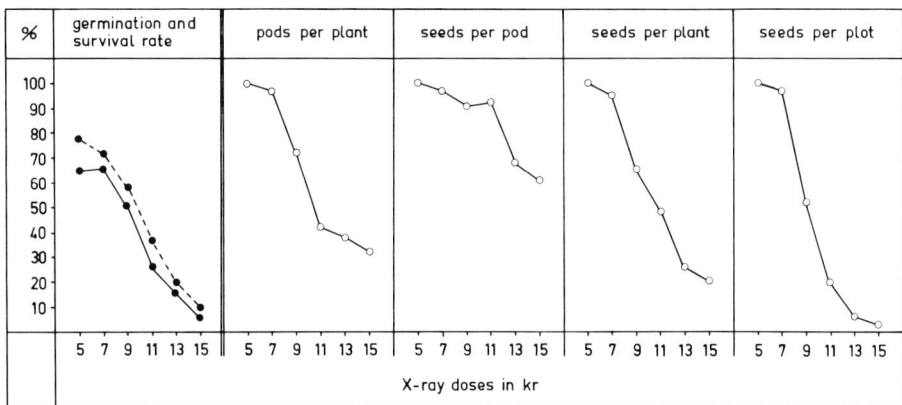

Fig. 3. *Left* Dependence of germination *(broken line)* and survival rate on the X-ray dose applied following irradiation of *Pisum* seeds. The values are related to the values of the nonirradiated control material. *Right* Dependence of different yielding criteria of *Pisum* M$_1$ plants on the dose following X-ray seed treatment. The values are related to the 5 kr values = 100%

a survival rate of 25%–30%, and a not too strong reduction in M$_1$ fertility. Nevertheless, the family size of the M$_2$ generation was small, resulting in all the negative consequences discussed above. Only those M$_2$ families were propagated in which segregations for mutants occurred. In the respective M$_3$ sibs, many new mutants were selected which were not present in the parental M$_2$ families. About 40% of the 800 pea mutants of our collection were selected in M$_3$. It can be concluded from these findings that a large number of mutant genes has been heterozygously present in the nonharvested M$_2$ families. The mean value for the total segregation of our segregating M$_2$ families was 6.4:1, demonstrating the high degree of M$_1$ chimerism. The most unfavorable segregations of single M$_2$ families were 14:1, 19:1, and even 45:1. In later generations, the mutant genes in question showed normal 3:1 segregations. Thus, the mutant sectors of the respective M$_1$ plants have been very small. Details on the chimerism in *Pisum* following mutagenic treatment have been published by Weiling and Gottschalk (1961), and by Blixt (1972).

The problem of chimerism can be avoided by treating pollen grains and by using them for pollinating emasculated flowers. This is a very circumstantial method which can only be used in exceptional cases. In *vegetatively propagated plants*, effective methods for avoiding the chimerical status of the M$_1$ plants are available (see Chap. 11).

3 The Selection Value of Mutant Genes

The selection value of mutant genes is an important theoretical criterion for judging the prospects of experimental mutagenesis in plant breeding. It represents a complex phenomenon of mutants and recombinants which is composed of a large number of different traits, most of them having not yet been studied in detail in the mutant material available. All those criteria which influence the competitiveness of the plants directly or indirectly, their vitality and fertility, should be considered in this respect. This holds true for:

- duration of seed germinability and speed of seed germination;
- speed of plant growth, particularly during early stages of ontogenetic development;
- metabolism and degree of the utilization of the nutrients available;
- begin of flowering and duration of the life cycle;
- resistance or immunity against pests and diseases as well as against unfavorable climatic conditions, and many other traits more.

Moreover, interactions between the mutants and other genotypes of the same species or with plants of other species living in the same biosphere, i.e., the behavior under distinct conditions of competition, are important factors. In cross-pollinating crops, the competitiveness of the mutant gene will relate not only to its homozygous but also to its heterozygous condition. A very important point is furthermore the adaptability of the mutant to the given and to altered ecological conditions. Also interactions between the mutant gene and the genotypic background have to be considered. The most important factor, however, is the relative fertility of the mutant, i.e., its seed production as related to that of the initial line. These factors and many others can positively interact with each other in different ways, resulting in a distinct reaction of the mutant which represents much more than merely its vitality. According to Gustafsson (1951), the selection value is the total sum of possible reactions of a genotype in varying environments. This definition emphasizes the important role of ecological factors for the efficiency and productivity of mutants. This point has not been considered by many authors working in the field of mutation research.

Let us now discuss the seed production of experimentally produced mutants and recombinants which is the most important part of the selection value of the genes involved; the reaction of mutants to altered ecological conditions are discussed in Chap. 18. Our collection of *Pisum* mutants, selected after X-ray and neutron irradiation, contains more than 800 mutants a small portion of them being heterozygous for reciprocal translocations. The whole material derives from the same mother variety. Only about 30% of the total number of our mutants have a fertility high enough for propagating them directly by means of seeds. The remaining 70% represent in about

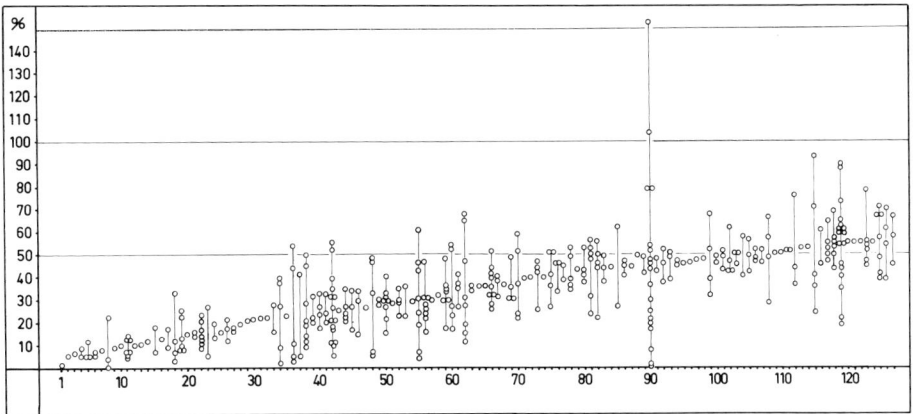

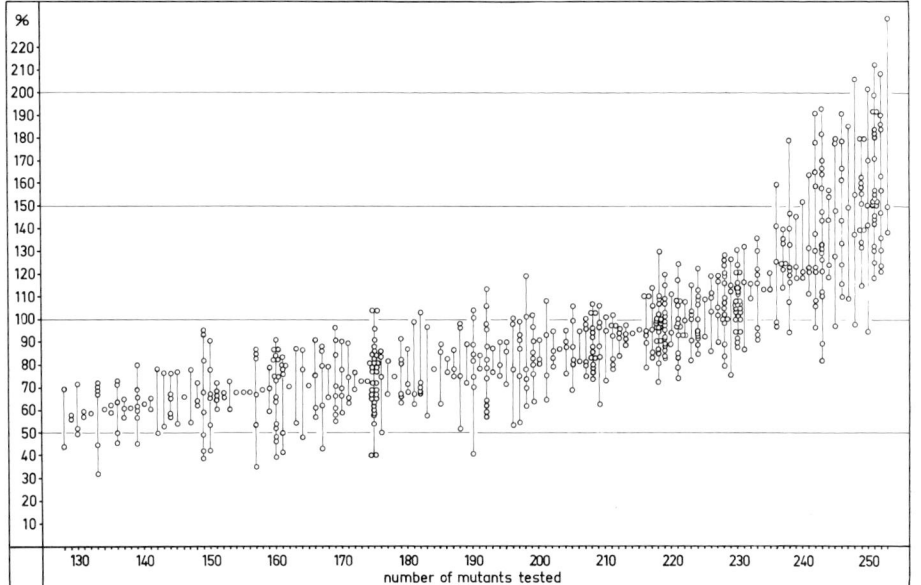

Fig. 4a, b. Seed production of 253 radiation-induced *Pisum* mutants. Each *dot* gives the mean value for the trait "number of seeds per plant" as related to the control value of the mother variety grown on the same field. Means of the same mutant, obtained in different generations, are connected by *vertical lines*

similar proportions lethal, sterile, and very weakly fertile genotypes, the propagation of which is only possible by means of segregations from plants heterozygous for the respective genes.

The seed production of 253 independently arisen mutants of our collection is graphically presented in Fig. 4. About 20 of them have arisen several times in our radiation genetic experiments; thus, the graph gives information on the seed production of about 230 different *Pisum* mutants. Reliable values on the productivity of a genotype can only be obtained by testing it in the form of yielding analyses with several replications over several generations. This cannot be done with so many genotypes because of lack of space and personnel. Nevertheless, most of the mutants were investigated in several years, those which might be of agronomic interest also in several replications per year. Their mean values for the trait "number of seeds per plant" are related to the control values of the mother variety grown in the same year at the same location. Altogether, about 80,000 plants of the 253 mutants and of the control material were evaluated. In this way, a relatively reliable survey on the capacity of the genotypes tested was obtained. More details on the selection value of mutant genes have been published elsewhere (Gottschalk 1962, 1971a).

The graph shows two regularities, both being of interest for the application of mutagenesis in plant breeding:

1. The relative seed production of a mutant can considerably differ from year to year.
2. The proportion of high-yielding mutants is very low.

The first regularity appears in almost each mutant tested over several generations. A particularly drastic behavior shows mutant 189 homozygous for three mutant genes. One of them — gene *ipc* — could become of interest for pea breeding because it increases the protein content of the seed flower considerably (Gottschalk and Wolff 1983). This genotype shows in general a reduced seed production from reasons not yet known so far. Structure of sex organs and course of meiosis are normal; nevertheless, the seed fertility was very low in most of the 18 generations tested so far (Fig. 4a). Surprisingly, the mutant outyielded its mother variety by about 50% in 1978, whereas no seeds at all were produced from a large number of completely healthy plants in 1976 and 1977. These extreme differences are certainly due to a specific reaction of the mutant to differences of distinct climatic factors in the three subsequent years. Similar findings were obtained in many other mutants, some of which being of direct agronomic interest because of their favorable yielding properties or of other useful characters. The seed production of the high-yielding fasciated mutants, for instance, is negatively influenced by rainfalls during the period of flowering and early ripening. Details on the reaction of early flowering genotypes and of mutants with bifurcated or fasciated stems are given in Chaps. 6 and 10. All these results show that many mutants and recombinants differ from their mother variety with regard to their adaptability to distinct ecological conditions. Moreover, our findings demonstrate that reliable conclusions on the productivity of mutants cannot be drawn if data of only one or two generations are available.

The second point mentioned above is of interest for the performance of treatments in mutation breeding. Only about 15% of the 253 mutants tested are equivalent or better than their mother variety with regard to seed production. If we relate this value to the total number of mutants selected (about 800), a proportion of 4%–5% of our material was found to have a seed production competitive or better than that of the initial line. This does, however, not mean, that all the mutants of this group are of value for pea breeding. Most of the high-yielding genotypes are too tall and are thus not suited for field cultivation; others are very late. The principally favorable yielding properties of some mutants are negatively influenced by the unstable penetrance or by the pleiotropic action of the respective genes. If we select those mutants which may really be of agronomic interest, a proportion of about 1% of the total material available may remain. This value, however, has been estimated only on the basis of seed production. It is possible that some genotypes with a slightly reduced seed yield have other favorable characters, such as a specific tolerance or resistance against pests or unfavorable climatic or soil conditions. This has not yet been studied in detail. Moreover, the true proportion of "useful mutants" may be somewhat higher if the "micromutants" are considered, the evidence of which is often difficult and time-consuming. The estimated frequency of about 1% of useful pea mutants is in principle in concordance with findings obtained by Gustafsson (1954) in *barley:* he found a frequency of 0.1%–0.5% of positive mutants. A more optimistic judgment of applied mutagenesis in *barley* gave Gaul et al. (1969) under consideration of their micro-mutants. But there is no doubt that the selection value of the great majority of mutant genes is negative. Thus, the use of mutants in plant breeding can only be done in a prospective way, if the mutation treatments are carried out on a very broad scale, i.e., if a very great number of mutants is selected in the M_2 and in later generations.

The reasons for the reduced seed production of most mutants are not yet fully understood. This negative behavior does not only appear in genotypes with reduced vitality; on the contrary, many mutants with normal vitality, undisturbed course of meiosis, and obviously with full physiological capacity show this peculiarity. Most of the mutant genes studied display two different actions:

— A specific action resulting in a distinct alteration of the mutant as compared to its mother variety. This effect is certainly due to the altered biochemical structure of the respective gene as each gene shows its specific behavior in the mutant condition.
— An inspecific action resulting in the reduced seed production of the mutant. As this effect is in principle similar in many genotypes homozygous for different mutant genes, it cannot be due to distinct changes of the nucleotide sequence of the DNA.

This behavior becomes understandable by assuming, that the presence of a mutant gene as such is in most cases a negative factor with regard to the seed production of the mutant irrespective of the question which gene of the genome is involved. The mutant gene can be regarded to be a foreign element which disturbs the genic harmony of the genome. This disturbance is a more or less similar effect of many mutant genes resulting in the inspecific reduction of the seed fertility of the mutants.

This hypothesis is supported by the fact that the accumulation of mutant genes in the same organism often leads to a stronger reduction of its seed production. Figure 5, however, shows that this behavior cannot be generalized. On the contrary, many recombinants of complicated genotypic constitution were found to show an unexpectedly high seed fertility. We have selected more than 600 different recombinant types after having crossed different mutants of our *Pisum* collection with each other. They were developed into pure lines and are homozygous for two up to more than ten mutant genes. This was possible as some genetically very similar fasciated mutants proved to have at least 16 mutant genes. They were used in many crosses as partners for other genotypes (details see p. 58ff.). Because of the high degree of heterozygosity of the F_1 hybrids, a large number of different recombinants arose in the F_2 families and the F_3 sibs. Many of them were studied over several generations with regard to their seed production. The yielding properties of 417 of them are given in a summarized form in

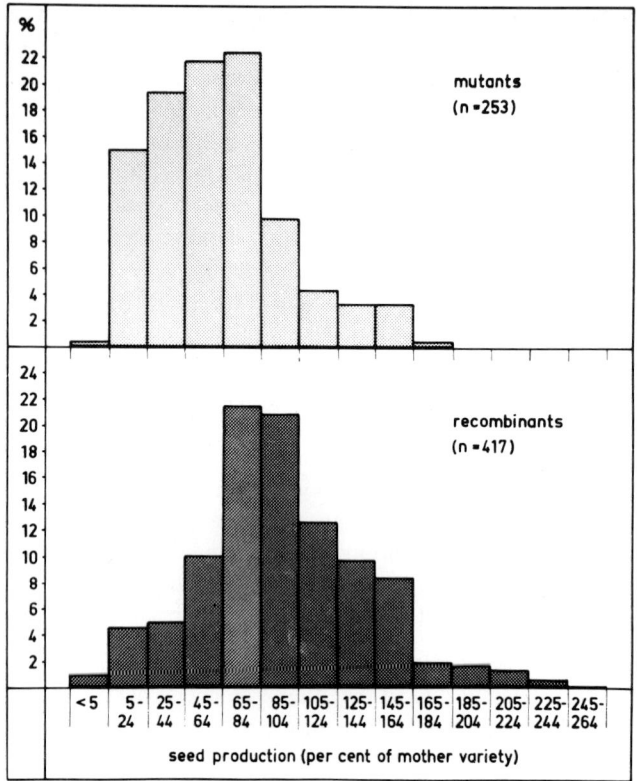

Fig. 5. Seed production of 253 different radiation-induced mutants and of 417 different recombinants of *Pisum sativum*. The mean values for the trait "number of seeds per plant" of the genotypes tested are related to the control values of the mother variety = 100%. Altogether, the yield of more than 128,000 pea plants is considered in the graph

Figure 5; moreover, they are compared with the 253 mutants already mentioned. The histogram of the *mutants* shows its maximum in fertility classes of 45%–85% of the control values of the mother variety; only a small proportion of the mutants tested has a better seed production. The curve of the *recombinants* is shifted more to the right-hand side of the graph, demonstrating thereby that the proportion of high-yielding genotypes is essentially higher than in the mutants.

This unexpected result needs some additional remarks. We would have certainly not obtained these favorable findings if we had used randomly chosen mutants for our hybridizations. This has insofar not been done as we were interested in improving some prospective mutants of our collection by combining specific useful genes with each other. Therefore, the high-yielding fasciated mutants and a low-yielding early flowering mutant were more frequently used as partners for the crosses than other mutants of our collection. But the graph demonstrates very clearly that the negative selection value of many mutant genes could be improved by means of recombination. This holds particularly true for gene *efr* for earliness, which is of direct value for pea breeding (for details see Sect. 10.1).

4 The Seed Production of Mutants and the Alteration of Quantitative Characters

4.1 The Alteration of Quantitative Characters

Many quantitative characters of our crops are controlled by polygenic systems. Under the influence of mutant genes these can easily be altered positively or negatively. Shifts of the genetically conditioned variability of these traits are so common that it is not possible to go into details in the present review paper. Some examples and references for a large number of different crops are given in Tables 2 and 3. The problem has been discussed by Brock (1965, 1970) and Scossiroli (1965) among others. In *cereals*, the action of the respective genes refers to the following characters:

- plant height,
- plant weight,
- date of heading,
- degree of tillering,
- number of productive tillers per plant,
- ear length,
- ear weight,
- ear density,
- number of spikelets per spike,
- number of kernels per spike,
- kernel weight,
- grain hardness,
- protein and carbohydrate content of the grain,
- grain yield,
- number of fertile ears per unit area of land.

Corresponding traits are influenced in dicotyledonous crops. In addition, genes causing better mechanical harvesting should be mentioned, for instance in *Haricot bean* (Khvostova 1967) and in *soybean* (Josan and Nicolae 1974). In many cases, only M_2 or M_3 findings are available which do not yet justify any conclusive statements. It is necessary to study the genotypes selected over a period of generations in order to ensure that the effects observed in earlier generations are really stabilized. This was for instance done in *Triticum aestivum* by Borojević and Borojević (1969). It may be mentioned that the increase in variation of quantitative characters for a given generation can depend on the genotype, the trait investigated, and the mutagen used (Rao and Siddiq 1977; *rice*). Details on genetically conditioned alterations of specific quantitative characters are given in the following chapters.

Table 2. Mutants of cereals with altered quantitative characters

Species	Authors	Mutagens	Quantitative characters altered
Hordeum vulgare	– Tavčar 1965	Gamma rays	Plant height, time of maturity, awn characters
	– Gaul 1966; Gaul and Ulonska 1967	X rays	Grain yield, length of growing period
	– Vasudevan et al. 1969	X rays	Plant height, spike length, grain yield
	– Gill et al. 1974	Gamma rays, EMS	Plant height, number of spikelets per spike
	– Ibrahim and Sharaan 1974a	Gamma rays	Plant height, spike length, spike density, number of kernels per spike, heading date
	– Morsi et al. 1977	Gamma rays, NMU	Heading date, grain yield
Oryza sativa	– Bilquez et al. 1965	X rays	Plant size, grain characters, heading date
	– Singh 1973		Higher values for several characters
	– Rao and Siddiq 1977	X rays, EMS	Tillering, grain size, grain yield
	– Okuno and Kawai 1978a, b	X rays, gamma rays, EMS, EI	Plant height, number of panicles per plant, number of spikelets per panicle, grain size, heading date
Triticum aestivum	– Tavčar 1965	Gamma rays	Plant height, ear length, number of kernels per ear
	– Borojević 1965, 1966; Borojević and Borojević 1969	X rays, neutrons	Plant height, number of kernels per spike
	– Goud 1967, 1968	Gamma rays, neutrons, EMS	Tillering, spike number, number of spikelets, grain weight
	– Trujillo-Figueroa 1968	EMS	Number of fertile ears per unit area, grain weight
	– Wagner 1969	X rays, EMS	Time of maturing, ear form, grain size, yielding ability, baking quality
	– Popovic et al. 1969	Gamma rays	Plant height, ear length
	– Mehta 1972	Gamma rays, EMS, NMU	Grain number per ear
	– Galal et al. 1975	Gamma rays	Stem and spike length, number of spikelets, number of kernels, grain weight
	– Larik 1975	Gamma rays	Plant height, culm diameter, grain weight, heading date
	– Sichkar et al. 1975; Mar'yushkin et al. 1977	EMS, NMU	Plant height, spike length, number of grains per spike, grain weight
	– Dhonukshe and Bhowal 1976	Gamma rays	Culm length, tillering, yield
	– Kumar 1977a	Gamma rays	Plant height, tillering, spike length, number of spikelets per spike, grain weight, grain hardness
Triticum durum	– Khamankar et al. 1978	Gamma rays, NMU, HA	Number of grains per ear, grain size
	– Bozzini et al. 1967	X rays	Plant height, number of fertile spikelets per spike, heading date

Avena sativa	– Frey 1965	Neutrons, EMS	Plant height, grain weight, heading date
	– Shevtzov 1969a	DES, EI, NMU	Plant height, heading date, yield
Sorghum subglabrescens	– Sree Ramulu 1974b	X rays, gamma rays, EMS	Plant height, ear weight, grain weight
Eleusine coracana	– Goud et al. 1971	EMS	Plant height, tillering, number of fingers per plant
Pennisetum typhoides	– Bilquez et al. 1965	X rays	Tillering, length of vegetative cycle

Table 3. Mutants of dicotyledonous crops with altered quantitative characters

Species	Authors	Mutagens	Quantitative characters altered
Arachis hypogaea	– Bilquez et al. 1965	X rays	Size of pods and seeds; oil content
	– Patil 1971, 1972	X rays	Seed size
Lens culinaris	– Uhlik and Urban 1976	Neutrons, ENU	Plant height and weight, seed weight, seed set
Phaseolus aureus	– Dahiya 1973	Gamma rays	Number of pods per plant, grain size, days to flower
Phaseolus vulgaris	– Swarup and Gill 1968	X rays	Number and size of pods, number of seeds per pod, seed size, seed yield
Vicia faba	– Ismail et al. 1976	EMS	Plant height, number of pods and seeds per plant, number of branches, seed weight per plant
	– Abdalla and Hussein 1977	Gamma rays, EMS	Plant height, tillering, number of pods and seeds per plant, seed weight, seed yield per plant
Medicago sativa	– Scossiroli and Sarti 1967	X rays	Increase for a few productive traits
Brassica campestris	– Kumar and Das 1977	Gamma rays	Plant height, number of branches per plant
Linum usitatissimum	– Rai and Das 1976	Gamma rays	Number of tillers, number of capsules per plant, number of seeds per capsule
Gossypium hybrids	– Siddiqui 1971	Gamma rays	Higher values for several characters
Corchorus olitorius, C. capsularis	– Ghosh and Sen 1974	X rays, gamma rays	Plant height, basal diameter, number of nodes

4.2 Mutants with Increased Seed Yield

One of the most important characters for judging the agronomic value of mutants is their yield as related to that of the mother variety. On the other hand, the seed production should not be overestimated. If a mutant has a specific desired peculiarity, for instance a distinct resistance behavior, it may be used in crossbreeding even if its yield is insufficient. In general, however, favorable yielding properties are a good prerequisite for the agronomic utilization of a mutant. According to the experiences made in *barley* and *peas,* only about 0.5% of all experimentally produced mutants were found to be of interest for breeding purposes. The true proportion may be higher if the "micromutations" are considered in an adequate way. These are mutations causing so small effects that the respective mutants cannot be discerned reliably in segregating families. The concept of micro- and macro-mutations and its importance for mutation breeding has been repeatedly discussed by Gaul and co-workers (Gaul 1965b, 1966, 1967; Gaul and Ulonska 1967; Gaul et al. 1969, 1971). According to these authors, it is relatively easy to select micro-mutants of *barley* having a yield potential 10% higher than that of the initial line. The presupposition for the selection of genotypes of this kind are tests over many generations and a careful statistical evaluation of the data obtained. This has been done so far only in a small number of cases. There is no doubt that many micro-mutants with good yielding properties are present in the collections which, however, have not yet been discerned. Investigations of Walther and Seibold (1978) in *barley* demonstrate that the chances for obtaining prospective strains are about equivalent in material obtained by normal hybridization and in genotypes arisen through micro-mutations. Most of the useful mutants utilized for breeding purposes belong, however, to the group of "macro-mutants" in Gaul's sense, which can reliably be distinguished from the nonmutated plants of segregating families because of their diverging characters. Other authors (for instance Scholz 1971) are more sceptical with regard to the genuine suitability of micro-mutations in plant breeding. Unfortunately, many desired macro-mutants show additional negative features besides the positive ones, which may have been caused by simultaneously induced micro-mutations. They can sometimes be eliminated by crossing the mutants with other genotypes.

In spite of the difficulties just mentioned, a large number of examples is given in the literature demonstrating an improved seed production of mutants in comparison to that of their mother varieties. We can give only a small selection. The usefulness of some of the mutants listed below has become evident insofar as they have been commercialized in the meantime. Some examples of high-yielding mutant strains are graphically presented in Figs. 7, 8, 9, 13, 25.

Hordeum vulgare

– Two Norwegian mutants yielded 7%–9% more than the mother variety (Aastveit 1966).
– In Hungary, about 1000 induced mutants were already available in the middle of the 1960's. Among 100 mutant lines examined, 18 surpassed the initial lines with regard to some agronomically important traits (Pollhamer 1966, 1967).

- In Russia, mutants with one or several valuable characters were selected (Preeatschenkoo and Greenvald 1967).
- Twelve neutron-induced Swedish mutants proved to be superior to the best control lines (Gustafsson et al. 1968).
- Several Austrian mutant lines are being tested in the state trials for becoming released varieties (Hänsel et al. 1972).
- A Danish dwarf mutant with a remarkably high degree of lodging resistance has very good grain-yield performance (Haahr and v. Wettstein 1976).

Oryza sativa

- In a voluminous Japanese collection of gamma-ray-induced mutants, containing 1405 genotypes, only a small number was found to be of agronomic interest with regard to yield. About 10% of the mutant lines showed an increased panicle weight per plant (Tanaka 1968).
- Thirty Indian mutants were isolated in gamma-ray treatments surpassing the mother varieties with regard to distinct characters. Some of them seem to be particularly suited for cultivation under conditions of high fertility (Saini and Sharma 1970).
- A high-yielding mutant was selected in Guyana (Pawar 1971).
- In North India, about 400 mutants were isolated, some of them showing a desired combination of favorable traits such as high yield and earliness or high yield and short straw causing lodging resistance (Kaul 1978).
- Three early ripening Egyptian mutants were found to have a high yield potential considering findings of M_5 and M_6 generations (Ismail and Ahmad 1979).

Triticum aestivum

High yielding mutants are available in Czechoslovakia (Jech 1966b), Russia (Khvostova 1967; Mar'yushkin et al. 1977; Sichkar et al. 1980), Pakistan (Siddiqui and Arain 1974), India (Sawhney and Sharma 1979). One of the Pakistani wheat mutants has been approved for release because of its favorable yielding properties and its shattering

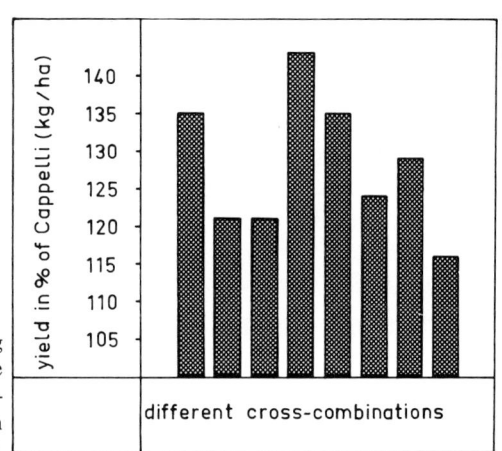

Fig. 6. Productivity of the best crosses among the *durum wheat* variety "Cappelli" and some of its mutants. Values (kg/ha) of eight different cross-combinations in F_5 generation (according to Scarascia Mugnozza et al. 1972)

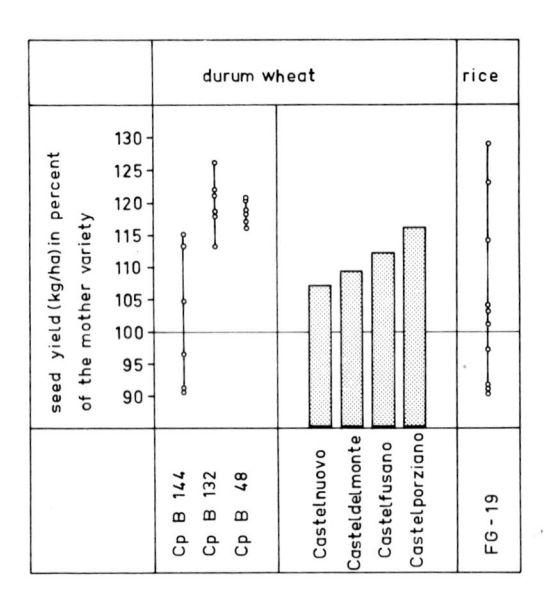

Fig. 7. Yield (kg/ha) of some *durum wheat* and *rice* mutants in comparison with that of the mother varieties. *Left* Three short-stemmed *wheat* mutants with improved lodging resistance tested at six localities in Italy. Each *dot* represents the mean value for one locality (according to Scarascia Mugnozza 1966a). *Center* Four Italian mutant cultivars of *Triticum durum* (Scarascia Mugnozza et al. 1972). *Right* A dwarf *rice* mutant tested at ten different localities in India (Reddy 1975)

resistance (Siddiqui et al. 1981). An Indian semidwarf wheat mutant shows a reduction with regard to the number of spikelets per spike, but a marked increase of the kernel size resulting in 56% higher yield than the control (Kumar 1977b).

Triticum durum

High-yielding *durum* wheat mutants were tested at different locations in Italy. Some of them were found to be superior to the mother varieties, particularly under conditions of high nitrogen level of the soil. The increased productivity was confirmed in several successive generations (Bozzini 1965; Scarascia Mugnozza et al. 1965, 1966; Scarascia Mugnozza 1966a, b, 1967; Pacucci et al. 1966; Mosconi 1967; Figs. 6, 7).

Avena sativa

One out of 160 EMS-induced mutant lines proved to have a positive deviation for grain yield as compared to the mother variety (Arias and Frey 1973).

Lycopersicon esculentum

Prospective mutants with very good fruit set and with relatively simultaneous ripening of the fruits have been selected in Bulgaria after gamma irradiation (Zagorcheva and Yordanov 1978).

Pisum sativum

— The fasciated mutants of our collection show a strongly increased seed production. Some other genotypes with deviating leaf shape, narrow leaves, long internodes, or stem bifurcation were found to be equal to or somewhat better than the mother variety, considering yield trials during 6–16 generations (Gottschalk 1966, 1977a, 1978c; Gottschalk and Hussein 1975; Figs. 8, 9, 13).

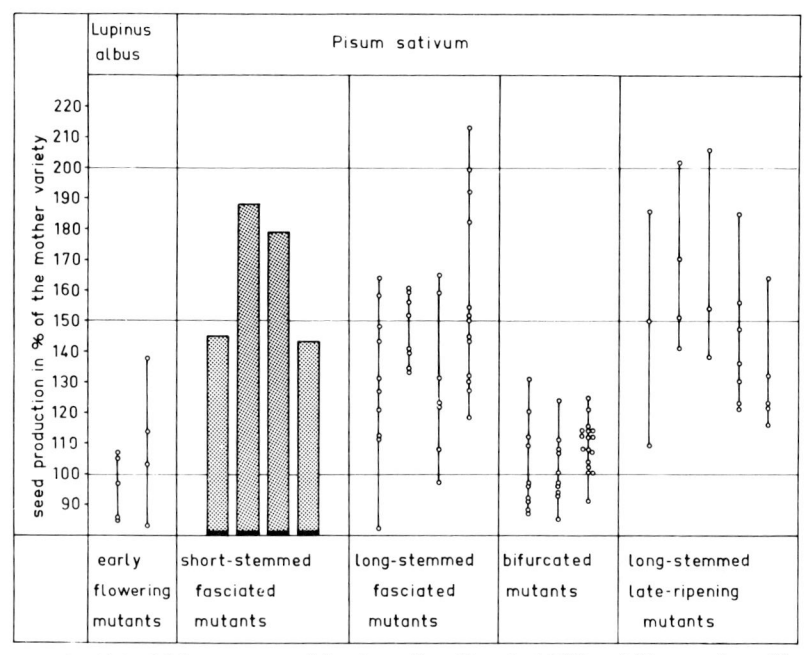

Fig. 8. Examples for high-yielding mutants of *Lupinus albus* (Porsche 1967) and *Pisum sativum*. The short-stemmed fasciated pea mutants according to Vassileva (1978); all other data own results. Each *dot* and each *column* represents the mean value for the character "number of seeds per plant" for one generation. Most mutants have been tested in a large number of subsequent generations. Mean values of the same mutant are connected by *vertical lines*

- A pea mutant with an about 20% increased seed yield is available in Russia (Sidorova et al. 1969).
- An Indian early ripening pea mutant is significantly better than its mother variety, but not better than the released varieties. Because of its earliness, however, it is a prospective partner in cross-breeding (Kaul 1977).

High-yielding mutants have also been selected in other pulses, for instance in:

- *Phaseolus vulgaris* (Swarup and Gill 1968; Rubaihayo 1975; Hussein and Disouki 1979),
- *Arachis hypogaea* (Patil and Thakare 1969; Patil 1980),
- *Glycine max* (Josan and Nicolae 1974; Rubaihayo 1976b),
- *Lathyrus sativus* (Chekalin 1977),
- *Cicer arietinum* (Kharkwal 1981),
- *Vigna trilobata, aconitifolia, radiata* (Subramanian 1980; Singh and Chaturvedi 1981b).

Some gamma-ray-induced mutants of Egyptian cotton *(Gossypium barbadense)* are promising because of improved quantitative characters, such as earliness, boll weight, lint yield, gin turnout, and fiber properties (Al Didi 1965; Khvostova 1967).

Unexpected findings, which cannot yet be interpreted in a satisfactory way, were obtained by Wellensiek et al. (1975) in *Pisum sativum*. A group of 26 mutants, obtained after EMS treatment, proved to be sterile in M_2. Four of these genotypes, however, were finally obtained in a fertile condition in later generations. They were obviously able to restore their fertility by means of a mechanism not yet known. We have made similar experiences with a flower mutant of the garden pea (Gottschalk 1978d).

4.3 Released or Approved Mutant Varieties

So far, almost 500 experimentally produced mutants have been developed into officially released or approved varieties, more than 250 of them belonging to the group of asexually propagated crops and ornamentals. Lists with the latest informations have been published by Sigurbjörnsson and Micke (1969, 1973, 1974); Sigurbjörnsson (1976); Broertjes and van Harten (1978); by the International Atomic Energy Agency (1972) and in the *Mutation Breeding Newsletter* edited by this organization. These lists contain:

— the name of the cultivar;
— place and date of release along with the institution, where the variety was developed;
— the kind of mutagenic treatment;
— the main improved character of the variety.

The most remarkable results in this field were obtained in *barley,* if we do not consider the big group of ornamentals showing a specific behavior which cannot be generalized. The first mutant barley variety, released in 1955 in the German Democratic Republic, was "Jutta," derived from an X-ray-induced mutant. According to information available to us, at least 61 mutant barley cultivars have been developed until 1981. This value is certainly still too low. The increasing importance of induced mutations for barley breeding becomes clear from Fig. 1. The graph shows that the number of released varieties increased rapidly since 1967. The main improved characters are as follows:

— early maturing;
— increased grain yield, in some cases especially with heavy fertilization;
— higher 1000-kernel weight;
— shorter and stiffer straw, often resulting in lodging resistance;
— increased winter hardiness
— mildew resistance or field tolerance;
— resistance to smut, stripe rust, leaf rust, yellow rust;
— drought resistance;
— shattering resistance;
— sprouting resistance;
— better threshability;
— better malting quality;
— increased protein content.

Some of the cultivars were withdrawn from the market because of their breakdown in mildew resistance.

The strong increase of the number of commercialized mutant varieties during the past years is valid for all groups of cultivated plants which are at all suited for the application of mutagenesis. Forty-five mutant *rice* varieties, 28 *bread wheat*, 12 *durum wheat*, 9 *oats*, and 7 *maize* varieties have been released. The Hungarian rice variety "Nucleoryza," obtained by neutron irradiation, has been released in 1976. Because of its excellent yielding properties during the past years, it covers now more than 80% of the Hungarian rice area (Simon and Sajo 1976, 1977). Values for the cereals and for all the sexually propagated crops treated in this way are given in Fig. 1. The methods of applied mutagenesis proved to be especially successful in ornamentals for developing commercial varieties within a very short period. In *Chrysanthemum*, at least 97 cultivars arose by means of X-ray and gamma-ray irradiation. The year of release of the majority of them is as follows:

— 1974: 12 varieties
— 1975: 14 varieties
— 1976: 27 varieties
— 1977: 5 varieties
— 1978: 13 varieties
— 1979: 10 varieties

Most of them were developed in the The Netherlands, India, and the Soviet Union. Similar examples could be given for other ornamentals, such as *Streptocarpus, Begonia, Dahlia* among others. Further details on mutant cultivars of asexually propagated crops and ornamentals have been published by Broertjes and van Harten (1978).

With regard to the mutagens used, the situation given in Table 1 is realized considering a total of 485 released mutant varieties of all the crops treated. The majority of them derives from mutants induced by physical mutagens. The X-ray values are so high because not only the X-ray-induced mutants are considered in the table, but also those cultivars which have been developed in crossbreeding by means of commercial varieties of X-ray origin. The differences between the total sums for the three different kinds of rays do not mean that there are true differences between X rays, gamma rays, and neutrons with regard to their usefulness in mutation breeding. In countries having potent sources of gamma rays, this kind of irradiation will be given preference to neutron irradiation. The low proportion of cultivars, originated from chemically induced mutants, is likewise not a criterion for the inefficiency of the mutagenic chemicals. Many prospective strains, derived from chemically induced mutants, are available in the collections and have a good chance for becoming commercialized in the near future.

5 The Utilization of Mutants in Crossbreeding

The experiences, obtained in mutation breeding, show that it is possible in principle to develop a commercial variety directly from an induced mutant. According to the results available in this field, it is, however, the opinion of many geneticists, that the utilization of useful mutants in crossbreeding programs is more prospective. In this way, negative features of some mutants can be eliminated and only the positive ones are utilized. Moreover, there may be a chance in specific cases that a pleiotropic spectrum of a mutant gene is altered in a favorable manner (for details see Chap. 16). Until 1976, 47 *barley* varieties of induced mutant origin have been officially released. Only 17 of them were directly developed from the respective mutants. The others derive from crossing mutants with varieties or strains (Sigurbjörnsson 1976). These data demonstrate the utilization of induced mutations in crossbreeding programs in a convincing way. The problem has been discussed by Scholz (1965, 1967, 1971, 1976) Borojević and Borojević (1972); Scarascia Mugnozza et al. (1972). Much additional information is found in the proceedings of a symposium organized by the International Atomic Energy Agency (1976b).

There are two different possibilities for using a mutant in a hybridization program:

— The mutant gene can be incorporated into the genome of a commercial variety or a prospective strain by crossing the mutant with the variety or the strain.
— Two or more mutant genes can be placed in the same genome by crossing the mutants with each other.

In both cases, recombinants have to be selected in F_2 and F_3 homozygous for the genes involved. They are developed into recombinant lines. Their breeding value can be tested by comparing them with the parental genotypes and the mother varieties.

5.1 The Incorporation of Mutant Genes into the Genomes of Varieties or Strains

An impressive example for the utilization of induced mutations in the improvement of a crop refers to the breeding work in *barley* carried out in Sweden (Hagberg 1967; Gustafsson et al. 1971). Since 1960, seven *barley* varieties, partially originating from three X-ray-induced mutants, have been officially released in the country:

— Pallas: Directly obtained from the *erectoides* mutant *ert-k*^{32} showing a better lodging resistance than the mother variety Bonus. Released in 1960. Widely cultivated in several European countries.

- Hellas: Improvement of Pallas with regard to the degree of lodging resistance by crossing Pallas with the variety Herta. Released in 1969.
- Visir: Adding mildew resistance to Pallas and Hellas from the primitive "Long glumes" barley. Released in 1970.
- Mari: Directly developed from the early maturing mutant *early-a^8*. Semidwarf habit with pronounced lodging resistance. Released in 1960.
- Kristina: Improvement of lodging resistance, yield, and malting quality by crossing Mari with the Norwegian variety Domen. Released in 1969.
- Mona: Introducing mildew resistance into Mari from the primitive variety Monte Christo. Released in 1970.
- Gunilla: The mutant giving rise to this variety arose already in 1939 in Gull barley and was originally an inferior genotype, particularly because of its reduced kernel size and bad malting quality, but it had an extreme straw strength. The mutant was included into a crossbreeding program resulting in the variety Gunilla released in 1970.

Some of these mutant varieties were crossed with other cultivars or with each other, and new cultivars were developed (Sigurbjörnsson 1976; International Atomic Energy Agency 1972–1982):

- Rupal (from Pallas)
- Senat (from Pallas and Hellas)
- Eva, Salve (from Mari)
- Hankkija's Eero (from Mari in Finland)
- Stange (from Mari in Norway)
- Atlanta (from Hellas in Canada)

Mari is one week earlier than its mother variety in South Sweden. Under more northern conditions, however, it is relatively later ripening. In order to utilize the *early-a^8* locus in Finland, Mari was crossed with the Finnish "Otra" barley ripening 9 days earlier than Mari. In F_4 and F_5, lines were available heading still earlier than Otra, obviously due to a transgressive effect of the two genes for earliness (Kivi 1967). In Argentina, Mari behaves as a drastic early mutant. The *early-a^8* gene was incorporated into the Castelar breeding material in order to study its reaction in different genetic backgrounds and to select prospective recombinants (Favret and Ryan 1966).

A similar intensive utilization of a mutant variety in crossbreeding programs was carried out in Czechoslovakia. The cultivar "Diamant," deriving from an X-ray-induced mutant in 1965, gave rise to the new Czechoslovakian varieties Ametyst, Favorit, Hana. By crossing Diamant with other mutants, the cultivars Rapid and Atlas arose. In the German Democratic Republic, the cultivars Trumpf and Nadja were developed by means of Diamant (Bouma 1977; Sigurbjörnsson 1976; International Atomic Energy Agency 1972–1982). In the United Kingdom, "Golden Promise" was used for developing the cultivars Midas, Goldspear, Goldmaster (Ruddick and Marsters 1977). "Luther" is utilized in the same way in the United States (Nilan et al. 1968), giving rise to the cultivar Boyer. This holds also true for a mutant *peanut* variety (Patil 1975).

The examples just given refer to mutant varieties which were successfully used in crossbreeding programs. This, of course, is also done with prospective mutant strains that have not yet reached the varietal level. Model experiments in this field were carried out by Gaul and Hesemann (1966) using the awn length of *barley* as a mutant character which can be altered to a great extent in different genotypic backgrounds. In the German Democratic Republic, some *erectoides* and early maturing barley mutants were improved by crossing them with varieties and by selecting efficient recombinant types (Scholz 1967, 1971). Similar work is done in Norway (Aastveit 1966; Ali et al. 1978). New genotypes of *Triticum durum,* showing an improvement with regard to yield, lodging resistance, earliness, and yellow berry frequency, were obtained in Italy by crossing mutant lines with varieties. Some of them show excellent yielding properties (Fig. 6; Pacucci et al. 1966; Scarascia Mugnozza et al. 1972). Moreover, mutant strains of the following crops are included into crossbreeding programs:

- *Oryza sativa* (Kawai 1968; Ismachin and Mikaelsen 1976). A Taiwanese semidwarf mutant was crossed with a variety giving rise to a popular strain (Li et al. 1971).
- *Triticum aestivum* (Ramirez et al. 1969; Borojević and Borojević 1972; Konzak 1972, 1976).
- *Triticum durum* (Bozzini 1974a, b).
- *Poa pratensis.* Some M_3 families with good dry and green matter yield could be directly developed into a variety. Some other mutants are used in crossbreeding (Svetlik 1967).
- *Arachis hypogaea* (Patil 1972, 1975).
- *Pisum arvense.* Some Polish mutants were crossed with varieties of *Pisum sativum* in order to utilize the mutant genes for pea breeding (Jaranowski 1977). A pea mutant of our collection, homozygous for gene *ion* responsible for a 30% increase of the number of ovules per carpel, proved to be a good general combiner with Indian varieties (Kumar and Agrawal 1981). Also some other mutants and recombinants of our material are utilized in Indian breeding programs. This holds particularly true with regard to an early flowering genotype showing a certain degree of drought resistance (Fig. 11).

In some other crops, investigations of this kind have just been started. The viability of 60 *tomato* mutants was studied by Butler (1977), and none of them was found to be superior to the initial line. Selection on different genetic backgrounds, however, resulted in some cases in an improved viability. The utilization of a large number of spontaneously arisen or induced *pea* mutants is being tested at Weibullsholm, Sweden, by means of a computer system (Blixt 1976b).

5.2 The Joint Action of Mutant Genes

In some crops, a large number of useful mutants is already available. One of the next steps of mutation breeding consists therefore in crossing different mutants with each other in order to combine their characters of agronomic value. So far, not many experiences are available in this field, but some findings show that the combination of mutant genes does not always lead to the combination of the mutant characters controlled by the genes involved. There are certain interactions between the genes which can influence the effect of one of them negatively. Many of the combinations tested are harmonious, but disharmonious combinations are also known, these being without any interest for breeding purposes (Kawai 1968, *rice;* Gottschalk 1972b, c; Jaranowski 1977, *peas*). It is not possible to make any predictions in this field. Each gene combination has to be tested individually. Moreover, the genotypic background can play an important role with regard to the harmony or disharmony of the combined gens (see below).

5.2.1 Negative Interactions

In Fig. 9, some examples for negative interactions between desirable genes of the *Pisum* genome are given. The X-ray- and neutron-induced fasciated mutants of our collection show a very high seed production, but they are too tall and too late; moreover, they are homozygous for gene *sg*, causing a considerable reduction of the seed size. The genetic peculiarity of all the mutants of this group consists in the fact that they are homozygous for 12 to 16 mutant genes, most of them being present in a hypostatic state (details see in Sect. 6.3). Theoretically, some of them could be used for improving these mutants. This holds true for three different genes for reduced internode length and for two genes causing earlier flowering and ripening. By crossing the fasciated mutants with nonfasciated genotypes, the respective epistatic genes can be removed and the hypostatic ones just mentioned can manifest their action. For these crosses, recombinant R 46C was used homozygous for genes *efr* for earliness, and *bif-1* for dichotomous stem bifurcation. Gene *efr* is not identical with the hypostatic genes for earlier flowering of the fasciated mutants. Its action is essentially stronger, because the first flowers are already formed at very low nodes of the stem, resulting in 7–10 days earlier flowering than the mother variety, and 10–18 days earlier flowering than the fasciated mutants when grown under German field conditions. By crossing the fasciated mutants with R 46C, it should be possible to produce a great number of different recombinant types having the following favorable traits:

- different kinds of stem fasciation,
- different kinds of reduced plant height,
- earliness,
- normal seed size.

If the genes involved cooperate normally with each other, at least some of these recombinants should have a better breeding value than the parental fasciated mutants, because some of their negative traits have been replaced by positive ones. The main question is whether it is possible to maintain the high seed production of the fasciated mutants in the presence of the other desirable genes just mentioned.

The seed production of 44 different fasciated recombinant types of our *Pisum* collection is compared with that of the fasciated mutants and the early flowering recombinant R 46C in Fig. 9. All the recombinants are homozygous for gene *efr* for earliness, furthermore for at least one of the three of four *fasciata* genes present in the fasciated mutants. They differ from each other with regard to the *Sg* alleles for seed size, and to hypostatic genes for different plant heights, which can display their action because the respective epistatic genes have been eliminated.

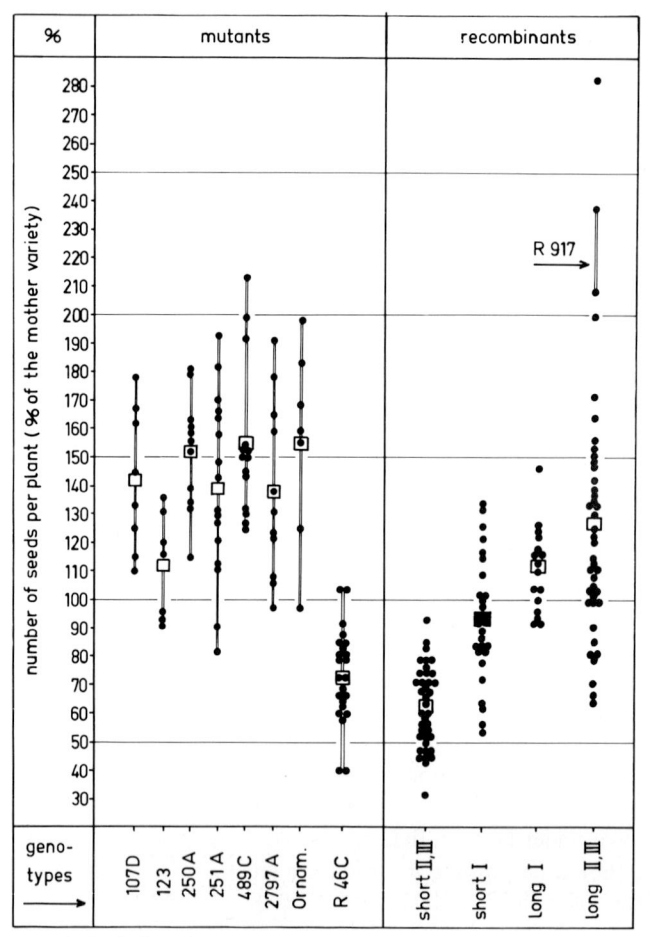

Fig. 9. *Left* Seed production of six fasciated *Pisum* mutants, the fasciated fodder pea variety "Ornamenta" and recombinant R 46C containing gene *efr* for earliness. *Right* The seed production of 44 recombinant lines homozygous for *efr*, for *fasciata* genes and for genes controlling different internode lengths. The material is subdivided into four groups according to the plant height. Each *dot* represents the mean value for the character "number of seeds per plant" for one generation as related to the control value of the mother variety. The *squares* are the total means of each genotype studied

The fasciated mutants outyielded the mother variety considerably in nearly all the generations tested, whereas the early flowering genotype R 46C was not able to reach the level of the mother variety (left-hand part of Fig. 9). The voluminous group of early flowering fasciated recombinants in the right-hand part of the figure has been subdivided into four subgroups according to their plant height. The genotypes of the first subgroup are homozygous for genes *"short III"* or *"short II"*, their stem lengths ranging between 20 and 50 cm. The graph shows that stem fasciation does not result in high seed production in our material, if the *fasciata* genes are combined with genes *"short III"* or *"short II"* for reduced internode length. This holds in principle also true

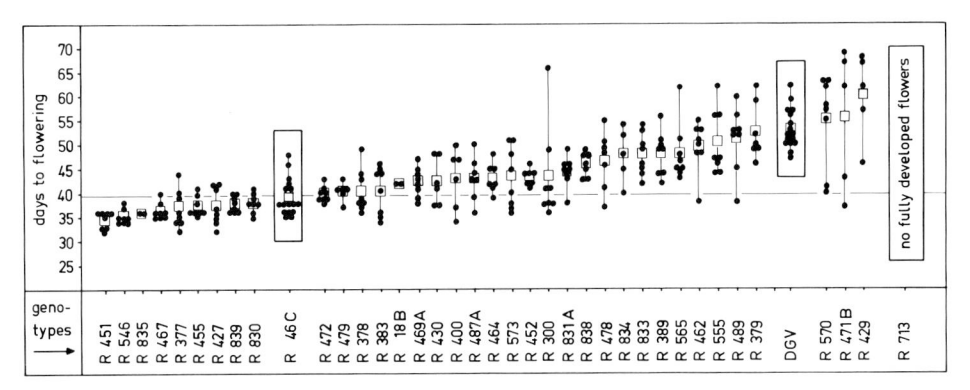

Fig. 10. Flowering behavior of 38 different *Pisum* recombinants homozygous for gene *efr* for earliness, under short-day phytotron conditions. Each *dot* represents the value for a single plant; the *squares* are the mean values for each genotype tested. The recombinants are compared to R 46C, the donor of gene *efr*, and to the later flowering mother variety "Dippes Gelbe Viktoria" (DGV), used as initial line for our irradiations

for the combination of the *fasciata* genes with gene *"short 1"* reducing the internode length only slightly (second subgroup). The recombinants of this subgroup are clearly better than those of the first subgroup, some of them are also better than R 46C, but they do not reach the yielding level of the fasciated mutants. In the third and fourth subgroup, the *fasciata* genes are joined with genes for long or very long internodes. Some of them may be competitive with the fasciated mutants in their yielding properties, but they are not suited for field cultivation because of their long internodes.

These examples demonstrate certain negative interactions between theoretically useful mutant genes which reduce the breeding value of the recombinants produced. Similar cases of a certain degree of disharmony have been found in many other recombinant types of our collection; they seem to be the rule rather than the exception. Gene *ion* for increased number of ovules per carpel or gene *efr* for earliness, for instance, are negatively influenced in their action by distinct other mutant genes or gene groups (Gottschalk 1972b, 1982a, d).

Of particular interest are those cases in which specific interactions between mutant genes are only discernible under distinct climatic conditions. This holds true for some of our *Pisum* genotypes grown in India. The plants of recombinant RM 849 are tall and linearly fasciated; moreover, they are homozygous for gene *efr* for earliness deriving from recombinant R 46C. Both these genotypes were grown in Kurukshetra (North India), and it should be expected that they show in principle the same flowering behavior because of their being homozygous for the same gene *efr*. This is, however, not the case. On the contrary, the plants of RM 849 began flowering 4 weeks later than those of R 46C, the donor of *efr* (Gottschalk and Kaul 1980). This delay can only be due to the negative influence of one of the other mutant genes of RM 849 on the action of gene *efr* which is not observed under the climatic conditions of Germany. Because of these interactions, RM 849 cannot be utilized agronomically in North India in contrast to R 46C. The plants of R 46C entered flowering period in Kurushetra 16 days earlier than the mother variety, but their seed production should be improved.

Stem fasciation is a good possibility for increasing the seed yield. Thus, the plants of the fasciated recombinant RM 849 should have good prerequisites for the union of the two desirable traits, earliness and high yield. One of the consequences of the specific gene combination of RM 849, however, consists in the loss of the trait "earliness," although gene *efr* is present in its genome.

Positive or negative interactions between mutant genes can be easily discerned under controlled phytotron conditions. The flowering behavior of 38 different *Pisum* recombinants is illustrated in Fig. 10. With the exception of the mother variety, all these genotypes are homozygous for gene *efr* for earliness, whereas they differ from each other with regard to other mutant genes or gene groups. Many of the recombinants studied show in principle the same flowering behavior as the parental genotype R 46C, the donor of gene *efr*. In these cases, neither positive nor negative interactions between *efr* and the other genes involved occur. Some recombinants, however, enter flowering period clearly earlier than R 46C, obviously due to positive interactions between *efr* and specific other mutant genes. Many recombinants tested, however, show the opposite reaction: they begin flowering later than R 46C due to a negative influence of distinct genes on the action of *efr*. Some of these genotypes are even later than the not early flowering mother variety. The plants of recombinant R 429, for instance, are homozygous for genes *efr* and "*short* I" (reduced plant height); they flowered 3 weeks later than R 46C. They formed – like R 46C – minute floral buds at very low nodes of the stem, but these buds were not developed into flowers. Functionable flowers arose only from buds which had been produced during an essentially later stage of ontogenetic development.

Of particular interest in this concern are those recombinants which are completely unable to form functionable flowers under short-day conditions. So far, four different genotypes, showing this behavior, were selected in our phytotron trials, one of them being considered in Fig. 10 (R 713). The tendency to produce tiny buds without further development in short-day is already observed in R 46C, but to a very weak degree. This tendency is strengthened under the influence of other mutant genes in a way that either lateness occurs or that the development of functionable flowers is completely suppressed. This happens, however, only under short- and not under long-day conditions. Further details on these problems have been published elsewhere (Gottschalk 1981c, 1982a).

The complexity of the interactions between different genes and different environmental factors becomes especially clear in recombinant R 20D of our *Pisum* collection selected after having crossed the strongly fasciated mutant 489C with the early flowering recombinant R 46C. The plants of this genotype contain the following mutant genes:

– *efr* for earliness (from R 46C),
– *bif-1* for dichotomous stem bifurcation (from R 46C),
– a gene for stem fasciation (from 489C),
– a gene for very long internodes (hypostatic in 489C),
– a gene for a slight reduction of the chlorophyll content (from 489C).

These plants, homozygous for five mutant genes, are very vigorous and highly productive. Their mean values for the trait "number of seeds per plant" ranged between 168%

and 269% of the corresponding means of R 46C in yielding analyses of five subsequent generations. In spite of this strong increase in yield, the recombinant R 20D does not represent an improvement of R 46C. First of all, the plants are too tall, therefore, they are not suited for field cultivation. But also with regard to the flowering and ripening behavior, R 20D is inferior to R 46C. Begin of flowering of the two genotypes is about equal, but in R 20D, almost no pods and seeds are produced from the lower flowers of the plants. Thus, R 20D is not able to realize its favorable genetic potentialities. This becomes particularly clear from the harvesting times. In 1981, the seeds of R 46C were harvested 5 to 6 days earlier, those of R 20D 2 to 3 days later, than those of the mother variety which is a delay of 7 to 8 days in relation to R 46C. Thus, these plants are genetically early due to the presence of gene *efr*, but in practice they are late, because *efr* is unable to manifest its normal action in the presence of the other mutant genes of R 20D. Besides this negative influence, however, a positive interaction between the genes involved occurs. A peculiarity of gene *bif-1* for stem bifurcation, deriving from R 46C, is its strongly reduced penetrance. All the plants of this recombinant line are bifurcated and the high seed production is mainly due to this gene action. The stabilization of its penetrance can only be caused by one of the other mutant genes present in the genome of R 20D.

But the situation is much more complicated. So far, only the interactions between the mutant genes present in R 20D were considered. For judging the general usefulness of the recombinant, we have also to consider its response to different climatic conditions. The photoperiod is a particularly important environmental factor influencing the flowering behavior of many genotypes specifically. Therefore, the two genotypes were grown under controlled long- and short-day phytotron conditions. In long-day (17 h light, 6 h darkness, twice half-an-hour dawn), the plants of R 46C started flowering about 10 days earlier than the mother variety, thus demonstrating a clear manifestation of the action of gene *efr* under the ecological conditions offered to it. Begin of flowering of R 20D, however, was 18 days later than that of R 46C, although both the genotypes are homozygous for gene *efr*. The induction of flowering was similar in the two recombinants, but they differed strongly from each other with regard to the formation of fully developed flowers. Under long-day field conditions, the plants of R 20D entered flowering period 2 days later than those of R 46C, under long-day phytotron conditions, however, 18 days later. Thus, the photoperiod is not the only environmental factor influencing the flowering behavior of the two genotypes. Another factor, probably temperature, plays an important role in this concern. Interestingly, the differences in the behavior of the two genotypes are even stronger in short-day (11 h light, 12 h darkness, twice half-an-hour dawn). Under these conditions, the plants of R 46C flowered 9 days earlier than the mother variety, whereas those of R 20D did not produce any functionable flowers. There were exclusively tiny floral buds which did not undergo any further development. If we consider the total reaction of the two genotypes, it becomes clear that R 20D has more disadvantageous characters than R 46C, in spite of its high seed production.

In order to eliminate one of the negative traits of recombinant R 20D, we have replaced the gene for long internodes by a gene for short ones, which is likewise present in the fasciated mutant 489C in a hypostatic state (gene *"short II"*). In this way, the recombinant R 20E arose being genetically identical to R 20D with the exception of the genes for plant height just mentioned. The seed production of these short plants is considerably lower than that of R 20D, but about twice as high as that of the early flowering recombinant R 46C used as one of the parents of the crosses. This relatively good yield may be mainly due to the fully stabilized penetrance of gene *bif-1*, which is not only observed in R 20D, but also in R 20E. In total, however, the plants of recombinant R 20E show the negative peculiarities of R 20D in an even stronger form. Under German field conditions, they started flowering 4 days later than R 46C, but seed harvesting was 12 days later. No pods and seeds were produced from the lower flowers of the stems. Under long-day phytotron conditions, begin of flowering was 26 days later than in R 46C and there was no formation of normally developed flowers in short-day.

These are two interesting examples for demonstrating both the interactions between different mutant genes and between genes and distinct environmental factors with regard to the manifestation of specific gene actions. Further problems of gene ecology are discussed in Chap. 18 of the present book.

5.2.2 Positive Interactions

So far, preferably negative interactions between mutant genes were discussed. This does not mean, however, that the combined action of mutant genes leads necessarily to unfavorable effects. On the contrary, an increasing number of findings is available demonstrating thereby that it is in principle possible to combine several mutant genes without decreasing the selection and the breeding value of the respective recombinants. Distinct *barley* recombinant lines, homozygous for two or three mutant genes, were found to exhibit the same average yield in several subsequent years as the parental mutants (Scholz 1965, 1967). Some Norwegian *barley* recombinants outyield not only the parental mutants, but even the best market varieties (Berg et al. 1976). Very favorable results were obtained in *Triticum durum* (Fig. 6, Scarascia Mugnozza et al. 1972). The efficiency of this method becomes evident through the fact that the Italian cultivar "Augusto" of *durum* wheat was developed using two neutron-induced mutants (Bagnara and Porreca 1977). Two *rice* mutants of a high-yielding Indian variety have fine grains. They were crossed with each other and a recombinant selected was found to be very promising because of overall better agronomic performance in combination with fine grains (Kaul 1981).

Some of our *Pisum* recombinants, selected after having crossed the fasciated mutants mentioned above with nonfasciated genotypes, were found to be superior to their parental mutants with regard to specific characters (for details see Gottschalk 1972a, c, 1977a, 1982d; Gottschalk and Hussein 1975, 1976; Hussein and Gottschalk 1976; Gottschalk and Bandel 1978; Bandel and Gottschalk 1978; Gottschalk and Müller 1979). Figure 11 shows in a summarizing way that we were able to develop quite a number of short-stemmed recombinant lines of *Pisum* which are equal to or better than the mother variety in seed production. The short-stemmed mutants, available in our collection, are inferior to the mother variety in this respect. Thus, distinct gene combinations resulted in an improvement of this category of genotypes. Positive interactions were also observed between gene *efr* for earliness and distinct other mutant

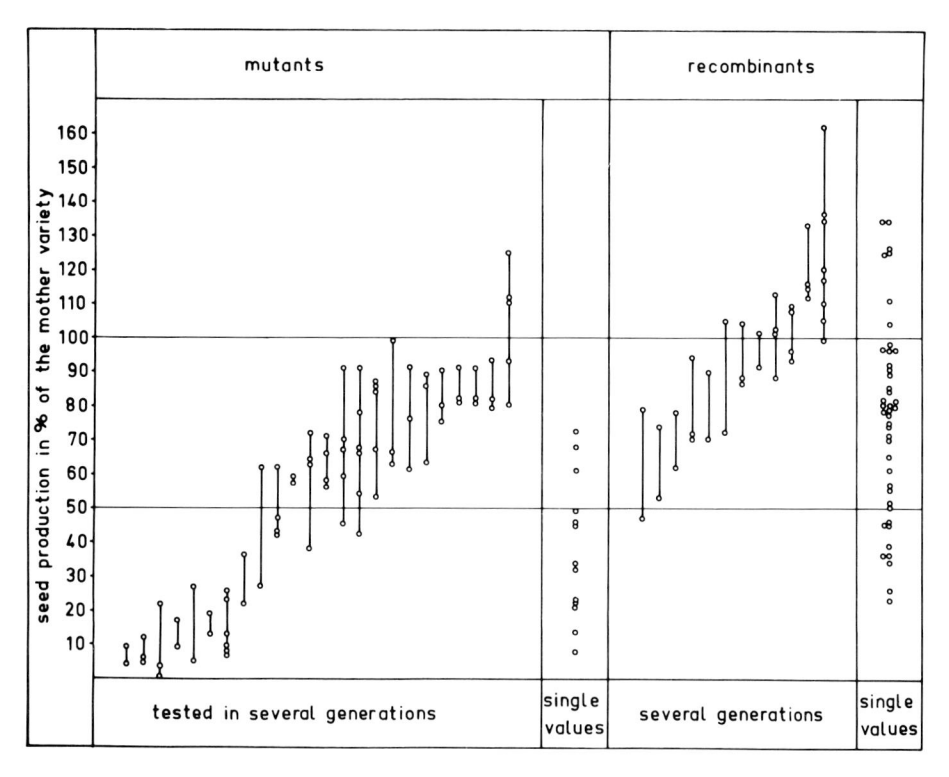

Fig. 11. Seed production of 37 short-stemmed mutants and of 55 short-stemmed recombinants of *Pisum sativum* as related to the corresponding values of the mother variety. Each *dot* represents the mean value for the character "number of seeds per plant" for one generation. Means of genotypes tested in several generations are connected by *vertical lines*. From the remaining genotypes, means of only one generation are available so far. The yield of the recombinants is essentially better than that of the mutants

genes of our collection (Gottschalk 1981c, 1982a, d). The stabilization of the penetrance of unstable genes under the influence of other mutant genes is likewise a positive effect. This phenomenon was mentioned in Sect. 5.2.1 for the *Pisum* gene *bif-1* for stem bifurcation resulting in a considerable increase of the seed production (recombinants R 20D, R 20E). Further details concerning this problem are given in Chap. 17.

An interesting example for the role of the genotypic background in this respect refers to the yielding capacity of a specific *Pisum* recombinant. By crossing the leaf mutants *acacia* (leaflets instead of tendrils, gene tl^w) and *afila* (tendrils instead of leaflets, gene *af*), the so-called *pleiofila* type is obtained having hundreds of tiny leaflets and no tendrils at all (Fig. 16). The plants are homozygous for tl^w and *af*. They differ so extremely from normal peas that they can hardly be discerned as belonging to the species *Pisum sativum* in the nonflowering condition. In the genotypic background used for the two genes in our experiments, the *pleiofila* recombinant has a very low seed production, much less than that of the two parental mutants (Gottschalk and Kaul 1973). The same recombinant was obtained in Poland by crossing the mutants *"viciaefolialis"* and *"cirrhifolialis"* of *Pisum arvense* (obviously identical with *acacia* and *afila*). Some of these recombinant types were found to be very vigorous and high-yielding, producing up to 480 seeds per plant (Jaranowski 1976, 1977). In spite of these favorable yielding potentialities, the *pleiofila* recombinant is not used in pea breeding because of the lack of tendrils, i.e., the lack of any possibilities for stem stiffness. It can, however, be assumed that a similar behavior is realized in certain recombinants of agronomic value. The differences in the effectiveness between the Polish and the German *pleiofila* material can only be due to the divergent genotypic backgrounds in which the two genes express their action.

6 The Alteration of the Shoot System by Means of Mutations

A very large proportion of genes of the genome of higher plants controls the organization of the shoot system. This becomes obvious from the large number of mutants with alterations in this plant organ which arise in mutation treatments. In our radiation genetic experiments in *garden peas*, about 30% of all genotypes selected show any kinds of anomalies in the stem. The reduction of the plant height results often in an improved lodging resistance and is therefore of considerable agronomic importance in *cereals*, to some extent also in *ornamental plants*. Specific alterations of the stem, such as bifurcation or fasciation, are interesting characters in distinct crops because they lead to an increased seed production.

6.1 Mutants with Reduced Plant Height: Erectoides Types, Semidwarfs, Dwarfs

Mutations with reduced plant height belong to the most frequently arising genotypes in mutation treatments. They do not represent a uniform group, neither morphologically nor anatomically or genetically. The reduced plant height of *Pisum* mutants, for instance, can be due to:

— The reduction of the number of internodes.
— The reduction of the length of internodes being either due to the reduction of cell length, while cell breadth is not altered, or to the reduction of cell number (Weber and Gottschalk 1973).

Moreover, the genetic background of these genotypes varies not only from crop to crop but even within the same crop. The mutant genes of short-stemmed genotypes can be dominant, recessive, or intermediate. Therefore, only a few generally valid remarks can be made. In *cereals,* the reduction in culm length is often combined with an improved straw stiffness resulting in an increased lodging resistance. This is the main advantage for the utilization of these mutants in plant breeding. There are so many details given in the literature that only some of them can be mentioned in this review paper. References and some remarks on the yielding properties of *erectoides* mutants, semidwarfs and dwarfs are given in Tables 4—9. A review on the usefulness of induced mutations for breeding lodging-resistant cereals has been published by Scarascia Mugnozza (1965). Further information on the use of dwarfy mutants are found in the proceedings of a symposium organized by the Insitute of Radiation Breeding in Ohmiya, Japan (1979b; *Gamma Field Symposia 16*).

Table 4. Examples for short-stemmed *barley* mutants

Mutant type	Authors	Mutagens	Remarks
Erectoides mutants	— Moës 1965	X rays	2 semidominant mutants with higher yield
	— Gustafsson et al. 1966; Gustafsson 1969	X rays, neutrons, DES, EI	
	— Hänsel 1966	X rays	5% less yield
	— Pollhamer 1966, 1967	X rays	Some *erectoides* mutants with higher yield
	— Gustafsson and Ekman 1967	X rays	Commercial variety "Pallas barley"
	— Hagberg 1967; Persson and Hagberg 1969		Cytogenetics of *erectoides* mutants
	— Scholz 1967, 1971	X rays, EMS	Normal yield, used in crossbreeding
	— Sethi et al. 1969	^{32}P, ^{35}S	
	— Donini and Devreux 1970	Gamma rays	
	— Gaul and Grunewaldt 1971; Grunewald 1974a		Studies on pleiotropism
	— Sethi 1975	^{32}P, ^{35}S, EMS	Normal yield
	— Haahr and v. Wettstein 1976	Neutrons	Lodging resistance, high-yielding dwarf
	— Abdalla et al. 1980	Gamma rays	Partially overdominance
Semidwarfs	— Yamashita et al. 1972	Gamma rays	
	— Kivi et al. 1974	Gamma rays	
	— Nettevich and Tzukanov 1976		Variety "Fakel"
	— Ali et al. 1978	Rays	Incorporated into crossbreeding
Dwarfs	— Kumar et al. 1967		23 dwarfs
	— Sethi et al. 1969	^{32}P, ^{35}S	
	— Chandola et al. 1971	Neutrons	Higher yield
	— Bansal 1972	Gamma rays, neutrons, EMS, NMU	Early flowering dwarfs with normal yield
	— Sethi 1974, 1975	EMS	Fully fertile
	— Yamaguchi et al. 1974	DES	
	— Haahr and v. Wettstein 1976	Neutrons	Very good grain yield
	— Konishi 1976	EMS	8 dwarf genes; 3 lines competitive to the mother variety in yield
	— Stephanov and Gorastev 1976; Stefanov et al. 1978	Gamma rays, EMS, DES	
	— Gorny 1978	NMU	

6.1.1 Barley

The huge collections of barley mutants available in many countries are intensively studied with regard to the genetics, morphology, and yielding ability of the short-stemmed mutants (Table 4). Mainly the *erectoides* mutants have to be mentioned in this context. In Sweden, more than 700 independently arisen *erectoides* mutants were analyzed genetically. At least 26 different loci of the barley genome were found to determine the *erectoides* habit of the plants (Gustafsson 1969; Persson and Hagberg 1969). Some of them show the tendency to react in a specific way to different mutagens with regard to their mutation frequency. All the *erectoides* genes display a pleiotropic action influencing the number and length of culm and spike internodes, the kernel length, the size of the sclerenchyma cells and the root system among others. The number of internodes is generally reduced. Moreover, the length of the basal internode is reduced, whereas that of the highest or the uppermost two internodes is increased. The spike internodes are shorter. The single traits of this pleiotropic complex can be altered to some extent when the *ert* genes are transferred into specific genotypic backgrounds (Gaul and Grunewaldt 1971; Grunewaldt 1974a). The morphology of a great number of *erectoides* mutants and their cytogenetic situation have been described by Persson and Hagberg (1969); some physiological aspects are discussed by Hänsel (1966).

The grain yield of most *erectoides* mutants is less than that of the mother varieties. Some genotypes of this group, however, are equal to or even better than the control material in respect of the lodging resistance (Moës 1965; Pollhamer 1967; Scholz 1971). An X-ray-induced Swedish *erectoides* mutant was developed into the commercial variety "Pallas barley" which is widely grown in Western Europe. It shows a pronounced lodging resistance and a high productivity (Gustafsson and Ekman 1967). Some other *erectoides* mutants are incorporated into crossbreeding programs (Scholz 1967). General questions of the usefulness of these genotypes for barley breeding have been discussed by Gustafsson (1963, 1969); Scarascia Mugnozza (1965); Hagberg (1967).

Besides the typical *erectoides* mutants, other groups of short-stemmed genotypes are obtained in mutation treatments. Some of them are not yet studied in detail and may belong to the *erectoides* group. Others, however, differ morphologically in a characteristic way from these types. Eight pleiotropic dwarf genes were obtained in Japan, causing a reduction of internode length but not internode number. Some of them improve, others impair the stability of the stem. Two of the respective mutant lines have the same grain yield as the initial line. Another one outyielded the standard under specific ecological conditions (Konishi 1976). In India, a fully fertile dwarf mutant is available having a longer peduncle than the initial line (Sethi 1974, 1975). A promising mutant feeding barley strain is available in Finland showing shortness, straw stiffness, profuse tillering, slow vegetative development, but rapid ripening. It is well adapted to the northern climatic conditions (Kivi et al. 1974).

Some short-stemmed barley mutants have been developed into commercial varieties:

- The variety "Pallas" in Sweden. It has successfully been used for developing some other commercialized varieties which are widely grown in North and Western Europe (Gustafsson et al. 1971; see Chap. 5).
- The variety "Gamma No. 4" in Japan (Matsuo and Yamaguchi 1967).
- The variety "Diamant" in Czechoslovakia (Bouma 1967). The plants of this variety show good lodging resistance even on irrigated soils of high fertility. Moreover, their grain has a good malting and brewing quality.
- The variety "Luther" in the United States (Nilan 1966, 1972). It has an improved winterhardiness besides the good lodging resistance and is the highest-yielding winter barley grown in the coastal areas of the northwestern states of the U.S.A. Moreover, it has successfully been used in crossbreeding, giving rise to even more improved strains.

Further promising dwarf mutants with good yielding properties and high resistance to lodging are available in Denmark (Haahr and v. Wettstein 1976) and Bulgaria (Stefanov et al. 1978). The latter have been selected in a crossbreeding program.

Further details on lodging-resistant barley mutants were published by Jech (1966a); Trofimovskaya and Zhukovsky (1967); Shevtzov (1969b); Chandola et al. (1971); Nettevich and Sergeev (1973); Krausse et al. 1974); Enchev (1976). A cytologically unstable barley mutant yielded a genetically heterogeneous progeny, giving rise to some strains with improved lodging resistance. One of them had normal grain yield (Ibrahim et al. 1966).

6.1.2 Rice

In *Oryza sativa*, a great number of mutants with reduced culm length is available, preferably in the collections of some Asiatic countries (Table 5). Some of them are similar to the barley *erectoides* mutants, but the group just mentioned is also heterogeneous in rice. The dwarfism or semidwarfism is often combined with a reduction of the internode number. Surprisingly, some short-culm mutants have an increased number of internodes (Kawai and Narahari 1971). The internode length is either generally decreased or this alteration is limited to specific internodes of the plants. This holds for instance true for a strong reduction of the basal internodes, which may be one of the main factors responsible for the improved lodging resistance (Joshua et al. 1969). A characteristic feature of many short-culm rice mutants is their increased degree of tillering, which often contributes to the increased seed production of the plants (Bose 1968; Miah and Bhatti 1968; Misra et al. 1971; Reddy and Reddy 1973a, b; Reddy et al. 1975; Fig. 7). The genetic basis of the group seems to be polymery, perhaps not to such a high extent as in barley. The dwarfism or semidwarfism of some genotypes is obviously jointly controlled by two recessive genes (Seetharaman and Srivastava 1971; Roy and Jana 1975). At least three nonallelic semidwarfing genes have been selected in irradiated Californian material (Mackill and Rutger 1979).

The seed yield of many dwarf or semi-dwarf rice mutants is lower than that of their initial lines, for different reasons. In a Japanese mutant, it is due to the reduction of the number of spikelets per panicle, while other factors of agronomic interest are not influenced by the gene (Narahari 1969). In other genotypes of this group the grain size is reduced. There are, however, some reports on short-culmed rice mutants which are not inferior to their mother varieties with regard to seed yield (Ram 1974; Kaul 1978; Okuno and Kawai 1979). In Japan, ten promising mutant lines with stiff straw, high yield, and excellent cooking quality of the grain were selected from a mother variety with low lodging resistance (Samoto 1975). In some mutants the dwarfism is combined with earliness (Hu 1973) or with an increased protein content (Misra et al. 1971). The best short-culm mutant of a French collection had a better lodging resistance and a 50% increased yield, but the plants were 6 days later and the milling quality of the grains was insufficient (Marie 1967). Of considerable interest are some Taiwanian semidwarfs. They have lost their photoperiodic sensitivity. Thus, it may be possible to grow one more crop within a year in the respective regions (Hu 1973). The Japanese rice variety "Reimei," developed from a gamma-ray-induced mutant, is short-culmed and more lodging resistant than its mother variety. Moreover, it is protected against damage by coolness due to low position of the ear primordia (Kawai 1967). Some

Table 5. Lodging resistant rice mutants

Authors	Mutagens	Remarks
– Kawai 1967; Matsuo and Yamaguchi 1967	Gamma rays	Commercial variety "Reimei"
– Marie 1967, 1970	X rays, gamma rays, EMS	Short culm, 50% increased yield, late
– Bose 1968	^{35}S	normal yield
– Li et al. 1968; Hu et al. 1970; Hu 1973	X rays, gamma rays, neutrons, EMS	Normal yield; included in crossbreeding, 35% shorter
– Miah and Bhatti 1968; Miah et al. 1970	Gamma rays	22 short-culm mutants; many of them with increased yield
– Siddiq and Swaminathan 1968; Swaminathan et al. 1970	EMS, NMU	Semidwarfs of economic value, dwarfs
– Tanaka 1968, 1969	Gamma rays	Many short-culm lodging-resistant mutants
– Joshua et al. 1969 Gopal-Ayengar et al. 1971	Gamma rays, fast neutrons	A completely nonlodging dwarf mutant with normal yield
– Majumder 1969	X rays, ^{32}P	Tolerant to high nitrogen doses
– Narahari 1969	X rays	Reduced yield
– Kawai and Narahari 1971	^{32}P	68 short-culm mutants
– Misra et al. 1971	EMS, DES, EI, NMU	Short-culmed, stiff-strawed mutant with higher yield
– Haq et al. 1971	Gamma rays	High-yielding semidwarfs
– Li et al. 1971	Rays, chemicals	Semidwarf successfully used in crossbreeding
– Pawar 1971	Gamma rays, fast neutrons	
– Reddy and Reddy 1971, 1973; Reddy et al. 1975; Reddy and Padma 1976; Padma and Reddy 1977	DES, EMS	Large number of dwarfs and semidwarfs, some with higher yield
– Ram 1974	Gamma rays	Increased yield under high-fertility conditions
– Roy and Jana 1975	EMS, EI	Controlled by duplicate genes
– Samoto 1975	Gamma rays	10 promising lines selected
– Rutger et al. 1976	Gamma rays	25 cm shorter, promising
– Okuno and Kawai 1977, 1979	X rays, gamma rays	6 short-culm mutants; yield comparable to mother variety
– Kaul 1978	Gamma rays, EMS, DES	Lodging-resistant dwarfs with high yield
– Dasgupta 1979	X rays	
– Mackill and Rutger 1979	Gamma rays	3 nonallelic semidwarfing genes
– Afsar Awan et al. 1980	Sodium azide	
– Sampath and Jachuk 1969	DES, EI	*Oryza rufipogon;* improved tillering capacity
– Misra et al. 1971		*Oryza rufipogon*

early ripening Indian semidwarf mutants can successfully be used for commercial cultivation because of their favorable peculiarities (Reddy and Reddy 1971; Reddy et al. 1975). A number of promising lines were developed in Taiwan from *erectoides* mutants after they had been crossed with local varieties (Li et al. 1968, 1971; Hu et al. 1970).

In wild rices, the so-called *Spontaneas,* belongig to *Oryza rufipogon,* some short-culm mutants with fine grains, erect leaves, and improved tillering capacity are available in Assam. They perform well in waterlogged areas which are less suitable for short types of *Oryza sativa* (Sampath and Jachuck 1969).

6.1.3 Bread and Durum Wheat; Other Gramineae

Semidwarfism of *Triticum aestivum* is a very complex phenomenon, the genetic basis of which has been discussed by Konzak (1976). It can be due to single recessive, dominant, or semidominant genes; moreover, epistatic interactions are known in this group of genotypes. The genes show in general a pleiotropic action which can be modified under the influence of specific genotypic backgrounds. There are at least several, perhaps many loci of the genome involved but the genetic analysis is still in its very beginnings. The agronomic advantage of the *erectoides* mutants of wheat consists in their improved lodging resistance, especially when the supply of nitrogen in the soil is raised. In contrast to barley, only a very small number of *erectoides* mutants is used in bread wheat, essentially in crossbreeding programs. This may mainly be due to the fact that six widely used genes of natural origin are available in *Triticum aestivum* which reduce the height of the culm in the desired manner. The usefulness of *erectoides* mutants in breeding for lodging resistance was reviewed by Scarascia Mugnozza (1965). Some details are given in Table 6.

As in barley and rice, semidwarf bread wheat mutants are often inferior to their mother varieties with regard to seed yield (as an example, see Djelepov 1973). Lodging-

Table 6. Examples for *erectoides* mutants in bread and *durum* wheat

Species	Authors	Mutagens	Remarks
Triticum aestivum	— Khvostova et al. 1965; Elshuni and Khvostova 1966; Khvostova 1967	Gamma rays, neutrons, EI	Lodging resistant with normal or higher yield
	— Jech 1966b	Gamma rays	7 lodging-resistant mutants
	— Borojević 1967	X rays, neutrons	
	— Goud 1967	Gamma rays, neutrons, EMS	
	— Woo and Konzak 1969	EMS	
	— Wagner 1969	X rays, EMS	
	— Djelepov 1973, 1978	Gamma rays, EMS	High productivity
	— Khan 1973	Neutrons	7 dwarf mutants
	— Kumar 1976, 1977b	Gamma rays	Some erectoides mutants with improved yield
	— Barabás 1977	X rays	Dwarf with high protein content
	— Bovkis 1978	X rays	
Triticum durum	— Scarascia Mugnozza 1966a, b, 1967; Scarascia Mugnozza et al. 1965, 1966; Pacucci et al. 1966; Mosconi 1967	X rays, neutrons, EMS, DES, EI	Many short-culmed mutants with increased lodging resistance
	— Konzak et al. 1968	EMS	
	— Bagnara et al. 1971	Scarascia's material	Used in crossbreeding
	— Rossi and Bagnara 1972	Neutrons, EMS	788 short-straw mutants
	— Bozzoni 1965, 1974a	—	Used in crossbreeding

resistant bread wheat mutants with normal or higher yield and other valuable characters were reported by Khvostova et al. (1965); Kumar (1976); and Barabás (1977). An Indian semidwarf mutant is normal with regard to tillering and number of grains per spike, but the number of spikelets per spike is markedly reduced. This disadvantage, however, is compensated by a significantly increased kernel size. In M_5, the mutant gave more than 50% higher yield than the control. Also the total protein content is higher (Kumar 1977b). In Bulgaria, some short-culm mutants are just developed into new cultivars (Djelepov 1978).

In *Triticum durum,* especially the work carried out by Scarascia Mugnozza and co-workers in Italy has to be mentioned (Table 6). By studying 788 short-straw mutants in the M_3 generation, it became obvious that most of these genotypes show a series of other alterations in morpholoy, in the shape, size, and color of spikes and grains, in the chlorophyll apparatus, and the fertility. These by-effects are the higher, the stronger the culm reduction is (Rossi and Bagnara 1972). In this respect, a similar situation is realized as in the early ripening *durum* wheat mutants mentioned in Chap. 10. Nearly 300 mutant lines were tested at two Italian locations in M_4–M_6 by analyzing their lodging resistance. Some shorter and more lodging-resistant genotypes were found to be of direct agronomic value because of their favorable yielding properties (Scarascia Mugnozza 1966a, b, 1967; Fig. 7). By incorporating some of them into

Table 7. Examples for lodging-resistant mutants in other cereals and in *Lolium*

Species	Authors	Mutagens	Remarks
Avena sativa	– Votava and Nasinec 1974 – Cummings et al. 1978 – Velikovsky 1978	X rays EMS X rays	Dwarfs with increased lodging resistance stiff straw, compact panicle, upright leaves
Secale cereale	– Kivi et al. 1974 – Muszynski et al. 1976 – Kuckuck and Peters 1979	Gamma rays Neutrons, NEU X rays	Short straw, good winter hardiness Many short-straw mutants From *S. cereale* × *S. vavilovii;* short-straw mutants with long ears, some of them dominant
Zea mays	– Popova 1975 – Bálint et al. 1977	EMS, NMU Gamma rays, neutrons, EMS	Short-stemmed mutants Mutants with increased lodging resistance
Sorghum vulgare	– Hadley et al. 1965 – Goud 1972	 Gamma rays	Lower yield than tall mutants obtained from dwarf mutants
Sorghum bicolor	– Singh and Drolsom 1977	DES	Reduced plant height, low seed production
Lolium multiflorum, Lolium temulentum	– Malik and Mary 1971, 1974	EMS, MMS, DES	

crossbreeding, hybrid lines with a favorable combination of semidwarfism, lodging resistance, earliness, and high yield were obtained (Bozzini 1974a). Moreover, high-yielding strains with drastically shortened culm without any lodging were selected after having intercrossed short culm mutants (Bagnara et al. 1971).

The increased lodging resistance in *Triticum durum* is not in all the cases combined with short culms. Some high-yielding mutants with improved resistance are equal to the mother variety with regard to plant height (Scarascia Mugnozza et al. 1966).

In some other cereals *(oats, maize, millets)*, dwarfy mutants have been selected in small proportions (Table 7). They could be of practical value in *winter rye* in Finland, because only long-strawed varieties show the winterhardiness necessary for good productivity. These varieties are, however, poorly adapted to mechanical harvesting. Short-strawed mutants with improved lodging resistance and good winterhardiness were obtained in gamma-ray treatments (Kivi et al. 1974). Dwarf mutants of *Sorghum vulgare* were used in mutagenetic treatments in the United States, giving rise to tall genotypes which outyielded their parental mutants because of producing more tillers and ears (Hadley et al. 1965).

Table 8. Examples for short-stemmed and/or lodging-resistant mutants in *Leguminosae* and *Solanaceae*

Species	Authors	Mutagens	Remarks
Pisum sativum	– Khvostova 1967; Sidorova et al. 1969	Gamma rays	High-yielding mutants with robust stem
	– Enken 1967	Gamma rays, EI	Different dwarf types
	– Monti and Scarascia Mugnozza 1970, 1972	DES	Several lines; shorter, earlier, and more productive
	– Gottschalk 1964; Gottschalk and Hussein 1975, 1976; Gottschalk and Bandel 1978; Hussein and Gottschalk 1976; Bandel and Gottschalk 1978	X-rays, neutrons	Many short-stemmed mutants and recombinants, some of the latter with high yield
Glycine max	– Khvostova 1967	Gamma rays	Lodging resistant
	– Enken 1967	Gamma rays, EI	Erectoides types
	– Rubaihayo 1976	Gamma rays	Reduction of the number of internodes
Vigna trilobata	– Subramanian 1980	Gamma rays	Bushy; richly branched, increased seed production
Cicer arietinum	– Athwal et al. 1970	Gamma rays	Seed setting only slightly reduced
Arachis hypogaea	– Patil and Thakare 1969	X rays	2 short mutants with normal yield
Lycopersicon esculentum	– Monti 1972	Gamma rays, neutrons	No tutoring necessary
Capsicum annuum	– Daskaloff 1978	X rays, gamma rays	

6.1.4 Dicotyledonous Crops

Short-stemmed and/or lodging-resistant mutants are of particular importance in cereals. But in other crops also, some genotypes of this group are of interest, whereas most of them cannot be utilized for breeding purposes because of their strongly reduced seed production. Examples are given in Table 8.

In *legumes,* short-stemmed and/or lodging-resistant mutants of *Pisum sativum* have been intensively studied. Short-stemmed strains with good yielding properties are available in Russia (Khvostova 1967; Sidorova et al. 1969) and Italy (Monti and Scarascia Mugnozza 1970, 1972). A great number of dwarfs and semidwarfs of our own collection were tested with regard to seed production over a period of several years. Figure 9

Table 9. Examples for short-stemmed mutants in other dicotyledonous crops and in fruit trees

Species	Authors	Mutagens	Remarks
Brassica juncea	– Nayar and George 1969	X rays	Short, early flowering, reduced yield
	– Narain and Prakash 1969	Gamma rays	Dwarf with higher yield
Ricinus communis	– Ankineedu et al. 1968	Gamma rays, neutrons	2 dwarf mutants with higher yield per unit area
	– Kulkarni 1969	X rays, neutrons	Dwarfs with normal yield
Corchorus olitorius	– Thakare et al. 1973	Gamma rays, neutrons	1 stiff-stem dwarf, nonlodging
Cryptomeria japonica	– Kukimura et al. 1975	Gamma rays	
Coffea arabica	– Moh 1973	Gamma rays	Shortened internodes; fruit clusters grow compactly along the fruit branches
Malus sylvestris	– Visser et al. 1971; Visser 1973	X rays	Shoots thicker; short internodes, compact growth
	– Nishida 1973	Gamma rays	
	– Zagaja and Przybyla 1973, 1976	Gamma rays	Trees with shortened internodes
	– Poll 1974	Gamma rays	Shorter shoots, thicker shoot diameter
	– Lacey and Campbell 1979	Gamma rays	Many dwarf or compact types; 24 listed for possible commercial introduction
Pyrus communis	– Visser et al. 1971; Visser 1973	X rays	
Prunus avium	– Visser 1973	X rays	
	– Donini 1976	X rays	
	– Milenkov 1978	Gamma rays	5 dwarfs of practical value
Prunus cerasus	– Zagaja and Przybyla 1976	Gamma rays	
Olea europaea	– Donini and Roselli 1972; Donini 1976	Gamma rays	Short internodes; loss of apical dominance

shows that only one of them reaches the capacity of the mother variety. The situation is, however, better in short-stemmed recombinants derived from mutants after they have been crossed with other genotypes (right-hand part of the graph). A bushy mutant of *Vigna trilobata* with a high number of branches was found to have a considerably better seed weight per plant than the control material. This holds also true for a pyramid-shaped genotype of *Vigna aconitifolia* (Subramanian 1980). A brachytic *tomato mutant* could be useful for intensive cultivation. The plants are only 35 cm in height and do not need any tutoring. Under conditions of close spacing, their fruit production is about equal to that of the mother variety (Monti 1972). In *Brassica juncea* a dwarf, having only about half the length of the initial line, was found to outyield the mother variety by 5% due to its lodging resistance and increased number of pods per unit area (Narain and Prakash 1969). Of particular interest are two Indian dwarf *castor* mutants showing a higher yield per unit area because of their lesser space requirements. One of them, having a stem height of only one-third of that of the mother variety, is in addition more resistant to the capsule borer and the gall fly (Ankineedu et al. 1968). A dwarfy, compact-branched *sesame mutant* showed 15%–20% increase in yield considering tests over 4 years. Because of its favorable properties, the mtuant has been released in Orissa/India (Panda 1981).

In *fruit trees*, compact mutants with shortened internodes giving the so-called spur types can be of great practical value because these characters facilitate fruit harvesting considerably. Such clones are available in *apple, pear, sweet,* and *sour cherry,* and in *olives* (Table 9). In *apple,* nearly 3000 dormant shoots have been treated with gamma rays in Poland, giving rise to 44 trees with significantly shortened internodes. From this material, six clones of practical value were selected (Zagaja and Przybyla 1973). A Dutch semicompact clone of the variety "Belle de Boskoop" with smooth-skinned, bright-red fruits is likewise of interest for apple breeding (Visser 1973). In England, 272 mutants of Cox's Orange Pippin apple were obtained by gamma irradiation, most of them being dwarfs or compact types without any practical value. Twenty-four of them, however, were listed for possible commercial introduction (Lacey and Campbell 1979). Of particular interest are certain *olive* mutants with short internodes. They have lost the apical dominance and show a pronounced growth of lateral shoots (Donini 1976).

A great number of further examples of this kind has been discussed and summarized by Broertjes and van Harten (1978). Dwarfy genotypes are of increasing interest in horticultural plant breeding. Potentially useful mutants of this group have recently been obtained in *Kohleria* (Parliman and Stushnoff 1979).

6.2 Mutants with Increased Plant Height

Increase of the plant height is likewise not a uniform phenomenon. It is in many crops fundamentally due to an initial increase in internode length, sometimes accompanied by an increase in internode number. Moreover, length of the cells is increased, in some genotypes also their number per unit area (Weber and Gottschalk 1973; *Pisum sativum*).

Tallness is generally an undesired trait, because the stability of the stem is negatively influenced. Therefore, tall mutants are only in exceptional cases of agronomic interest. Some of them show an increased productivity. This holds for instance true in *peas* (Enken 1967; Chen and Gottschalk 1970). In our material, the tallness is combined with lateness. These mutants could not be grown on a large scale in the field because of their susceptibility against lodging. For private use in gardens along wire fences, however, they could be of some interest because of their favorable yielding properties (Fig. 8). A tall mutant of *Sorghum bicolor* shows several positive peculiarities, such as an increase of the number of leaves and productive tillers. An about 20% increase in grain weight was obtained in successive generations, mainly due to the increased number of grains in the panicles (Bhaskara Rao and Reddi 1975). In *jute,* tall mutants are of direct agronomic interest because of the longer fibers. Twenty-seven tall genotypes out of 136 viable mutants were selected in India and are being tested in different regions of the country with regard to fiber yield and fiber quality (Thakare et al. 1973).

Occasionally, a better lodging resistance appears in mutants having an abnormally long stem due to an increased stability of the stem. A *rice* mutant of this kind was selected in India, having a combination of favorable and unfavorable characters. The plant height is increased by about 10%, but the panicle length is reduced. Moreover, the number of grains per panicle and the tillering capacity are reduced. The plants mature 10 days later than the initial line and have a reduced yield although their 1000-grain weight is enormously increased. They might possibly be useful in crossbreeding programs (Bhan and Kaul 1974). The long culm of rice mutants can be due to either recessive or dominant genes (Okuno and Kawai 1978b).

6.3 Mutants with Altered Stem Structure

The organization of the stem of a crop plays an important role with regard to the productivity of the plant. In some cases, genetically conditioned alterations of the shoot structure lead to an improvement. The degree of branching can influence the yielding properties of a crop positively. In other cases, the lack of lateral branches proves to be a favorable character. In exceptional cases, finally, stem fasciation or stem bifurcation result in an increase of the seed production and thus represent characters of agronomic value. In the following, some examples are given covering this field.

Alterations of the shoot structure in vegetatively propagated crops or ornamentals are sometimes of direct agronomic or horticultural value. Improved plant types are known in mutants of *Achimenes* (Broertjes 1972a), *Kalanchoë* (Broertjes and Leffring 1972), *Rosa* (Dommergues et al. 1967), *Chrysanthemum indicum* (Satory 1975), and *Dianthus caryophyllus* (Badr and Etman 1977) among others. An X-ray-induced blackcurrant mutant proved to be important for the improvement of *Ribes nigrum* due to its erect growing habit. This trait — favorable with regard to mechanical harvesting — is not available in the natural genetic variability of the species. The mutant arose already in 1952 as the result of a bud mutation and was developed into the commercial variety "Westra" (Bauer 1974).

6.3.1 Branching, Tillering

The increase of the degree of branching leads to an increase of the number of fruits in a mutant of *Brassica juncea* (Nayar and George 1969). Bulgarian *pea* mutants with highly branched stems and narrow leaves show a good seed set and seem to be suited for cultivation in dry areas. Another strongly branched *pea* mutant was found to have an increased protein content (Vassileva 1978). In *cotton,* multiple alleles control the type of branching. Extremely branched mutants were obtained in Russia (Fursov and Konoplia 1967). An interesting combination of the experimental methods for inducing gene and genome mutations refers to the improvement of *Brassica campestris* var. *toria* in India (Swaminathan 1965, 1969a, b). From this important oil plant, tetraploid strains are available since 1939. Because of their poor branching ability, they are inferior in yield to the diploid initial material. By means of X rays, mutants with an increased number of secondary branches were obtained one of them outyielding the best diploid control.

In *cereals,* high tillering can be a favorable character which has already been mentioned in the previous part of this chapter. Further mutants showing this trait are known in *rice* (Mallick 1978), *millets* (Goud et al. 1970), and *barley* (Sethi 1975) among others. A promising Egyptian *rice mutant* has about 15 tillers per plant instead of 9 tillers of the mother variety (Ismail and Ahmad 1979). Occasionally, mutants of this type show an improvement with regard to the number of spikes per unit area, this being a favorable prerequisite for an increased seed yield (Ibrahim et al. 1966, *barley*). The released Czechoslovakian *barley* variety "Diamant" has an increased tillering and shorter straw resulting in high lodging resistance. Its yield is very high, especially in soils of good fertility, high manuring and sufficient water supply (Bouma 1966). The corn-grass mutants of *Zea mays,* obtained in Hungary after application of physical and chemical mutagens, belong to the category of "systematic" mutations deviating very strongly from normal maize plants. They have a bushy habit with many narrow, grasslike leaves. Moreover, they form 20 to 50 tillers of similar length per plant which produce a relatively large number of ears. Because of this high degree of tillering, they are of interest for breeding strains of silage maize. Furthermore, a high degree of heterosis is obtained when they are crossed with normal inbred maize (Pásztor 1978, 1979).

In some cases, poor branching was found to be a favorable character under specific conditions of cultivation. A branchless mutant of Egyptian *cotton* outyielded the initial line in boll weight and lint percent; moreover, it is somewhat earlier. An additional advantage of this genotype is its suitability for mechanized harvesting (Bishr et al. 1973). Also, a single stemmed *tomato* mutant can be harvested easier by machines than the branched cultivars (Alexander et al. 1971).

6.3.2 Stem Bifurcation

A specific kind of alteration of the branching system of a crop was found in *Pisum sativum.* Normal pea plants have a monopodial shoot structure. In many varieties, the lateral branches participate in seed production either not at all or only to a small extent. In these cases, dichotomous branching should result in an improvement with regard to seed yield. Some bifurcated mutants and a great number of bifurcated recombinants are available in our collection, some of them being of interest for pea breeding. The agronomic value of many of these genotypes is negatively influenced by the unstable penetrance of some of the genes involved. The penetrance problem and its importance in mutation breeding is discussed in Chap. 17.

6.3.2.1 Bifurcated Mutants

The two polymeric genes *bif-1* and *bif-2* were found to alter the monopodium of the species *Pisum sativum* into dichotomy; the plants show a stem bifurcation in the upper part of the shoot (Fig. 12). Theoretically, this should be a positive trait because the number of pod-bearing stems per plant is increased. Therefore, an increase of the number of pods should be expected to result in a general increase of the seed production. This holds, however, only true to a certain extent. Four mutants of this group are available in our collection, showing the following genetic relations to each other:

− mutant 1201A: homozygous for gene *bif-1^{1201}*
− mutant 239CH: homozygous for gene *bif-1^{1201}*
− mutant   37B: homozygous for gene *bif-1^{37}*, an allele of *bif-1^{1201}*
− mutant 157A: homozygous for gene *bif-2*, polymeric to the *bif-1* alleles.

Thus, the phenomenon of stem bifurcation in *Pisum* is controlled by polymeric genes as well as by multiple alleles. Unfortunately, the genes *bif-1^{1201}* and *bif-2* have an unstable penetrance which reduces the agronomic value of the respective mutants. This disadvantage, however, can be overcome by transferring the genes into specific genetic backgrounds (Gottschalk 1965a, 1966, 1978c; Gottschalk and Chen 1969).

Fig. 12. Apical part of a bifurcated *Pisum* mutant with two corresponding stems due to the action of gene *bif-1^{1201}*

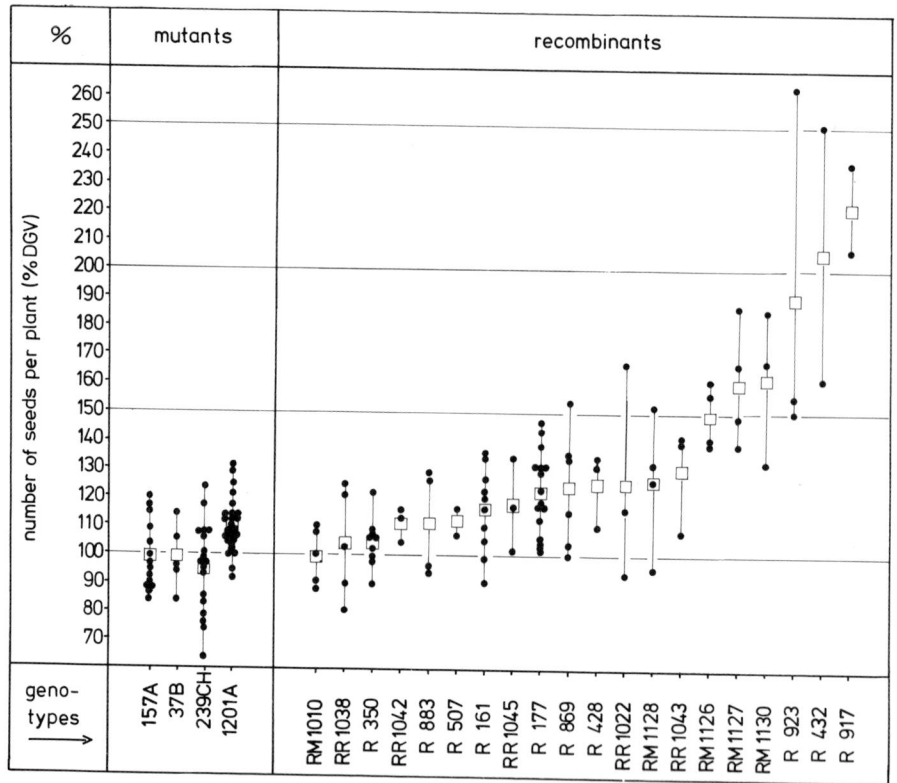

Fig. 13. Seed production of four bifurcated *Pisum* mutants and 20 bifurcated recombinants as related to the control values of the mother variety "Dippes Gelbe Viktoria" (DGV). Each *dot* gives the mean for one generation; the *squares* represent the mean values for the whole material studied of each genotype

With regard to the seed production, only mutant 1201A of the group was found to exceed its mother variety (Fig. 13). The mean values for the trait "number of seeds per plant" varied between 92% and 131% of the control values of the initial line, considering yield analyses over 22 generations. The total mean for the whole material studied is 110.1%. Thus, mutant 1201A is a useful genotype in spite of the reduced penetrance of the $bif\text{-}1^{1201}$ allele. The mean values of the other three bifurcated mutants are lower ranging around the control values of the mother variety.

For judging the breeding value of the bifurcated *Pisum* mutants, some additional information is necessary. The seed production of mutant 1201A is clearly higher than that of mutant 157A. This difference would be easily understandable, if gene *bif-2* would have a lower penetrance than $bif\text{-}1^{1201}$ (for details see Chap. 17). Such a regularity, however, was not found if we consider all the generations studied so far. But there is no doubt that the selection value of gene $bif\text{-}1^{1201}$ is better than that of its polymeric gene *bif-2*. Mutants 1201A and 239CH are homozygous for the same gene $bif\text{-}1^{1201}$, but the yield of 239CH is essentially lower than that of 1201A, about 15% less if we consider the whole material tested in 14 or 22 generations, respectively. The main reason for the reduced yield is obviously the fact that mutant 239CH is homozygous for a second gene which influences the flower structure of the plants.

This gene does not cause a direct reduction of the functionability of the flowers, but it has possibly an indirect effect on the seed fertility resulting in the lower yield observed. The allele $bif\text{-}1^{37}$, present in mutant 37B, finally, has full penetrance in contrast to the genes $bif\text{-}1^{1201}$ and $bif\text{-}2$. If we consider the relations between penetrance and seed production, realized in the other three mutants of the group, the stabilized penetrance should lead to an increase in yield (see Chap 17). This is, however, not the case; on the contrary, the seed production of mutant 37B is clearly lower than that of 1201A. The reasons for this unexpected behavior are not yet known, but the results demonstrate that there are some additional difficulties preventing that mutant 37B can display its positive potentialities in an optimal way.

In many plants of the genotypes discussed, the position of bifurcation is so high at the stem that it cannot become effective with regard to an increase of the seed production. Its influence on the yield would be more favorable if this point would lie lower, enabling the shoot to form two well-developed corresponding stems. Plants of this shoot structure were selected in M_2 and M_3 generations following X-ray and neutron treatments, but this trait proved to be due to modifications and not to mutations. Some recombinant lines of our *Pisum* collection, however, show this anomaly as a constantly appearing character resulting in a considerable increase in yield.

6.3.2.2 Bifurcated Recombinants

The bifurcated *Pisum* mutant 1201A was crossed with our fasciated mutants and with some other genotypes, resulting in a large number of different recombinant types. Most of them are homozygous for gene $bif\text{-}1^{1201}$; some others contain a similar gene derived from the fasciated mutants. These recombinants are available as pure lines and have been under study for several generations with regard to both their yielding potentialities and the penetrance behavior of the *bif* gene. The seed production of 20 of them is illustrated in Fig. 13. The graph shows that it was not difficult to develop bifurcated recombinants exceeding the bifurcated mutants considerably in yield.

Many of the high-yielding recombinants of this group have genes for long or very long internodes. Therefore, they are not suited for field cultivation in the presently existing form. We shall try to replace these genes by genes for shorter internodes, hoping that the favorable yielding properties can be maintained at least in some genotypes. This aim has already been reached to some extent. The plants of recombinant R 161 for instance, selected after having crossed the fasciated mutant 489C with the bifurcated mutant 1201A, are homozygous for at least four mutant genes. They have the following characters:

— short I (from 489C),
— linear stem fasciation (from 489C),
— later flowering and ripening than the mother variety (from 489C),
— dichotomous stem bifurcation (gene *bif-1* from 1201A).

The three 489C genes are present in the fasciated mutant in a hypostatic state. The recombinant R 161, however, does not contain the respective epistatic genes; therefore, the hypostatic ones are effective. The seed production is about equal to that of mutant 1201A and lower than that of 489C, but the plants are shorter and not so late.

Of particular interest in this concern is recombinant R 177 deriving from the same cross. The plants are homozygous for three genes as follows:

- weak stem fasciation (from 489C),
- reduced grain size (gene *sg* from 489C),
- stem bifurcation (gene *bif-1* from 1201A).

In this combination, gene *bif-1^{1201}* shows full penetrance, obviously due to the presence of gene *sg*. The mean value for the number of seeds per plant of the whole material tested in 16 generations lies 22% higher than the control value of the mother variety, whereas that of mutant 1201A lies only 10% higher. The increased production seems to be mainly due to the stabilized penetrance of the *bif* gene. The reduced seed size, however, reduces the breeding value of the recombinant. Therefore, gene *sg* was replaced by a gene for somewhat bigger seeds (recombinant RM 1010). Interestingly, the stabilized penetrance of gene *bif-1^{1201}* could be maintained, being combined with about normal seed size. The plants of this genotype are nonfasciated, but all of them are bifurcated. Unfortunately, the point of bifurcation lies so high at the stem that this kind of abnormal shoot structure does not lead to the increased seed production expected. On the contrary, the yield is only similar to that of the mother variety and clearly lower than that of mutant 1201A and recombinant R 177.

The two bifurcated recombinants R 507 and RM 1128 are likewise short-stemmed, the latter having in addition to stem bifurcation an increased number of seeds per pod. Their yield is considerably better than that of the mother variety. The high seed production of many long-stemmed bifurcated recombinants is partly due to the stabilized penetrance of gene *bif-1^{1201}*. Most of them have only been studied over two or three generations. That is not enough for judging their yielding properties reliably. The values available, however, demonstrate the high physiological capacity of these genotypes in a convincing way. They are certainly interesting cross-partners for developing further prospective recombinant lines. Details on the yielding properties of our bifurcated genotypes are given in Fig. 13.

6.3.3 Stem Fasciation

Genetically conditioned stem fasciation is widespread in higher plants. Usually, it is a monstrosity having no practical value. The upper part of the main stem, in addition sometimes also that of the lateral branches, is broadened band-like. This anomaly is often combined with the formation of a very great number of floral buds and an increased number of flowers accumulated in the top region of the plants. A review paper about early findings on fasciated plants has been given by White (1948). In plant breeding, stem fasciation is only rarely utilized.

6.3.3.1 Fasciated Mutants

In *Celosia argentea*, spontaneously arisen fasciated lines have been used in horticulture for many decades. Similar genotypes, obtained in mutagenic treatments in India, are available in *Amaranthus hypochondriacus* and could become of economic interest, as in the cockscomb (Behera and Patnaik 1979). Among cultivated plants, fasciated genotypes are of some importance for breeding purposes only in *Pisum sativum*. These genotypes are well known since Mendel's experiments *("Pisum umbellatum")*. They have often been studied for both theoretical and practical reasons (Table 10). A review on the older literature has been given by Gottschalk (1971a). The genetic

Table 10. Fasciated and bifurcated *Pisum* mutants and tall genotypes of some other crops

Divergent character	Species	Authors	Mutagens	Remarks
Stem fasciation	*Pisum sativum*	Gottschalk 1966, 1977a, 1981a; Chen and Gottschalk 1970	X-rays, neutrons	Higher seed production
		Enken 1967	Gamma rays, EI	
		Sidorova 1970	–	
		Jaranowski 1976	Gamma rays	
		Vassileva and Milanova 1977; Vassileva 1976, 1978	Gamma rays	Normal or increased seed production
		Scheibe 1965, 1968, 1971, 1977; Kaschi and Zoschke 1968	Spontaneous	Commercial varieties "Ornamenta" and "Rosakrone"
		Gottschalk 1970, 1976d, 1977a; Gottschalk and Milutinović 1973a,b; Gottschalk and Wolff 1977	X rays	Strong heterosis effect in crosses with nonfasciated genotypes
		Gottschalk 1972a, b, 1977c; Gottschalk and Hussein 1975, 1976; Gottschalk and Wolff 1977; Gottschalk and Bandel 1978; Hussein and Gottschalk 1976; Bandel and Gottschalk 1978	X rays	*Fasciata* genes used in recombination lines
Stem bifurcation	*Pisum sativum*	Gottschalk 1965, 1966, 1968c; Gottschalk and Chen 1969	X rays	Increased seed production; unstable penetrance
Tallness	*Pisum sativum*	Enken 1967	Gamma rays, EI	High seed production; late
		Chen and Gottschalk 1970	Neutrons	High seed production
		Enken 1967	Gamma rays	
	Glycine max	Mallick 1978	EMS	
	Oryza sativa			
	Sorghum bicolor	Bhaskara Rao and Reddi 1975	Gamma rays	21% increase in grain weight
	Corchorus olitorius	Thakare et al. 1973	Gamma rays, neutrons	Longer fibers

basis of stem fasciation in *Pisum* is not yet completely clear in spite of the tremendous amount of work done in this field. It is considered to be due to one single gene (Scheibe 1954; Marx and Hagedorn 1962), two polymeric genes (Lamprecht 1952), three major genes supplemented by several modifiers (Sidorova 1970), or to a series of three multiple alleles (Rod and Vagnerova 1970). Our own investigations in a great number of radiation-induced fasciated mutants show that three or four different genes are responsible for the fasciation in the material tested; moreover, multiple alleles participate in realizing this anomaly. These interpretations are not necessarily in contradiction to each other. On the contrary, certain observations indicate that the fasciated pea mutants are not a uniform group. They are heterogeneous in their morphology as well as in their genetic constitution.

At least four morphologically and genetically different types of fasciated mutants are known in *Pisum sativum:*

- Strong fasciation. Plants tall; apical internodes extremely shortened, all the flowers and pods accumulated in the top region of the stem; high yield (Fig. 14).
- Linear fasciation. Plants tall; apical internodes not shortened; number of pods per node increased; high yield (Fig. 15).
- Very weak fasciation. Apical internodes not shortened; stem fasciation not in each plant of the genotype clearly discernible; reduced seed production.
- Different kinds of fasciated dwarfs with varying degrees of seed production.

Fasciated *Pisum* mutants have favorable and unfavorable characters. They are — at least in some intensively studied mutants of the group — not due to pleiotropic effects of the *fasciata* genes but to the action of whole groups of independently functioning mutant genes. The main advantages are as follows:

- The large number of flowers per plant. As a consequence of the stem fasciation, a large number of floral buds is present in the top region of the stem. Only a small proportion of them develops into pods; nevertheless the number of pods per plant is considerably higher than in nonfasciated genotypes, resulting in a strongly increased seed production. This holds true for the first two groups just mentioned (Fig. 9).
- The accumulation of tendrils in the top region of the stem due to the shortening of the apical internodes and the increase of the number of foliage leaves as a consequence of the fasciation. It leads to a high amount of lodging resistance when the material is grown in the field without utilizing fences.

Fig. 14. Apical part of a strongly fasciated *Pisum* mutant. All the flowers and pods are accumulated in the top region of the plant due to strongly shortened internodes

- The short flowering period. The plants of the strongly fasciated pea mutants flower 5–12 days only, whereas the flowering period of nonfasciated strains is 3–6 weeks.

The main disadvantages of the first three groups mentioned above are as follows:

- The tallness and lateness of the plants.
- The strongly reduced seed size annulling the positive effect of the increased number of pods per plant in many cases.
- A negative response to specific climatic conditions preventing flower formation.

The good yielding properties of some fasciated pea mutants are graphically presented in Figs. 8 and 9. A spontaneous mutant of this type was used by Scheibe (1965, 1968, 1971, 1977) for developing the commercial fodder pea varieties "Ornamenta" and "Rosakrone." He made more than 1500 crosses in order to eliminate some negative traits of the original mutant. About three decades were required before the two varieties were on the market. Their seed production can be improved by cultivating the peas together with oats or *Sinapis* (Kaschi and Zoschke 1968; Scheibe 1971). Scheibe's "Ornamenta" is identical with one of our X-ray-induced mutants with regard to the genes for stem fasciation. Four fasciated mutants, selected after gamma-ray treatment in Bulgaria, are short-stemmed and show likewise very good yielding properties (Vassileva 1978). The stems of two others, however, are very unstable. They break below the inflorescences reducing the seed yield (Vassileva and Milanova 1977).

Fig. 15. Apical part of a linearly fasciated *Pisum* mutant. The uppermost internodes are long; very often, the nodes bear more than one inflorescence

For judging the suitability of the fasciated mutants for pea breeding we have to consider their complicated genotypic constitution. Six of our fasciated mutants have been studied in detail. They are homozygous for 14 to 18 genes which have mutated more or less simultaneously during the X-ray or neutron irradiation. Mutant 489C of this group, studied since almost 20 generations, contains genes controlling the following traits:

— three, possibly four genes responsible for different kinds of stem fasciation;
— a gene for apical stem bifurcation;
— six genes for different internode lengths;
— three genes for different flowering and ripening times;
— gene *sg* for reduced grain size;
— a gene causing a slight reduction of the chlorophyll content;
— a gene for slightly reduced leaflet breadth;
— gene *fis* for long-day requirement with regard to flower formation.

Only 4 out of these 17 or 18 genes are visible in the mutant; all the others are hypostatic. Their action becomes discernible only in distinct recombinants selected in F_2 and later generations, after having crossed the mutant with nonfasciated genotypes. In this way, the epistatic genes could be removed and the hypostatic ones can display their actions. The other five fasciated mutants of this group proved to be identical to 489C with regard to most of the mutant genes listed above. Moreover, they are identical with corresponding genes present in that spontaneous mutant which was used for developing Scheibe's "Ornamenta" (Gottschalk 1977a, 1981a, f).

Another characteristic feature of our fasciated pea mutants consists in the high heterosis effect which occurs in F_1 hybrids when the mutants are crossed with nonfasciated genotypes. Moreover, they show a specific reaction to distinct ecological factors which influences their breeding value negatively. These problems are discussed in Chaps. 12 and 18.

One should expect that similar positive effects, due to homologous mutations, are also known for other pulses. This is, however, not the case. So far, the garden pea is the only leguminous crop in which the phenomenon of stem fasciation is agronomically utilized. A fasciated mutant of the *soybean*, available in Korea, shows a very high seed production. It is, however, too early for judging its productivity reliably, because yielding analyses have not yet been carried out (Won, pers. commun.). Fasciated plants of green gram *(Vigna radiata)* are without any practical value because of having abnormal flowers and an increased pollen sterility (Singh 1981).

6.3.3.2 Fasciated Recombinants

The complicated genotypic constitution of our fasciated *Pisum* mutants is a considerable advantage for pea breeding because these genotypes contain several genes for useful traits. This holds not only true for all the genes for stem fasciation causing high seed production or for earlier flowering and ripening, but also for genes reducing the plant height. Most of them are hypostatic. This disadvantage, however, can be easily overcome. If we cross the fasciated mutants with other useful mutants or with nonfasciated varieties or strains, F_1 hybrids with a high degree of heterozygosity arise showing a tremendous segregation in F_2 and in later generations. In this way, hundreds of recombinant lines were developed at our institute. In many of them, the negative traits of the fasciated mutants were eliminated, because the epistatic genes could be replaced by the respective hypostatic ones responsible for positive traits. Stem fasciation was combined with reduced plant height, normal seed size, normal or even earlier flowering and ripening. The agronomic usefulness of most of these genotypes is reduced because of negative interactions between the genes involved, but some of them outyield the mother variety considerably. The whole group is under intensive study and it is certain that prospective genotypes with favorable combinations of useful traits will be selected in near future.

The mutagenic situation of our fasciated *Pisum* mutants is very uncommon not only because so many different genes have simultaneously mutated, but especially because the same big group of genes has mutated several times in the initial cells of different embryos. This behavior would be more easily understandable if all the genes involved would be closely linked, thus demonstrating that a distinct region of a specific *Pisum* chromosome shows in total a reduced stability against mutagenic treatments. The *dim*-segment is such a susceptible chromosome region (for details see Chap. 16). This is, however, not the case in our fasciated mutants. On the contrary, it can be concluded from the findings available that the 14–18 mutant genes of these genotypes are more or less randomly distributed over the seven chromosomes of the complement. They have not yet been localized, but this becomes clear from the fact that they could be easily separated from each other and that a great number of different recombinants is available, in which these genes are recombined with each other in almost all combinations theoretically conceivable. Most of the fasciated recombinants, however, have small seeds. It can therefore be assumed, that the three or four genes for stem fasciation are closely linked with each other and that these genes are relatively closely linked with gene *sg* for reduced grain size. The simultaneous mutation of such a big group of specific genes is certainly not a matter of chance; on the contrary, it is obviously a regularity not yet understood in detail. It could be possible that the primary event in the respective seven *Pisum* embryos was a mutation in a kind of a "mutator gene," causing subsequently the mutations of a specific group of other genes of the genome as secondary events (Gottschalk 1981a).

Details concerning the utilization of fasciated pea mutants for producing a great variety of different recombinant types are found in a series of papers (Gottschalk 1972a, c, 1977a, 1979c; Gottschalk and Hussein 1975, 1976; Hussein and Gottschalk 1976; Gottschalk and Wolff 1977; Gottschalk et al. 1978; Gottschalk and Bandel 1978; Bandel and Gottschalk 1978; Gottschalk and Kaul 1980). The yielding properties of different recombinants, selected after having crossed fasciated mutants with nonfasciated genotypes, are found in Fig. 9. The reaction of some fasciated recombinants to specific ecological conditions are discussed in Chap. 18; interactions between *fasciata* genes and other mutant genes are described in Sect. 5.2.1.

6.3.4 Mutations in Fiber Plants

In some fiber plants, mutants suitable for practical utilization are known. In jute *(Corchorus olitorius, C. capsularis)*, radiation-induced mutants with 15%–40% increased fiber yield and with improved fiber quality have been selected. A new commercial variety was developed in *Corchorus olitorius* after having crossed two mutants with each other. Moreover, high-yielding long-stemmed hybrids and recombinants were obtained by crossing two dwarfs (Singh 1971; Singh et al. 1973; Sinhamahapatra and Rakshit 1981). A chlorophyll-deficient mutant of *Corchorus olitorius* was found to have a 26% increase in fiber length as well as an improved fiber quality with regard to fineness and tenacity. An M_6 inbred population from this mutant yielded 13% more fiber than the mother variety (Hossain and Sen 1978a). Mutants of *Corchorus capsularis* with higher fiber yield were selected by Shaikh et al. (1980) in gamma-ray treatments using a quicker screening method. Similar results were obtained in a mutant of *Linum usitatissimum*, it having an increased yield of long fibers in the stem (Pospisil 1974).

7 Alterations of Flower Shape, Color and Function

Many flower mutants are available in the collections but most of them are without any agronomic interest. In the majority of these cases, the genetically conditioned flower anomaly results in sterility or a drastic reduction of fertility of the plants. A short-styled *tomato* mutant was found to be especially suited for greenhouse cultivation. Greenhouse tomatoes must be artificially pollinated, which is much easier in short-styled flowers as compared to the normal ones. In this way, the high costs for pollination can be strongly reduced (Alexander et al. 1971). A serious problem of *red clover* cultivation in Middle Europe is the low seed set due to the fact that the *Trifolium* flowers cannot be fertilized by bees because of their long corolla tube. A mutant with short tubes was isolated by Scheibe and Bruns (1953). Unfortunately, the bees are not able to distinguish between short- and long-tubed red clover flowers when the pollination experiments are made in isolation chambers. In the field, they avoid visiting the flowers of the short-tubed genotypes. Thus, they cannot be utilized for solving the problem of a better seed production of *Trifolium pratense* (Bugge 1970). Experimental mutagenesis, however, plays an important role with regard to the problem of male sterility in certain crops and for obtaining new flower colors and shapes in ornamentals. In these fields, plenty of findings are available in the literature.

Certain mutations alter the breeding system of the respective species in a desired manner. This holds true for the "advanced stigma mutant" of *Phaseolus aureus,* an important pulse crop of India (Raghuvanshi et al. 1978). The calyx of the flowers is enlarged and the style protrudes in the bud condition. In this way, a tendency toward cross-pollination is realized, which is a favorable peculiarity in hybridization programs. A morphologically similar mutant is known in *Physalis ixocarpa*. Already the very young floral buds are open due to a very slow growth rate of the corolla resulting in a short tube. The style projects out of the bud. In spite of normal pollen fertility, the mutant shows poor fruit setting when the flowers are isolated. The seeds obtained from these flowers do not germinate. When crossed with the mother variety, normal seeds are obtained and a 3:1 segregation occurs. The original mutant was obviously heterozygous for a dominant gene, causing formation of open buds. Homozygosity for the gene results in lethality. This genetic mechanism is useful in *Physalis* breeding as it guarantees cross-fertilization (Mahna and Singh 1975).

7.1 Flower Shapes and Flower Colors in Ornamentals

The genetically conditioned alterations of many ornamentals are due to somatic mutations. Therefore, the new character can often only be propagated vegetatively. A voluminous survey on this field has recently been given by Broertjes and van Harten (1978). It is therefore not necessary to go into the details. Some examples of flower alterations are given in Tables 11 and 12. With regard to flower shape, the alterations can refer to the size or to the doubling. In *Compositae,* the shape of individual florets can be altered. With regard to flower color, some new colors, not yet existing in the respective species, were obtained. The flower pigments of the mutants of *roses* were biochemically analyzed by Dommergues et al. (1967). In some genotypes, quantitative

Table 11. Alterations of flower shape and size under the influence of mutant genes in different ornamentals

Species	Authors	Mutagens	Remarks
Roses	– Heslot 1966	Gamma rays, EMS	
	– Streitberg, 1966a, 1967	X rays	
	– Dommergues et al. 1967	Gamma rays, EMS	
	– Nakajima 1970	Gamma rays	
Chrysanthemum morifolium	– Bowen 1965	Gamma rays	Changes in size and shape of individual florets
	– Broertjes 1966; Broertjes et al. 1976	X rays, neutrons	
	– Yamakawa 1970	Gamma rays	Smaller inflorescences; tubular florets
Chrysanthemum carinatum	– Rana 1965	X rays	Dissected rays, tubular rays; seeds irradiated
Chrysanthemum indicum	– Satory 1975	Gammay rays	
Dahlia variabilis	– Broertjes and Ballego 1967	X rays	4 mutants commercialized
Rudbeckia	– Tamrazian 1968	EI	Seeds treated
Tithonia rotundifolia	– Kaicker et al. 1971	Gamma rays	
Streptocarpus	– Broertjes 1969a,b, 1972c	X rays, neutrons	5 mutants commercialized
Achimenes	– Broertjes 1972c, 1976a,b	X rays, neutrons	
Kalanchoë	– Broertjes and Leffring 1972	X rays	
Azalea	– Streitberg 1966b, 1967	X rays	
Begonia x *hiemalis*	– Doorenbos and Karper 1975	X rays	
Portulaca grandiflora	– Gupta 1969	Gamma rays	
Dianthus caryophyllus	– Badr and Etman 1977	Gamma rays	
Hibiscus	– Das 1969	Gamma rays	
Canna	– Nakornthap 1965	Gamma rays	
Iris	– Hekstra and Broertjes 1968	X rays	

Table 12. Alterations of the flower color under the influence of mutant genes in different ornamentals

Species	Authors	Mutagens
Roses	– Heslot 1966	Gamma rays, EMS
	– Streitberg 1966a, 1967	X rays
	– Dommergues et al. 1967	Gamma rays, EMS
	– Nakajima 1970	Gamma rays
	– Swarup et al. 1971	X rays
	– Kaicker and Swarup 1972	Gamma rays
	– Lata 1980	Gamma rays
Chrysanthemum morifolium	– Bowen 1965	Gamma rays
	– Broertjes 1966; Broertjes et al. 1976, 1980	X rays, neutrons; repeated irradiations
	– Yamakawa 1970	Gamma rays
	– Gupta and Shukla 1971	Gamma rays
Chrysanthemum indicum	– Satory 1975	Gamma rays
	– Jung-Heiliger and Horn 1980	Gamma rays, EMS
Dahlia variabilis	– Broertjes and Ballego 1967	X rays
Tithonia rotundifolia	– Kaicker et al. 1971	Gamma rays
Streptocarpus	– Broertjes 1969a, b, 1972c	X rays, neutrons
Achimenes	– Broertjes 1972c, 1976a	X rays, neutrons
Kalanchoë	– Broertjes and Leffring 1972	X rays
Azalea	– Streitberg 1966b	X rays
Dianthus caryophyllus	– Buiatti et al. 1965	Gamma rays
	– Badr and Etman 1977	Gamma rays
Hibiscus	– Das 1969	Gamma rays
Canna	– Nakornthap 1965	Gamma rays
Lilium	– Broertjes 1972c	X rays, neutrons
Gladiolus	– Buiatti and Tesi 1969	Gamma rays
	– Moës 1969	Gamma rays
Alstroemeria	– Broertjes and Verboom 1974	X rays
Petunia	– Khalatkar and Kashikar 1980	Sodium azide
Kohleria	– Parliman and Stushnoff 1979	Gamma rays

alterations in flavonols and anthocyanidins have been found. Frequency of these alterations is very high. In *roses*, for instance, 14.9% out of 2650 buds treated had bud variations and 5.8% showed new characters (Streitberg 1967). Whole groups of mutants of this kind have been commercialized within a very short period after the experiments had begun. This holds true for new varieties of *Dahlia* (Broertjes and Ballego 1967), *Streptocarpus* (Broertjes 1969b), *Alstroemeria* (Broertjes and Verboom 1974) among others.

7.2 Inflorescences

In some cases, an increase of the seed production is reached by an increase of the number of flowers per inflorescence or the number of flowers per plant in general. In a Russian *pea* mutant, the number of pods per inflorescence is increased. At first the mutant was inferior to its initial line, but later it outyielded the mother variety by more than 20% (Sidorova et al. 1969). A similar effect, due to a somatic mutation, is known in *Iris*. The increased number of flowers per stalk represents an improvement of the cultivar as it enhances its ornamental value (Hekstra and Broertjes 1968). The plants of *Sesamum indicum* form two extrafloral nectaries per leaf axil. These organs are transformed into normal functioning flowers in a Japanese mutant. They form capsules with normal seed fertility (Kobayashi 1965). The inflorescences of normal *castor* plants have some male flowers interspersed among the female ones. In an Indian mutant, the number of male flowers is increased from 10 to 100; this is a desirable character (Kulkarni 1969). *Maize* is one of the few cross-pollinating crops in which applied mutagenesis is carried out. Multi-eared mutants, selected in Hungary, show an increased number of ears per stalk resulting in an increase in yield. A unisexually female maize mutant may be of interest for the production of hybrid seed (Pásztor 1978, 1979).

A new flower shape of *snapdragon (Antirrhinum majus)*, interesting from a horticultural point of view, was obtained by crossing some spontaneously arisen mutants. The combination of *hemiradialis, centroradialis,* and *divaricata* results in plants having a peloric terminal flower and a picturesque short and rounded inflorescence (Knapp 1967). This is a homozygous recombinant which can be propagated sexually.

Awned mutants of awnless varieties proved to be of agronomic interest in *rice* (Majumder 1966) and grain *Sorghum* (Sree Ramulu 1974a, b). Awnedness is insofar a desirable character as it reduces damages of the crops caused especially by birds.

7.3 Genetic Male Sterility

Male sterile mutants are of increasing interest for hybrid breeding and for the utilization of the heterosis effect. The problem has been recently discussed repeatedly with regard to the development of hybrid programs in *barley* (Künzel and Scholz 1973; Foster 1976; Hagberg et al. 1976). Practical experiences, however, are available only in *Capsicum annuum* so far. A group of male sterile pepper mutants was tested under natural conditions of pollination in Bulgaria with regard to yield of hybrid seeds, number of fruits per plant, and number of seeds per fruit. The hybrid seed production of certain male sterile genotypes proved to be quite satisfactory when tested on a large scale (Daskaloff 1976). There are obviously good chances for utilizing these mutants in heterosis breeding. A general survey on the isolation and utilization of male sterility in crops was given by Driscoll and Barlow (1976); male sterility in *cotton* has been reviewed by Meyer (1969).

In principle, there are two different types of genetic male sterility. An obviously very small group of mutant genes in higher plants causes a misdifferentiation of the growing points in such a way that the stamens are either completely lacking or that they are present in a rudimentary form. The female sex organs are fully functionable. Such a situation is for instance realized in a *tomato* mutant (Dolgih 1969). Moreover, male sterility can be caused by genes realizing a transformation of the stamens into carpel-like organs. Examples from older literature are given by Gottschalk and Kaul (1974) and by Gottschalk (1976c). The same effect, finally, appears if the pollen grains are inaperturate, if they do not possess any pores for the germination of the pollen tube. A spontaneously arisen mutant of this type has been selected in *barley* (Ahokas 1975).

The typical *ms*-genes, however, become effective in otherwise normal flowers. Their main effect consists in causing the breakdown of microsporogenesis in a distinct meiotic stage characteristic for each single gene of the group. Afterwards not only the chromosomes, but the whole living contents of the pollen mother cells degenerates. In many male sterile mutants the degeneration occurs at the end of the second meiotic division or during the postmeiotic stages. In others, this process occurs already in the beginning of microsporogenesis or at pachytene. Male germ cells are not produced in the anthers of these genotypes, whereas the female sex is normally functioning. More than 400 male sterile mutants of this type are known in higher plants. In most cases, only the existence of male sterility is known, but meiosis has not been analyzed in detail. About 100 mutant *ms*-genes of 48 species belonging to 12 different families, however, are studied with regard to their action on microsporogenesis. They have been reviewed by Gottschalk and Kaul (1974) and by Gottschalk (1976c).

It is obvious that the genetically conditioned male sterility is not a uniform phenomenon. So far, four groups of male sterile mutants in higher plants are known, due to

- the action of single recessive *ms*-genes;
- the action of single dominant *Ms*-genes;
- the joint action of several recessive *ms*-genes in a polymeric manner;
- the cooperation of *ms*-genes with a specific type of cytoplasm.

In *Triticum aestivum*, 12 male sterile mutants have been selected after EMS treatment. In 11 of them, male sterility is controlled by single recessive genes, three of which were found to be allelic. In the 12th mutant of this group, a dominant gene becomes effective (Sasakuma et al. 1978).

The role of the tapetal tissue with regard to the degeneration of the pollen mother cells is not yet definitely clarified.

Some male sterile mutants of a large number of different crops are listed in Table 13. In *rice*, they are considered to play an important role for a backcross breeding system (Hiraiwa and Fujimaki 1977), whereas a *durum wheat* mutant of this type is already used for hybrid seed production (Bozzini 1974b). They are of particular interest in typical outcrossing species, such as *Brassica napus* (Takagi 1970) or *Helianthus annuus* (Plotnikov 1973). In *Capsicum annuum*, male sterile mutants are commonly grown with varieties of good combining ability in Bulgaria. Some of them show a high heterosis effect for earliness; the hybrids were found to be somewhat superior in yield (Daskaloff 1968, 1978; Stoilov and Daskaloff 1976). Thirteen radiation-induced male sterile *Pisum* mutants are available in our collection which have been analyzed meiotically in detail (Gottschalk 1968b; Klein 1969; Gottschalk and Baquar 1972; Gottschalk and Klein 1976). They cannot be used in pea breeding because of the cleistogamous pollination system of the species. Pea breeders are, however, interested in developing cross-fertilizing strains. With these prerequisites, the male sterile mutants could be of particular interest.

Table 13. Male sterile mutants in cereals and dicotyledonous crops

Species	Authors	Mutagens	Remarks
Hordeum vulgare	– Tavčar 1965	Gamma rays	
	– Favret and Ryan 1966	X rays, EMS	2 male sterile mutants
	– Mian et al. 1974	Spontaneous	5 male sterile lines
	– Ahokas 1975	Spontaneous	Pollen grains inaperturate
	– Sethi 1975; Sethi and Bhateria 1977	^{32}P, ^{35}S, EMS	
	– Sharma and Reinbergs 1976		3 new recessive *ms*-genes
Oryza sativa	– Pavithran and Mohandas 1976	EMS	
	– Hiraiwa and Fujimaki 1977	Gamma rays, EI	No pollen grains formed
	– Singh and Ikehashi 1981	EI	4 male sterile mutants
Triticum aestivum	– Kleijer and Fossati 1976	X rays	Gene ms^{a1} located on chromosome 4A
	– Sasakuma et al. 1978	EMS	12 male sterile mutants; one gene dominant
Triticum durum	– Bozzini 1974b	Neutrons	Used in a breeding program
Zea mays	– Popova 1975	EMS, NMU	
Sorghum bicolor	– Anrews and Webster 1971	Gamma rays	ms_7; no pollen produced
Pennisetum typhoides	– Burton 1974	Neutrons	
Lycopersicon esculentum	– Rick and Boynton 1967		
	– Dolgih 1969	Gamma rays	Devoid of stamens
	– Yamakawa 1969	Gamma rays	4 male sterile mutants
	– Zagorcheva and Yordanov 1978	Gamma rays	
Capsicum annuum	– Daskaloff 1968, 1973, 1976, 1978; Stoiloff and Daskaloff 1976	X rays, gamma rays	6 male sterile mutants; *ms-3* to *ms-8*
	– Shifriss 1973	Spontaneous	Small and shrunken anthers; no pollen grains ms_1
Glycine max	– Brim and Young 1971		
Lathyrus sativus	– Chekalin 1977	Gamma rays and chemicals	
Lens culinaris	– Sharma and Sharma 1978a	Gamma rays, NMU	
Pisum sativum	– Gottschalk 1968; Klein 1969; Gottschalk and Baquar 1972; Gottschalk and Klein 1976	X ray, neutrons	13 male sterile mutants; microsporogenesis analyzed
Vigna radiata	– Singh and Chaturvedi 1981a	Gamma rays	Morphological anomalies
Brassica napus	– Takagi 1970	Gamma rays	Abnormal development of the tapetum
Helianthus annuus	– Plotnikov 1973	DMS	
Corchorus capsularis	– Mitra 1977	X rays	Several undesirable characters
Gossypium hirsutum	– Weaver and Ashley 1971	?	1 dominant gene

In a Canadian mutation breeding program in fruit trees, some self-compatible mutants of the self-incompatible *sweet cherry* were obtained. They were included into crossbreeding resulting in the first self-compatible sweet cherry cultivar "Stella" (Lapins 1973).

8 Leaf Mutants of Agronomic Interest

In *ornamentals*, certain alterations of the leaf color can improve the ornamental character of the plant and may thus be a new character. This is the case in *Iresine herbestii*, a tropical foliage ornamental which normally has purple-colored leaves. Mutants with bronze-colored foliage represent an increase of the genetically conditioned variability within the species which can be utilized for horticultural purposes (Das 1969). Similar findings were obtained in *Begonia* (Doorenbos and Karper 1975), *Canna* (Nakornthap 1965) and *Polyanthes tuberosa* (Abraham and Desai 1976). In the last-mentioned two species, some mutants show different types of leaf variegation which make the plants very attractive already during the vegetative period and enhance their ornamental value. Alterations of the leaf color and leaf shape are also known in mutants of *Kalanchoë* (Broertjes and Leffring 1972) and *Dianthus caryophyllus* (Badr and Etman 1977). All these alterations are due to somatic mutations and can be maintained only by means of vegetative propagation.

In *sugarcane*, mutants with glabrous leaf sheath proved to be favorable when the material has to be harvested by hand. The respective clones are not inferior to their initial lines with regard to yield (Jagathesan and Sreenivasan 1970; Jagathesan 1976). Similar problems exist in *rice*. It is therefore one of the aims of rice breeding to replace the pubescent varieties by glabrous ones in order to avoid skin injuries of farmers. A gene for glabrousness was only known in *indica*, but not in *japonica* varieties. Eight glabrous mutant strains are now available in Japan, arisen after gamma irradiation and EI treatment. Some of them show good yielding properties (Hiraiwa and Tanaka 1979). But also the opposite effect can be useful. An Indian *cotton* mutant shows an increase of hair density on the underside of the foliage leaves and is therefore less damaged by jassids (Swaminathan 1965). *Pea* mutants with a considerable increase in leaf size, available in Bulgaria, have a higher yield (Vassileva 1978), and a *multifoliata* mutant of *Phaseolus aureus* is so leafy that it may be utilized as a green manure crop (Santos 1969). In *Morus latifolia, alba,* and *bombycis*, some mutants with high leaf yield, having a 10% increase in 2nd and 3rd year of cultivation, have been selected in Japan. This improvement seems to be due to both an increase in leaf thickness and in the number of leaves per branch (Nakajima 1973). *Rice* mutants with broad leaves are available in India; the mutant character is dominant (Roy and Jana 1975). The trifoliate leaves of *red clover* (*Trifolium pratense*, an allogamous species) are altered into multifoliate leaves composed of 4–7 leaflets in some mutants. By crossing them with other genotypes, recombinants with different leaflet shapes and sizes and with strongly divergent plant habit were obtained. The seed production of these genotypes is not negatively influenced (Jaranowski and Broda 1978).

Some leaf mutants and recombinants of *Pisum sativum* are of interest for discussing problems of leaf evolution in the *Leguminosae* (for details and references see Gottschalk 1971a). Two of them have been studied with regard to their usefulness for pea breeding. The *acacia* mutant, already known since 1910, has been repeatedly obtained in mutation treatments in different countries, for instance in Poland (Jaranowski 1976). Some of these genotypes, selected in Bulgaria, outyield the mother varieties by 20%–30% in grain production (Vassileva 1978). They are homozygous for gene tl^w, which causes formation of leaflets instead of the tendrils. Thus, the leaf area available for photosynthesis is somewhat increased. The loss of the tendrils, however, is such a negative effect that the gene will hardly be utilized in pea breeding (Fig. 16).

The opposite situation is realized in the *afila* mutants of *Pisum sativum* having branched tendrils instead of the leaflets (Fig. 16). Spontaneous mutants of this genotype were selected in Finland (Kujala 1953), Argentina (Goldenberg 1965) and Russia (Solovjeva 1958; Khangildin 1966); moreover, they were experimentally obtained in Poland (Jaranowski 1976) and in Bulgaria (Vassileva 1978). Their seed yield is mostly reduced; nevertheless, the genes *af (afila)* and *st* (reduced stipules) have been utilized in the United Kingdom, Poland, and Italy for developing "leafless" or "semileafless" peas (Snoad 1974, 1975, 1981; Snoad and Hedley 1981; Monti and Frusciante 1978). Their genetic constitution is as follows:

— semileafless peas: *af af* / *ST ST*
— leafless peas: *af af* / *st st*.

They represent an entirely new crop. Their main advantages are:

Fig. 16. Leaves of different *Pisum* genotypes. *Left acacia* mutant homozygous for gene tl^w; *middle afila* mutant homozygous for gene *af*; *right pleiofila* recombinant homozygous for genes tl^w and *af*

- a very high standing ability because of the tremendous increase of the number of tendrils per plant;
- uniformity of ripening;
- less attack by pathogens and insect pests;
- easier drying.

With regard to yield, distinct Italian lines are better than their mother variety due to a higher number of ovules per carpel. The English material has only about 50% seed yield of that of the conventional pea on a single-plant basis. Regarded as a crop, however, the recombinants of this genotype have so many advantages that they have been developed into commercial varieties ("Filby" in England, developed by means of Goldenberg's spontaneous *afila* mutant; "Hamil" in Poland, developed by means of the induced *afila* mutant mentioned above). Their good physiological capacity is insofar surprising as the leaf area of the plants is considerably reduced. We have crossed Goldenberg's *afila* mutant with some of our genotypes and a great number of different recombinants are available at our institute homozygous for gene *af* and distinct other mutant genes. The seed production of most of these recombinants is bad. It is obvious that the genotypic background, in which gene *af* becomes effective, is an important factor with regard to the productivity of the strains tested. It should be mentioned that the combination of the genes *af* and *st* does not lead to really "leafless peas," because the stipules are still present in a reduced form. By combining *af* with the *cochleata* gene, plants arise which do not have any leaflike organs; they consist exclusively of stems and tendrils. Their seed production is so low that they are without any agronomic interest (Gottschalk 1973; Gottschalk and Kaul 1973).

The alteration of leaflets into tendril-like organs is also known for two mutants of *Lens culinaris* which, however, cannot be used for breeding purposes because of sterility or strongly reduced fertility (Sharma and Sharma 1978b).

9 Mutations Affecting the Root System

Many thousands of mutants of a large number of different crops are known that show alterations in the shoot system in the broadest sense of this term, hundreds of them being of direct interest for realizing specific aims of breeding. It can be assumed that a relatively large proportion of genes of a genome controls the development and functionability of the root system of a plant. Mutations of these genes should realize distinct alterations of this plant organ. Pleiotropic effects influencing both shoot and root in a corresponding way may exist in exceptional cases. In general, however, it can be expected, that the characters of the shoot and the root system of our crops are controlled by different genes which mutate independently from each other. The small number of findings available in this field confirm this assumption. It is conceivable that the effectiveness of the root system with regard to water and nutrient uptake could be improved under the influence of some mutant genes. In this way it could be possible to improve the physiological capacity of a crop to some extent.

Unfortunately, this branch of mutation research has been completely neglected. No consequent selection of "root mutants" has been carried out and mutants of this type are not available in our collections. The main root of some *barley erectoides mutants* remains shorter and begins earlier to form primary and secondary lateral roots. Three weeks old plantlets of the mutant *ert-a^{23}*, for instance, have already many secondary lateral roots, whereas they are not yet present in the control material (v. Wettstein 1954). Dwarfy *barley* mutants with a strongly developed root system were also selected by Górny in Poland (1978). Some of our *Pisum* mutants show likewise significant differences in the structure of their root system already during the earliest stages of ontogenetic development (Quednau 1972). In older plants, genetically conditioned differences were found in:

— length of the main root,
— length of the lateral roots,
— number of the lateral roots.

Moreover, there are clear differences between different genotypes with regard to the speed of the mobilization of the storage substances stored in the cotyledons. In this way, certain advantages of distinct mutants during the earliest ontogenetic stages can be achieved (Gottschalk and Hasenberg 1977).

In all these cases, the root system of mutants, selected for other traits, has been investigated. The alterations observed are probably due to the pleiotropic action of the respective genes. It should, however, be tried to select genotypes with alterations in the root system specifically. This has not yet been done. Some gamma-ray-induced *mulberry* mutants were found to have an increased rooting ability in comparison to that of the mother variety. This is insofar an agronomically important character as *Morus* is multiplied by cuttings (Kukimura et al. 1975).

10 The Alteration of Flowering and Ripening Times

Flowering and ripening are quantitative characters in crops and are controlled by many genes. Therefore, a great number of genotypes can be expected in mutation treatments differing from each other in this respect. Earliness is an important aim of breeding for many crops grown under distinct ecological conditions. Very often it is combined with reduced seed production. Lateness is less desired as a mutant character. In some cases, however, late ripening genotypes are more favorable than their earlier ripening mother varieties.

10.1 Earliness

Earliness is one of those characters of a crop that can be obtained reliably in mutation experiments. A large number of early flowering and/or ripening mutants has been selected in many crops after having used different physical or chemical mutagens. Examples are given in Tables 14–18 along with some characteristics of the respective genotypes. Review papers with many references were published by Micke (1978, 1979).

The causes of the genetically conditioned earliness may often lie in a changed photoperiodic reaction. This holds true for instance for the early ripening *barley* mutant $ea\text{-}a^8$ giving rise to the commercial variety "Mari" (Hagberg 1967; Gustafsson et al. 1971). Some mutants of *Gossypium hirsutum* show photoinsensitivity. As a consequence of this behavior, the cultivation of cotton could be extended both in time and space. These genotypes take 2–2.5 months less for maturity (Raut et al. 1971). Similar genotypes are available in *jute* (Hossain and Sen 1978b). In other cases earliness is due to forming the first flowers at very low nodes of the stem. This is the case in some mutants of *Pisum sativum* (Gottschalk 1960a; Wellensiek 1965; Monti and Scarascia Mugnozza 1967, 1970; Chen and Gottschalk 1970) and *Corchorus olitorius* (Basu 1966). In *Pisum*, the position of the first flowers is controlled by a series of four multiple alleles as follows (Uzhintseva and Sidorova 1979):

- No^h first flower on 18. to 20. node
- no^m first flower on 11. to 14. node
- no first flower on 8. to 9. node
- no^l first flower on 5. to 7. node.

Table 14. Early flowering barley mutants

Authors	Mutagens	Remarks
– Tavčar 1965	Gamma rays	2 mutants with increased yield
– Gustafsson et al. 1966	X rays, neutrons, DES, EI	
– Dormling et al. 1966		Reaction of "Mari" in the phytotron
– Hagberg 1967		
– Gustafsson 1969		
– Dormling and Gustafsson 1969		Phytotron experiments
– Jech 1966a		
– Hänsel 1966	X rays	
– Favret and Ryan 1966	X rays	"Mari" from Sweden used in crossbreeding in Argentina
– Pollhamer 1966, 1967	X rays	
– Kivi 1967	X rays	"Mari" used in crossbreeding in Finland
– Scholz 1967, 1971	X rays, EMS	Used in crossbreeding
– Trofimovskaya and Zhukovsky 1967	EMS	
– Sethi et al. 1969	^{32}P, ^{35}S	
– Donini and Devreux 1970	Gamma rays	
– Bansal 1971, 1972	Gamma rays, neutrons, EMS, NMU	20–25 days earlier; normal yield 5 days earlier, increased yield
– Yamashita et al. 1972	Gamma rays, EI	38 early mutants
– Devreux et al. 1972	Gamma rays	25 days earlier blooming
– Ibrahim and Sharaan 1974b	Gamma rays	25–35 days earlier heading
– Yamaguchi et al. 1974	DES	
– Sethi 1975	^{32}P, ^{35}S, EMS	30 days earlier; very poor yield 7 days earlier, normal yield
– Stephanov and Gorastev 1976	Gamma rays, EMS, DES	6–12 days earlier
– Enchev 1976	Gamma rays, neutrons	
– Hussein et al. 1979, 1980	Gamma rays	25 and 38 days earlier
– Ukai and Yamashita 1980	Physical and chemical mutagens	150 early flowering mutants

The degree of earliness varies considerably in the different genotypes. In exceptional cases an extraordinarily high shift of the flowering and/or ripening periods is realized under the influence of a mutant gene. Heading of a Japanese *rice* mutant, for instance, is about 60 days earlier than that of the mother variety (Tanaka 1968). A mutant of *Corchorus capsularis,* in spite of maturing 40–50 days earlier, shows an increased fiber yield (Singh 1971). Two mutants of *Corchorus olitorius,* flowering about 60 days earlier, on the other hand, have a longer flowering period than the mother variety. Some other early flowering jute genotypes exceed the control material with regard to seed production (Basu 1966). An astonishing example refers to *castor (Ricinus communis).* The Indian varieties are in general late taking 220–270 days for maturity. One of the mutants, obtained after neutron irradiation, ripens within 120–150 days and shows a very high seed yield. It was released under the name "Aruna" (Kulkarni 1969; Swaminathan 1969a, b). A gamma-ray-induced *sugarcane* mutant proved to be very vigorous, showing a growth rate about 50% higher than that of the mother clone. Its

Table 15. Early flowering rice mutants

Authors	Mutagens	Remarks
— Majumder 1965, 1969	^{32}P	9 days earlier; reduced yield
— Kawai 1967	Gamma rays	Variety "Reimei"; increased yield
— Marie 1967, 1970	X rays, gamma rays, EMS	
— Ree 1968	X rays	8 days earlier; normal yield
— Viado 1968	X rays, gamma rays	14–24 days earlier
— Miah and Bhatti 1968	Gamma rays	
— Siddiq and Swaminathan 1968	Gamma rays, EMS, NMU	
— Tanaka 1968, 1969	Gamma rays	30 + 60 days earlier heading; some of the mutants with normal or increased yield
— Takenaka 1969	Gamma rays	Up to one month earlier; some mutants with normal or higher yield
— Kawai 1969; Kawai and Narahari 1971	^{32}P	7 early heading mutants; some of them normal or higher yield
— Miah et al. 1970	Gamma rays	39 lines up to 9 days earlier heading; reduced yield
— Basu and Basu 1970	^{32}P	Early ripening with short grains
— Haq et al. 1971	Gamma rays	10 and 21–25 days earlier
— Li et al. 1971	Rays, chemicals	
— Miah and Awan 1971	Rays, chemicals	Earliness with normal yield; 30–35 days earlier
— Mikaelsen et al. 1971	Neutrons	2–3 weeks earlier maturing; good yield
— Misra et al. 1971	EMS, DES, EI, NMU	26 early maturing mutants; up to 15 days earlier
— Pawar 1971	Gamma rays, neutrons	
— Reddy and Reddy 1971, 1973; Reddy 1975	DES	Semidwarf, 10 days earlier maturing; 15–20% more yield
— Hu 1973	X rays	
— Ram 1974	Gamma rays	
— Sharma et al. 1974	EMS	11 early maturing mutants; 17–30 days earlier flowering; higher yield 1 mutant 30 days earlier, higher yield, fine grain quality
— Reddy et al. 1975	EMS	One week earlier maturing
— Roy and Jana 1975	EMS, EI	One week earlier; normal yield
— Ismachin and Mikaelsen 1976	Gamma rays	Many mutants about 30 days earlier maturing; some of them normal or higher yield
— Rutger et al. 1976	Gamma rays	2 mutants; reduced yield
— Simon and Sajo 1976	Neutrons	Released variety
— Ismachin Kartoprawiro 1977	Gamma rays, EMS	107 early maturing mutants; some of them resistant to bacterial leaf-blight
— Mallick 1978	EMS	One mutant 14–23 days earlier maturing
— Kaul 1978	Gamma rays	Earliness combined with fine grain and high yield
— Dasgupta 1979	X rays	
— Ismail and Ahmad 1979	Gamma rays	3 early mutants with high yield
— Afsar et al. 1980	Sodium azide	

Table 16. Early flowering mutants of other *cereals*, of *Lolium* and *Saccharum*

Species	Authors	Mutagens	Remarks
Triticum aestivum	— Khvostova 1967	Gamma rays	
	— Sawhney et al. 1971	Gamma rays	An early ripening mutant with good yield
	— Mehta 1972	Gamma rays, EMS, NMU	18 mutants; 3–15 days earlier
	— Kalashnik et al. 1972		1 dominant mutant
	— Khan 1973	Neutrons	Strain "Rageni"; 3 weeks earlier maturing; recommended for commercial sowing
	— Djelepov 1978	Gamma rays, EMS	2 mutants, 4–5 and 10 days earlier
Triticum durum	— Scarascia Mugnozza 1966b, 1967	X rays, neutrons, chemicals	10 days earlier heading
	— Bozzini 1971		
	— Pacucci and Scarascia Mugnozza 1974	Neutrons	Variety "Casteldelmonte"
	— Rossi 1972	Neutrons, EMS	Up to 1 week earlier
Zea mays	— Popova 1975	EMS, NMU	
Sorghum bicolor	— Singh and Drolsom 1974	DES	3 weeks earlier flowering
Sorghum vulgare	— Goud et al. 1970; Goud 1972	Gamma rays	
Saccharum sp.	— Jagathesan and Ratnam 1978	Gamma rays	60 days earlier with higher yield
Pennisetum typhoides	— Burton 1974	Neutrons	
Panicum miliaceum	— Jashovsky and Golovchenko 1967	Gamma rays	
Lolium multiflorum	— Malik and Mary 1971, 1974	EMS, MMS, DES	10 days earlier flowering; allogamous species
Lolium temulentum	— Malik and Mary 1971, 1974	EMS, MMS, DES	

vegetative growth ends 60 days earlier. The plants produce a higher shoot population resulting in an increased number of millable canes at maturity. Other economically important traits such as sucrose content and juice purity are not negatively influenced (Jagathesan and Ratnam 1978).

In most cases, the early ripening mutants are not competitive to their later ripening mother varieties with regard to yield. This is often due to negative by-effects. They can be caused by the pleiotropic action of the mutant genes or by the presence of whole groups of different, independently mutated genes. Moreover, there are certain indications that the mutant gene may act as a kind of foreign element within the genome, disturbing its genic balance which results in a reduced vitality or seed production of the plants. These relations are not yet analyzed genetically, physiologically, and biochemically. But it is clear that the reduced yield of the respective genotypes is not equivalent to a reduced fertility. The meiosis is normal in these mutants and enough functional male and female germ cells, necessary for normal seed production, are available. The reduced yield is not a cytological but a physiological problem.

Table 17. Early ripening and/or early flowering mutants of different legumes

Species	Authors	Mutagens	Remarks
Pisum sativum	– Wellensiek 1965	Neutrons	
	– Fierlinger and Vlk 1966	Gamma rays	
	– Enken 1967	Gamma rays, EI	Up to 10 days earlier
	– Monti and Scarascia-Mugnozza 1967, 1970, 1972	DES	Very early; used in cross-breeding
	– Sidorova et al. 1969; Sidorova 1970; Sidorova and Khvostova 1972	Gamma rays, EMS, EI	7–9 days earlier
	– Chen and Gottschalk 1970; Gottschalk and Patil 1971; Gottschalk 1972, 1976a, 1981c, 1982a; Gottschalk and Hussein 1975; Gottschalk and Wolff 1977; Gottschalk and Bandel 1978; Bandel and Gottschalk 1978; Gottschalk et al. 1978; Müller and Gottschalk 1978; Gottschalk and Müller 1979	X rays, neutrons	10–14 days earlier flowering; reduced yield; tested in different climates; combined with other mutant genes
	– Vassileva 1976		10–12 days shorter vegetative period
	– Kaul 1977	Gamma rays	14 days earlier; increased seed size
	– Kaul 1980	X rays	31 and 40 days earlier; 2 nonallelic genes
	– Uzhintseva and Sidorova 1979	?	12 early flowering mutants
Psophocarpus tetragonolobus	– Kesavan and Khan 1977	Gamma rays, EMS	
Lupinus albus	– Porsche	X rays	Many early mutants; some of them with normal yield
Lupinus mutabilis	– Pakendorf 1974	Gamma rays	
Vigna radiata	– Prasad 1976	EMS	Higher yield, drought tolerant
	– Shakoor and Haq 1980	Gamma rays	10–15 days earlier maturing; higher yield
Phaseolus vulgaris	– Swarup et al. 1971	X rays	Cultivar "Pusa Parvati", high yield
Phaseolus aureus	– Dahiya 1973	Gamma rays	Only M_2 findings
Glycine max	– Enken 1967	Gamma rays	
	– Zacharias 1967	X rays	Low yield
	– Josan and Nicolae 1974	Ionizing radiations	
	– Rubaihayo 1976	Gamma rays	
	– Kotvics 1976	Gamma rays	
	– Baradjanegara 1980	Gamma rays, neutrons	3–7 days earlier
Medicago polymorpha	– Brock et al. 1971	Gamma rays	Up to 3 weeks earlier; normal vigor
Lathyrus sativus	– Nerkar 1976	Gamma rays, EMS, NMU	
	– Prasad and Das 1980	Gamma rays	Only M_2 findings

Table 18. Early flowering mutants in other dicotyledonous crops and in some vegetatively propagated species

Species	Authors	Mutagens	Remarks
Lycopersicon esculentum	– Khvostova 1967	Gamma rays	Variety "Luch 1"; high yielding
	– Poláček 1967	Gamma rays	
	– Yordanov et al. 1977	Gamma rays	6–10 days earlier ripening
	– Zagorcheva and Yordanov 1978		
Sesam indicum	– Kobayashi 1965	X rays, neutrons, gamma rays	10–15 days earlier
Ricinus communis	– Kulkarni 1969; Swaminathan 1969	Neutrons	4 months earlier
Corchorus olitorius	– Basu 1966	X rays, ^{32}P	80 days earlier, higher seed production
Corchorus capsularis	– Singh	X rays	40–50 days earlier; increased fiber yield
	– Hossain and Sen 1978b	Gamma rays	50 days earlier
Brassica juncea	– Nayar and George 1969	X rays	13 days earlier; reduced seed yield
Brassica campestris	– Kumar 1972	Gamma rays, neutrons	
Helianthus annuus	– Remussi and Gutierrez 1965	X rays	30 days earlier flowering; allogamous species
Vegetatively propagated species			
Malus sylvestris	– Nishida 1973	Gamma rays	
Prunus armeniaca	– Lapins 1973, 1975	Gamma rays	
Prunus persica	– Donini 1976	Gamma rays	22 days earlier
Prunus cerasus	– Zwintscher 1967	X rays	10 days earlier flowering
Citrus sinensis	– Spiegel-Roy and Kochba 1975	X rays	
Achimenes	– Broertjes 1972a	X rays, neutrons	2 early flowering mutants commercialized

Some examples may elucidate this widely realized situation. An early ripening Italian *pea* mutant shows an increased degree of branching but its seed production is very low. After having backcrossed this mutant with the initial line, two very early genotypes were selected. The plants were highly branched and formed three and a half times more flowers, but the number of seeds per pod was strongly reduced (Monti and Scarascia Mugnozza 1967). Similar observations were made in *durum wheat* (Rossi 1972). Early flowering mutants of this species showed certain negative "modifications" concerning the chlorophyll content, the whole vegetative habit, the spikes and caryopses, and especially the fertility. The earlier the mutants, the higher the frequency and the degree of these by-effects. Two X-ray-induced early maturing *barley* mutants outyielded the mother variety by 2.9% and 8.8%, respectively, due to an increased kernel size. They are nevertheless not of direct agronomic value because the plants break down heavily when in an overripe stage, this leading to the loss of much seed material (Hänsel 1966). The mutant "appressed pod" of *Brassica juncea* flowers about

13 days earlier than its initial line. The number of branches and fruits per plant is increased and the pods are less shattering. Unfortunately, these favorable traits cannot be utilized for breeding purposes because of the reduced seed size and the seed yield of the mutant (Nayar and George 1969).

Occasionally, early ripening mutants are competitive with or even superior to their mother varieties with regard to vitality and seed production. Examples are known for *Lupinus albus* (Porsche 1967), *Pisum sativum* (Kaul 1977), *Oryza sativa* (Sharma et al. 1974; Ismachin and Mikaelsen 1976), and *barley* (Bansal 1972). Tanaka (1969) selected 1400 *rice* mutants after gamma irradiation, among them 112 early ripening genotypes. Some of them had shorter culms, resulting in lodging resistance, and had higher yield particularly under the conditions of increased plant density. Two EMS-induced *rice* mutants, 10 days earlier maturing than the mother variety, were tested at 15 different locations in India. They exceeded the control in yield by 15%–20% (Reddy 1975). *Green gram (Vigna radiata)* is an important pulse crop for the drylands of Rajasthan in the western parts of India. A mutant strain, obtained in EMS treatments, is superior to its mother variety under varying conditions of rainfall due to a favorable combination of earliness, drought resistance, and increased seed production (Prasad 1976). An Indian *cotton* mutant seems to be promising for breeding because of a combination of favorable traits. The early ripening plants are short-branched with a clustering habit. Thus, they occupy only about one third of the space required by the control plants (Raut et al. 1971). Some mutants of *Medicago polymorpha*, flowering 2–3 weeks earlier, proved to be comparable to the controls in vigor (Brock et al. 1971).

The usefulness of some early ripening mutants becomes obvious from the fact that some of them have been developed into released commercial varieties. This holds true for the following genotypes:

- The Japanese *soybean* variety "Raiden" (Kawai 1967).
- The Indian cultivar "Pusa Parvati" of *Phaseolus vulgaris* (Swarup et al. 1971). It is an early ripening bush type with round meaty pods. The green pods and the mature seeds can be consumed.
- The Russian *tomato* variety "Luch 1" (Khvostova 1967).
- The Indian variety "Aruna" of *castor* already mentioned (Kulkarni 1969; Swaminathan 1969a, b).
- The Swedish *barley* variety "Mari" (Dormling et al. 1966; Hagberg 1967). It is about 8 days earlier than its mother variety "Bonus" and is grown on large areas in North and West Europe. Moreover, it is widely used in crossbreeding programs.
- The Italian variety "Casteldelmonte" of *Triticum durum*. It is well adapted to irrigated conditions and to high nitrogen levels (Pacucci and Scarascia Mugnozza 1974).
- The Pakistani mutant "Rageni" of *Triticum aestivum* maturing 3 weeks earlier than its mother variety and having an increased protein production per unit area. The mutant is recommended for commercial sowing (Khan 1973).
- The Hungarian variety "Nucleoryza" of *Oryza sativa* heading 3 to 4 weeks earlier and maturing 2 to 3 weeks earlier than its mother variety and having excellent yielding properties (Mikaelsen et al. 1971; Simon and Sajo 1976).

— An early ripening Indian semidwarf *rice* mutant being lodging resistant due to its stiff culms and yielding about 20% more than the control. The strain can directly be used for commercial cultivation (Reddy and Reddy 1971).

Other early ripening mutants are incorporated into crossbreeding programs and may be utilized indirectly (*barley:* Hagberg 1967; *peas:* Monti and Scarascia Mugnozza 1970). A gene for earliness of the *Pisum* genome was combined with many other mutant and nonmutant genes. In this way, early flowering recombinant strains with improved traits were developed exceeding not only the original mutant but even the mother variety (Gottschalk and Hussein 1975; Gottschalk and Bandel 1978; Bandel and Gottschalk 1978). Similar results have been obtained in *Lupinus albus* (Porsche 1967). The mutant *barley* variety "Mari," developed in Sweden, was tested at Castelar in Argentina. It was found to behave as a drastic early mutant. Another one has an even shorter life cycle. The respective two genes for earliness have been transferred into Castelar breeding material with the aim to develop lines enabling the increase of the number of generations per year (Favret and Ryan 1966).

Early flowering mutants are interesting genotypes for genecological studies. Investigations of this kind have been carried out in *barley* (Dormling et al. 1966; Dormling and Gustafsson 1969), *durum wheat* (Pacucci and Scarascia Mugnozza 1974), *soybeans* (Zacharias 1967) and *peas* (Gottschalk and Patil 1971; Sidorova and Khvostova 1972; Gottschalk 1976a, 1978b; Gottschalk and Wolff 1977; Kaul 1977). Experience drawn from these experiments shows that not only the degree of earliness but also the yielding ability of early ripening mutants can strongly vary under the influence of specific environmental factors. Thus, the ecological reaction of the mutants and their adaptability to climatic or soil conditions are important factors for judging their agronomic value. These problems are discussed in detail in Chap. 18.

10.2 Lateness

The best *rice* mutant, obtained in mutation treatments in South France, is a short-strawed genotype with improved lodging resistance and an about 50% increase in yield, but it is 6 days later than the mother variety (Marie 1967). An Indian *peanut* mutant, having 20% more yield, shows likewise a delayed maturity and requires either prolonged rains or additional irrigations for being able to exhibit its genetically conditioned advantages (Patil and Thakare 1969). These are two characteristic examples apart from many similar ones. In our X-ray and neutron experiments with *Pisum sativum*, more than ten genotypes were selected, outyielding the initial line considerably but ripening 1–2 weeks later. Details on their seed production are graphically presented in Fig. 8.

Under specific ecological conditions, however, lateness is a desired character in some crops. A late-maturing *rice* mutant, obtained in India, was found to have a high yielding potential and could be of great value for cultivation in the typical monsoon areas of the country (Misra et al. 1971). A high-yielding Indian *Sorghum mutant*, on the other hand, is able to escape the late rains at the time of harvest because it is 20 days later than the initial line. In this way, grain spoilage can be avoided (Kajjari et al. 1969).

10.3 Changes of the Photoperiodic Reaction

The photoperiodic reaction of crops can be altered under the influence of mutant genes. This has been demonstrated very clearly by Gustafsson's group testing *barley* mutants in the Stockholm phytotron. Similar experiences were made in Japan by Yamashita et al. (1972). They obtained 38 early ripening *barley* mutants after gamma-ray and EI treatment. With regard to their response to photoperiod, these genotypes can be subdivided into three different groups:

— nonsensitive mutants: emerging 10 days earlier;
— slightly sensitive mutants: emerging 7 days earlier;
— highly sensitive mutants: emerging 3 days earlier than the initial line.

In *rice,* the sensitivity against photo- and thermoperiods can be altered rather easily by mutations (Hsieh and Chang 1975). Some Taiwanian rice mutants have lost the sensitivity to photoperiod, unlike their mother varieties. As a consequence of this altered physiological behavior, it seems to be possible to grow one more crop within a year in distinct areas (Hu 1973). A similar situation is realized in certain Indian *cotton* mutants showing photo insensitivity: they mature 2–2.5 months earlier than their initial lines (Raut et al. 1971). *Gossypium hirsutum* ssp. *mexicanum* is a highly wilt-resistant perennial wild species which fructifies only under short-day conditions. An EI-induced Russian mutant was found to fructify under the conditions of normal day length (Egamberdiev and Payziev 1977). The earliness of a dominant *soybean* mutant is likewise due to a changed photoperiodic sensitivity (Lee Choo Kiang and Halloran 1977).

In some vegetatively propagated species, mutants of this category are of direct agronomic value. The Indian *potato* varieties are typical short-day plants with regard to tuber formation. After having treated dormant buds (eyes) of freshly harvested tubers with NMU, day-neutral mutants were selected giving very good tuber yields within 90 days. The control material had hardly produced any tubers under these conditions (Upadhya and Purohit 1971, 1973; Upadhya et al. 1974, 1976). The opposite situation was found in *Streptocarpus.* The commercial varieties of this ornamental plant form their flowers only under long-day conditions. An X-ray-induced mutant flowers under both long- and short-day conditions (Roy Davies and Hedley 1975). Because of its all-year-round flowering, it shows a very desirable improvement of the commercial varieties.

The *Pisum* variety, used for our radiation genetic experiments, reacts day-neutral; the plants are flowering under both long- and short-day conditions. Some high-yielding fasciated mutants of our collection, discussed in detail in Sect. 6.3.3, do not flower in Egypt, Brazil, and India. In Ghana, they were extremely late and produced only a few seeds. When grown under controlled phytotron conditions, the flowering behavior of these genotypes was found to depend highly on the photoperiod. Under **long-day phytotron conditions** (6 h darkness), they began flowering 5 to 7 weeks later than the mother variety, whereas they are only 6 to 10 days later under the long-day field conditions of West Germany. The phytotron plants were normally developed and healthy; nevertheless, their flowering was very poor, in strong contrast to the field plants. The delayed and poor flowering of these genoypes cannot be a photoperiodic effect because there was long-day in the field as well as in the phytotron. The differences observed are obviously due to the temperature which was considerably higher in the phytotron than in the field. Under **short-day phytotron conditions** (12 h darkness), four mutants of this group did not flower at all. This holds also true for a strongly fasciated mutant of Mrs. Vassileva's collections in Bulgaria and for the fasciated fodder pea variety "Ornamenta." Even at an age of 93 days after sowing the apical growing points of all the plants of these six genotypes produced exclusively minute foliage leaves; they were purely vegetative. After the phytotron conditions had been changed into long-day, the plants had small flower buds about 2 weeks later. Thus, these mutants require long-day for flowering. They are not day-neutral; on the contrary, they are long-day plants, being homozygous for gene *fis,* which suppresses the initiation of flower formation in short-day (Gottschalk 1981c, h). The plants of a nonfasciated "micromutant" of our collection show a similar behavior. When grown in the field, they differ from the initial line so slightly that they cannot be identified reliably in segregating families. Under long-day phytotron conditions, their internodes are strongly shortened and they flower extremely late. In short-day, they do not flower. It is not yet clear, whether they contain likewise gene *fis* for long-day requirement.

The effect of the *fis* gene of the *Pisum* genome is of interest in connection with the performance of mutational treatments. If such experiments are carried out in countries with short-day, the action of gene *fis* and similar mutant genes becomes immediately discernible because of the nonflowering of the respective plants in distinct segregating M_2 or M_3 families. We did not discern this gene in our X-ray and neutron experiments, because it is not able to express its action in the long-day climate of Middle Europe. On the contrary, the fasciated pea mutants carrying this gene are of direct importance for pea breeding because of their favorable yielding properties, whereas they are completely useless in India or other countries with short-day climate.

In exceptional cases, the photoperiod can influence qualitative characters. A distinct *Pisum* recombinant shows a pronounced stem fasciation in Germany. In Kurukshetra and Varanasi (North India), the plants were nonfasciated when sown at the usual sowing time. Sown 4 weeks later, all the plants of this genotypic constitution were fasciated, obviously due to gradual changes of the photoperiodic conditions during this period (Gottschalk and Kaul 1975).

11 Mutations in Vegetatively Propagated Crops and Ornamentals

Some vegetatively propagated crops, especially certain ornamentals, proved to be well suited for the application of mutagenesis in order to improve their breeding value. The problems related to this special field of mutation breeding have been discussed and reviewed by Nybom and Koch (1965); Privalov (1968); Broertjes et al. (1968); Broertjes (1969b, c); Ohba (1971). Moreover, volumes of symposia organized by the International Atomic Energy Agency in Vienna (1973a, 1975a, 1976a) and the Institute of Radiation Breeding in Ohmiya-machi, Japan (1973), are available. Further details are found in the *Manual of Mutation Breeding* published by the IAEA (1977b). An early list of released varieties was given by Sigurbjörnsson and Micke (1973). A voluminous literature review has recently been published by Broertjes and van Harten (1978) containing so many details that it is not necessary to repeat all the results obtained in this field in the present book. The methods used are carefully described. Moreover, the results available in *potato, garlic*, in tropical and subtropical crops such as *cassava, yam, sweet potato*, and other root and tuber crops, are reviewed and the respective references are given. This holds also true with regard to ornamentals, fruit crops, woody plants, and some other vegetatively propagated species which are utilized agronomically. The progress, made within a period of 20 years, becomes evident from the fact that about 250 mutants of vegetatively propagated species have been developed into officially released varieties. Some examples have been mentioned in the various chapters of the present book. Moreover, we want to discuss some general problems of this branch of mutation research which deviate characteristically from mutation breeding in sexually propagated crops.

For inducing the mutations, the mutagenetic agents can affect different plant organs such as:

— freshly cut leaves *(Streptocarpus, Achimenes)*
— leaf stalks *(African violet)*,
— tubers *(potato, Dahlia)*,
— young rhizomes *(Alstroemeria, Cynodon)*,
— bulbs *(Iris)*,
— dormant buds *(fruit trees, roses, grapes)*,
— cuttings *(cherries)*,
— dormant stolons *(peppermint, Bermuda grass)*.

Even tissue cultures are treated for obtaining mutants. A review on results obtained with this method has been given by Skirvin (1978).

A very effective method, important in practical aspects, with regard to the performance of mutation breeding in vegetatively propagated species, is the so-called adventitious bud technique (Broertjes 1969a,c, 1972b). Many plant species can be stimulated to form adventitious buds on isolated leaves; a voluminous list was published by Broertjes et al. (1968). These buds originate very often from a single meristematic cell. If mutational events have been induced in this cell, a plant having the same genotypic constitution in all its organs arises and is not a chimera. This is a great advantage. If mutated and nonmutated cells are present in a chimeric M_1 plant, diplontic selection occurs. Very often, the mutant cells are not fully competitive with the nonmutant ones. This behavior results in a low frequency of mutants and a narrow mutation spectrum. This unfavorable situation is avoided in many ornamentals derived from a single mutant cell. In this way, very high mutation frequencies as well as a wide genetic variability are obtained. Examples for such a situation are *Streptocarpus, African violet, Achimenes, Kalanchoë* (Broertjes 1969a,b, 1972a; Broertjes and Leffring 1972), and *Begonia* (Doorenbos and Karper 1975) among others. Moreover, there are good prospects to avoid chimerism in clonally propagated plants by treating tissue cultures (Skirvin 1978). In *Chrysanthemum,* mostly chimeras arise, having developed from several cells, when in vivo material is treated. In in vitro material, however, the treatment results in nonchimeral M_1 plants (Broertjes et al. 1976).

Some findings in this field are not yet fully understood. In *Alstroemeria,* for instance, M_1 plants without chimerism were obtained after X-ray treatment of young rhizomes, although the buds have probably multicellular apices (Broertjes and Verboom 1974). In *potatoes,* a high degree of chimerism is obtained in EMS treatments, which, however, is considerably lower after X-irradiation (Miedema 1973). Finally, it should be mentioned that the mutation frequency, obtained after X-ray and neutron irradiation of *Achimenes,* was found to be 20–40 times higher in autotetraploid material as compared to the diploid initial material. The reasons for this unexpected behavior are not yet known, but this is an interesting aspect for utilizing this material in mutation breeding (Broertjes 1976a,b).

An additional advantage of mutagenesis in many ornamentals consists in the fact that even mutations from the dominant to the recessive state of a gene become already discernible in M_1 plants. Recent findings of this kind have been obtained in *carnation* (Badr and Etman 1977). This is due to the fact that many ornamentals are highly heterozygous. After application of mutagens, the homozygous recessive condition is realized in some of these gene pairs. In this material, the selection for mutants can already begin in M_1.

Mutagenesis in some crops is the only possibility for any kind of agronomic improvement. This holds true for sterile cultivars which can be propagated only vegetatively and in which the conventional breeding methods cannot be applied. An interesting example is the *turf* and *forage Bermuda grasses.* These plants are sterile triploid hybrids between 4n *Cynodon dactylon* and 2n *C. transvaalensis.* After gamma-ray treatment of dormant stolons, a very high mutation rate was obtained. These results are due to the highly heterozygous condition of the treated initial material. More than 300 mutants are available; among them are some agronomically interesting genotypes. The genes involved influence leaf size, internode length, spreading rate, herbicide and nematode resistance among others (Burton 1974, 1976; Powell et al. 1974; Powell 1976). Because of the high mutation frequency, applied mutagenesis is now a recommended method for the improvement of these grasses. In *Bahia grass (Paspalum nota-*

tum) and *Kentucky bluegrass (Poa pratensis),* mutants with increased seed set and with improved disease reaction were selected (Powell 1976). A similar situation is realized in the very small group of *haploid cultivars* utilized in horticulture. Recessive mutations are immediately discernible in this material. *Thuja gigantea gracilis* is haploid with 2n = 1x = 11 chromosomes. Many chlorophyll mutants and genotypes with abnormal shoot structure have been selected due to spontaneous mutations (Pohlheim 1972). It would certainly be possible to induce experimentally a broad genetically conditioned variability and to select mutants of horticultural interest.

In some species, which can be propagated sexually as well as vegetatively, the methods of mutation breeding seem to have better chances than the conventional methods. *Brachiaria brizantha,* for instance, is a pasture grass well adapted to Ceylon's dry zones. Its seed set and viability are low and could not be improved by conventional breeding methods. After gamma irradiation of seeds and cuttings, a prospective mutant was isolated showing short internodes, profuse tillering, erect growth habit, reduced pubescence, and rapid regrowth (Ganashan 1970).

The improved characters of the vegetatively propagated plants, controlled by the mutant genes, cover a range of traits. An increase of the genetic variability was obtained with regard to:

— flower color and flower shape in many *ornamentals;*
— earliness, sometimes also lateness in almost all the crops treated;
— shortening of internodes in *fruit trees* and *ornamentals;*
— alterations of the plant type in *ornamentals;*
— improvement of the resistance behavior;
— desirable biochemical alterations in some *fruit trees*, in *tea plants*, and in other crops.

Some details on the influence of the respective genes are given in the various chapters of the present book. Distinct genes, influencing fruit shape and fruit color, may be of interest in *apple* breeding (Gröber 1967; Lapins 1973; Ikeda 1974). This holds also true for the reduced amount of russeting in fruit skin (Lapins 1973).

The effectiveness of the methods used in vegetatively propagated species may be demonstrated by means of a few examples. The number of mutants selected within a surprisingly short time is as follows:

— 1650 mutants in *Streptocarpus* (Broertjes 1969b);
— more than 300 mutants in *Ribes nigrum* (Bauer 1974);
— about 160 flower color mutants in bulbous *Iris* (Hekstra and Broertjes 1968);
— more than 300 mutants in *Bermuda grass* (*Cynodon* sp.; Powell 1976).

12 Heterosis

In a few cases, hybridization between mutants and their initial lines or between different mutants results in a more or less pronounced heterosis effect. The problem has been discussed with a review of the literature by Stoilov and Daskaloff (1976) stating that the combined use of mutations and heterosis seems to be promising in both cross- and self-fertilizing species.

Let us at first give some examples in *cross-fertilizing crops*. In *maize,* the general combining ability of 150 Bulgarian mutant lines of the M_4-M_6 generation was tested. Of these lines 15% were found to have high combining ability, and four lines revealed a 30% higher F_1 yield as compared to that of the standard. They were included into a diallel cross program in order to test their specific combining abilities (Stoilov and Daskaloff 1976). Similar experiments were carried out in Russia on an even larger scale and hybrids with improved productivity were obtained (Shkvarnikov and Morgun 1974). A high degree of heterosis is observed when the corn grass mutants of maize, selected in Hungary, are crossed with normal inbred lines (Pásztor 1979). Some mutant strains of *Mentha piperita,* grown in the United States for peppermint oil and showing a moderate wilt resistance, exhibit a genetically conditioned heterosis for herbage weight, some other ones for oil yield (Murray 1969).

The *sweet clover (Melilotus albus)* is a wild allogamous plant with very low demand for soil quality. Because of the favorable protein composition of its leaves, it could become a forage and green manure crop for poor soils. This disadvantage of the presence of bitter coumarins has already been overcome by isolating radiation-induced mutants with low glucoside content. They are, however, inferior to the initial lines in green-matter production and show a yield depression of about 30%. They can be considerably improved by crossing different low-glucoside sister mutants with each other. After having tested many strains of this category, a pronounced heterosis effect was obtained in 73 out of 91 cases. Full viability was restored and the yield was increased over that of the wild type. There are obviously good prospects for developing nonbitter mutant hybrids with high yield (Micke 1969a, b, 1974, 1976; Römer 1973; Römer and Micke 1974). The heterosis appears not only by crossing distinct mutants with each other but also by crossing low-glucoside mutants with the initial line. Some results covering the data of two subsequent years are graphically given in Fig. 17.

In some *self-fertilizing species,* cases of monohybrid heterosis are known due to the presence of specific mutant genes in the heterozygous condition. This is the case in *Pisum sativum.* Four chlorophyll mutants, out of 22 mutants tested in Russia, were found to exhibit heterosis with regard to seed yield in F_1 (Shumny et al. 1970, 1971; Vershinin et al. 1976, 1979; Shumny 1978). Similar observations with a large group of other mutants were made by Glazacheva and Sidorova (1973) and Sidorova (1975,

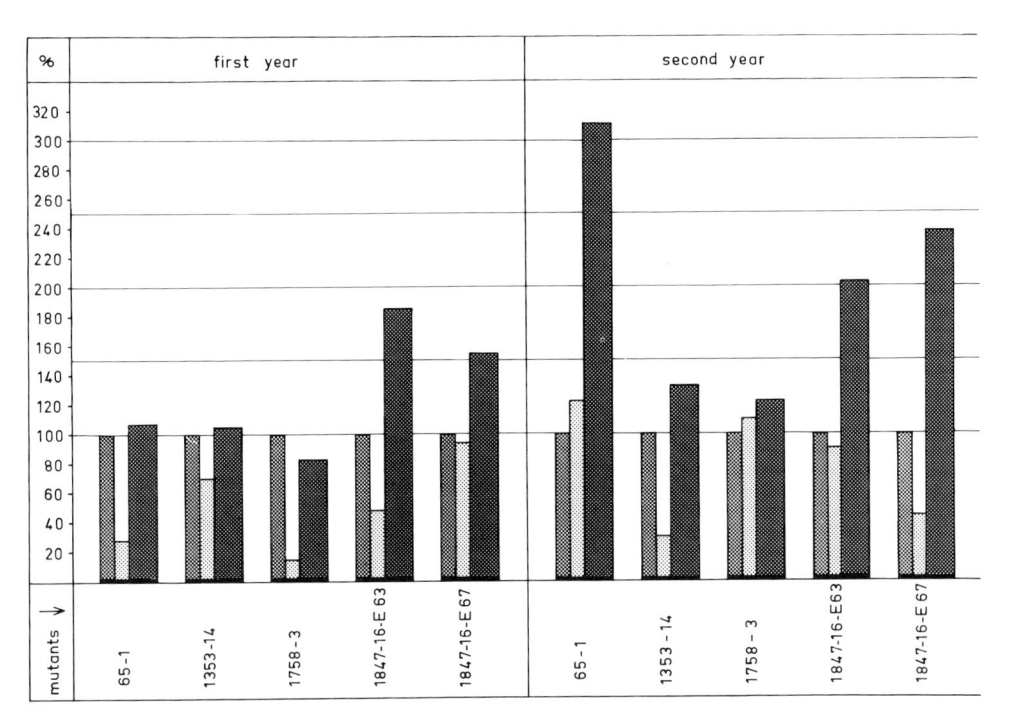

Fig. 17. Heterosis in *Melilotus albus* with regard to plant weight under field conditions investigated in two subsequent years after having crossed different low glucoside mutants with the initial line. *Left column* of each group: initial line; *middle column:* mutant; *right column:* F₁ hybrid. All values are related to the control values of the initial line = 100% (according to data published by Micke 1976)

1981a). In many hybrid combinations a high degree of heterosis, preferably with regard to seed production, was observed which, however, was found to depend highly upon environmental conditions. An extraordinarily high degree of heterosis occurs regularly when fasciated pea mutants of our collection are crossed with non-fasciated genotypes. So far, 17 different cross-combinations of this group were tested. The hybrid vigor refers to both vegetative and reproductive traits. With regard to the plant weight, the F_1 hybrids surpassed the initial line by 100%–280%, with regard to the number of seeds per plant in some cross-combinations by more than 300%. The real degree of heterosis is even higher because it appears also with regard to the seed size. In most cases, the seed production of the hybrids is not only essentially better than that of the nonfasciated mother variety but even better than that of the high-yielding fasciated mutants used for the crossings. This holds also true for the commercial pea variety "Ornamenta" developed by means of a spontaneous fasciated mutant (Gottschalk 1970, 1972a, 1976d, 1977a; Gottschalk and Milutinović 1973a, b). These findings were originally interpreted as cases of monohybrid heterosis. This interpretation, however, has to be revised. It was shown by Lönnig (1982) that the "heterotic effects" just mentioned are due to the rare phenomenon of recessive epistasis. The fasciated mutants of our collection contain a dominant gene for increased number and length of internodes which, however, is hypostatic and is thus not able to manifest its action in the mutants. By means of crosses the respective epistatic gene can be eliminated and the hypostatic gene for strongly increased plant height becomes effective. This trait is not restricted to the heterozygous condition of the F_1 hybrids; on the contrary, it is a permanent character of those strains which have lost the epistatic gene for reduced plant height. The recessive epistasis of our material is probably caused by one of the *fasciata* genes present in the fasciated mutants.

Cases of monohybrid heterosis are known in *tomato* (Popova and Mikhailov 1973; Mikhailov and Popova 1977) and in *oats* (Ahokas 1976). Heterosis in combination with the use of male sterile mutants for hybridization was observed in *Capsicum annuum* (Stoilov and Daskaloff 1976). Some *barley* mutants were studied in the Stockholm phytotron with regard to their reaction to specific photo- and thermoperiods. Heterosis was very common in some genotypes. Its degree was found to depend upon the environmental factors tested. Details have been published by Gustafsson and Dormling (1971); Gustafsson et al. (1973a–c, 1974, 1977). Heterotic effects were also observed after having backcrossed three early flowering Egyptian barley mutants with their mother variety (Hussein et al. 1980).

It is still too early to judge whether heterosis in combination with induced mutations may be a prospective tool in plant breeding. The number of clear examples in this field is very low. Doll (1966) has tested 49 different mutant genes of the *barley* genome in the heterozygous condition evaluating the trait "number of kernels per plant." He was not able to find any clear cases of overdominance. Moreover, some criticism has been uttered with regard to the justification to interpret the findings obtained in the sense of heterosis (Jaranowski 1977). Most of the mutants available in the collections show a reduced selection value. As the negative influence of the recessive mutant genes cannot become effective in the F_1 hybrids, it is clear that these plants are superior to their parental mutants. These are not cases of heterosis. If the hybrids, however, are significantly superior to the initial line with regard to specific vegetative or reproductive traits, I would not hesitate to designate this behavior as "heterosis." This is also valid in those cases in which hybrids between very vital, high-yielding mutants and their mother varieties show a higher yield level than both parents. As an example of this category, the hybrids between fasciated *pea* mutants and the initial line or other useful mutants can be mentioned.

13 Disease Resistance

13.1 Resistance Against Fungi, Bacteria, and Viruses

In the voluminous collections of mutants of different crops, many genotypes showing different kinds of resistance or tolerance against plant diseases have been isolated. These problems are discussed intensively in some symposia organized by the International Atomic Energy Agency in Vienna. The results and many references are found in the respective symposia volumes (1971, 1974a, 1975a, 1976c, 1977a) and the proceedings of a symposium held at the Institute of Radiation Breeding in Ohmiya, Japan (1978). General aspects of resistance breeding by means of mutants were discussed by Favret (1965, 1971, 1976). Recent review papers were published by Konzak et al. (1977; stripe rust in *wheat*) and by Campbell and Wilson (1977; resistance in *fruit plants*).

Some examples may demonstrate the effectiveness of mutagenesis for obtaining resistant mutants. In *rice*, 100 blast-resistant M_2 strains were found in Japan out of 2376 strains tested (Yamasaki and Kawai 1968). In Czechoslovakia, 140 *wheat* lines with improved disease resistance were selected since 1960 (Haniš 1974; Haniš et al. 1976, 1977). In *barley*, 46 mutants with changed reactions to powdery mildew resistance were obtained in EMS treatments in Germany (Röbbelen et al. 1977; Abdel-Hafez and Röbbelen 1979, 1981). Similar results have been obtained in Czechoslovakia (37 mutant barley strains with improved resistance; Haniš et al. 1977). Most of these mutants cannot be used for resistance breeding because of their low yielding ability and some other negative features. Some others, however, have been developed into commercial varieties or are being incorporated into crossbreeding programs. The experiences available so far in this field show that induced mutations can be successfully utilized in breeding for disease resistance in many crops. References on disease-resistant mutants of different crops are given in Tables 19–24.

13.1.1 Barley

The most important barley disease is **powdery mildew** caused by the fungus *Erysiphe graminis*. Plenty of efforts have been made to isolate resistant mutants after having applied physical and chemical mutagens. So far, several hundreds of mutants of this group are available in the collections but only a very small proportion of them can be utilized in barley breeding. References and some additional details are given in Table 19.

The genetic basis of mildew resistance has intensively been studied by Jörgensen (1971a, b, 1974, 1975, 1976, 1977). The resistance is mainly conditioned by the *ml-o*

Table 19. Disease-resistant barley mutants

Disease and pathogen	Authors	Mutagens	Remarks
Powdery mildew *(Erysiphe graminis f.sp. hordei)*	– Hänsel 1966	X rays	Mutant 3502 fully resistant to all *Erysiphe* races; another mutant developed into the variety "Vienna"
	– Pollhamer 1966, 1967 – Jech 1966a	X rays	Some mutants higher yield
	– Schwarzbach 1967	EMS	Resistant to 12 races of *Erysiphe*
	– Hagberg 1967		Used in backcross programs
	– Hänsel 1971	X rays	Resistant to all known races of the pathogen
	– Jörgensen 1971a,b, 1974, 1975, 1976, 1977	Rays and chemicals	
	– Wiberg 1973, 1977	Different mutagens	51 resistant mutants
	– Fuchs et al. 1974	EMS	
	– Haniš 1974	X rays, gamma rays, neutrons, EMS	10 resistant mutants
	– Einfeld et al. 1976	EMS	15 resistant mutants
	– Moës 1977	X rays	Resistant to biotypes A, B, C, D
	– Hentrich 1977, 1979		95 resistant mutants; multiple alleles of the *ml-o* locus
	– Röbbelen et al. 1977; Abdel-Hafez and Röbbelen 1979, 1981	EMS, NaN$_3$	46 mutants with lower or higher susceptibility; not *ml-o* locus
	– Parodi and Nebreda 1977	Gamma rays	
	– Haniš et al. 1977	Gamma rays	37 resistant lines
	– Yamaguchi and Yamashita 1979	EI	7 mutants; three nonallelic
	– Reinhold 1980a, b	EMS	9 mutants, two of them different alleles of the *ml-o* locus
Loose smut *(Ustilago nuda)*	– Pollhamer 1967	X rays	
Helminthosporium sativum	– Pollhamer 1967	X rays	
Yellow dwarf virus	– Parodi and Nebreda 1977	Gamma rays	
Yellow mosaic virus	– Ukai and Yamashita 1979, 1980	Gamma rays	One resistant early mutant

locus of the barley genome located on chromosome number 4. No mildew race was able so far to overcome this resistance. Crosses between ten independently induced resistant mutants show, that the ten genes are noncomplementing alleles of the *ml-o* locus. They are designated *ml-o/1* through *ml-o/10*. Some of the *ml/o* alleles have mutations in different sites within the locus. Multiple allelism for *ml-o* mutants was also found by Hentrich (1978, 1979) and by Reinhold (1980a, b). The alleles studied show quantitative differences in their reaction against mildew; moreover, they have pleiotropic effects. One of the alleles isolated by Hentrich *(ml-o/1)* seems to be of special interest because it deviates functionally and structurally from all the other alleles known so far. The recessive *ml-o* alleles cause almost immunity. But there are still other genes causing partial mildew resistance. They are dominant and are associated with the *ml-o* locus (Wiberg 1977). Some genotypes isolated by Abdel-Hafez and Röbbelen (1979) and by Reinhold (1980a, b) after EMS treatment were found to be due to mutations other than those in the *ml-o* locus. These new sources, which cause partial resistance, have not been described before and are of direct interest for barley breeding.

The following examples may demonstrate that a considerable amount of work is necessary for screening the mildew-resistant mutants:

- Jörgensen (1975, 1976) screened 951,000 M_2 plants resulting in 5 M_1 plants which produced powdery mildew resistant M_2 seedlings.
- Hentrich (1977) selected 95 resistant mutants after having tested more than 2.5 million M_2 plants.
- Yamaguchi and Yamashita (1979) selected 7 resistant mutants from 1,200,000 M_2 plants following gamma-ray or EI treatment. In 4 out of these 7 mutants, the genes for resistance were allelic to each other, in the remaining 3 they were not allelic.

The mildew resistance of barley is not a uniform phenomenon. On the contrary, at least four different groups of mutants with regard to their reaction to *Erysiphe graminis* were found by Einfeld et al. (1976) in EMS treatments:

- Highly susceptible mutants.
- Highly resistant mutants.
- Mutants showing resistance during the early stages of ontogenetic development but becoming susceptible in later stages.
- Mutants susceptible in the seedling stage but being resistant at maturity of the plants.

Similar findings have been obtained by Abdel-Hafez and Röbbelen (1979, 1981). Moreover, the mutants can show a divergent reaction to the pathogen under different ecological conditions. The dominant barley mutant 5455, induced by Hoffmann and Nover (1959) in Germany, was tested at Gembloux in Belgium. It proved to be resistant in the field but hypersusceptible in the greenhouse at 20°C (Moës 1977).

Most of the mildew-resistant barley mutants have a reduced grain yield. They show necrotic or chlorotic flecking on the leaves, effects which are obviously due to the pleiotropic action of the *ml-o* alleles. Out of ten mildew-resistant mutants isolated in Czechoslovakia, none was agronomically satisfactory (Haniš 1974). In Hungarian treatments, some genotypes with increased productivity were found (Pollhamer 1966). An X-ray-induced Austrian mutant, showing resistance against mildew race A_2, was developed into the variety "Vienna," outyielding the mother variety by about 8% due to its increased grain size. The variety has a very good malting quality but the beer produced by "Vienna" malt proved to be not suitable for a long cool storage. Another mutant, resistant to all known races of *Erysiphe,* has a reduced grain size and was therefore not developed into a commercial variety (Hänsel 1966, 1971). In Sweden, mutants of this category were used in a backcross program. In this way, "pure" mildew-resistant mutants were isolated which are obviously free from other background mutations influencing negatively the breeding value of the original mutants. They have normal yield (Hagberg 1967).

Among about 1000 mutants, obtained by X-irradiation in Hungary, some genotypes were selected showing resistance to *Helminthosporium sativum* and partial resistance to *Ustilago nuda* causing loose smut (Pollhamer 1967).

The *barley yellow mosaic virus* is only found in Japan. Out of 150 early ripening mutants, obtained by gamma irradiation, a single genotype was found to be completely resistant to the virus. The resistance is caused by one recessive gene. The mutant is 6 days earlier than its mother variety, due to shortening of the critical day length for heading and a certain reduction in vernalization requirement (Ukai and Yamashita 1979, 1980).

13.1.2 Rice

Rice is, besides barley, that crop to which the methods of experimental mutagenesis are applied particularly intensively, resulting in very comprehensive collections of mutants, preferably in some Asiatic countries.

One of the most important rice diseases is **blast** caused by the fungus *Pyricularia oryzae*. Considerable efforts have been made in order to select blast-resistant mutants. According to Kawai (1974), mutations of this kind occur only in very low frequencies. The expenditure may be seen from the following examples:

— In Japan, 888 viable mutants, obtained by X rays, were tested giving rise to one moderately blast-resistant strain (Kawai 1974). In a gamma field, three blast-resistant mutants were obtained from 51,530 M_2 plants tested (Yamasaki and Kawai 1968).
— Seven resistant lines and 15 individual plants were obtained in Korea from 9,760 M_3 lines tested (Kwon 1974).
— In India, 50 out of 1500 lines originally selected bred true for blast resistance (Kaur et al. 1977).
— In France and Italy, 2000 lines of a gamma-ray program were initially selected giving rise to 178 lines with improved blast resistance in M_6. But only one of them was multiplied for utilizing it as an improved cultivar in Italy (Marie and Tinarelli 1974).

These experiences show that only a very small proportion of blast-resistant mutants is suitable for rice breeding. An interesting mutant was obtained in Bangladesh by gamma irradiation, showing a combined resistance against blast, brown leaf spot, bacterial leaf blight and leaf streak. But also this genotype is not superior to its initial line (Haq et al. 1974).

Resistance against **bacterial leaf blight**, caused by *Xanthomonas oryzae*, is obviously controlled by a polygenic system (Nakai and Goto 1977; Padmanabhan et al. 1977). Examples for resistant mutants are given in Table 20. The yield of fully resistant strains is mostly low, whereas some moderately resistant lines are competitive with their mother varieties in this respect (Padmanabhan et al. 1976, 1977).

Some mutants show resistance to other rice diseases (Table 20). In French material, obtained by gamma irradiation from a [^{60}Co] source, a certain amount of resistance against *Sclerotium oryzae* and/or *Sclerotium hydrophilum* was found. Four M_6 lines proved to be clearly superior to the control, at least one of them being of agronomic value (Bernaux and Marie 1977).

Table 20. Disease-resistant rice mutants

Disease and pathogen	Authors	Mutagens	Remarks
Blast (*Pyricularia oryzae*)	– Yamasaki and Kawai 1968	Rays and chemicals	
	– Takenaka 1969	Gamma rays	Dominant mutation
	– Dasananda and Khambanonda 1970	Rays, EMS	
	– Haq et al. 1974	Gamma rays	Combined resistance to blast, brown leaf spot, bacterial leaf blight and leaf streak
	– Kawai 1974	X rays	
	– Kwon 1974	Gamma rays	7 lines and 15 individual resistant plants selected
	– Marie and Tinarelli 1974	Gamma rays	178 lines with improved blast resistance
	– Nayak and Padmanabhan 1974	Gamma rays, EMS	
	– Ram 1974	Gamma rays	
	– Ree 1971	X rays, neutrons	4 mutants resistant
	– Ree et al. 1974	Neutrons	
	– Woo and Ng 1974	EMS	1 mutant: blast resistant, late maturity, erectoides, lodging resistant
	– Kaur et al. 1976, 1977	EMS	
	– Kwon and Oh 1977	X rays	17 resistant lines
Bacterial leaf blight (*Xanthomonas oryzae*)	– Singh and Rao 1971	Gamma rays, EMS, NMU	Some resistant mutants
	– Haq et al. 1974	Gamma rays	Combined resistance to blast, brown leaf spot, bacterial leaf blight and leaf streak
	– Ram 1974	Gamma rays	Moderate resistance
	– Padmanabhan et al. 1976, 1977	EMS	Some moderately resistant lines with normal yield potential
	– Ismachin and Mikaelsen 1976		
	– Ismachin Kartoprawiro 1977	Gamma rays, EMS	Some early maturing mutants with full and moderate resistance
	– Nakai and Goto 1977	Neutrons, gamma rays, EI	Some resistant mutants
Sheath blight (*Rhizoctonia oryzae*)	– Ismachin Kartoprawiro 1977	Gamma rays, EMS	Some early maturing mutants resistant
Leaf spot (*Helminthosporium oryzae*)	– Haq et al. 1974	Gamma rays	Combined resistance to blast, brown leaf spot, bacterial leaf blight and leaf streak
	– Ram 1974	Gamma rays	Moderate resistance
(*Sclerotium oryzae, Sclerotium hydrophilum*)	– Marie 1967	EMS	
	– Bernaux and Marie 1977	Gamma rays	
Bacterial leaf streak (*Pseudomonas* spec.)	– Haq et al. 1974	Gamma rays	Combined resistance to blast, brown leaf spot, bacterial leaf blight and leaf streak
Tungro virus	– Ram 1974	Gamma rays	Moderate resistance

13.1.3 Bread and Durum Wheat

In *Triticum aestivum,* breeding for resistance to the different rusts and to powdery mildew is an important problem. Table 21 shows that plenty of work has been done in the last years for contributing to solve this problem by means of mutations. Additional results were published by Khvostova (1967); Khvostova et al. (1965); Raut et al. (1974). The frequency of mutants resistant to stripe rust, leaf rust, and mildew was found to vary between 0.005 and 0.16 mutations per 100 M_2 plants in Czechoslovakian treatments (Haniš 1973). The majority of these mutants had additional undesirable characters, but some genotypes are available showing a yielding level equal to or higher than that of the mother variety (Jech 1966b; Haniš et al. 1976; Skorda 1977). Others are used in crossbreeding programs (Haniš et al. 1977). This holds true also for some mutants showing resistance not only to one but to several rusts (Mohamed et al. 1965).

The **stem rust resistance** is controlled by a group of largely recessive genes showing small additive effects. As a consequence of this polymeric situation, it is unlikely that mutants arise in a mutation program having full resistance to *Puccinia graminis* (Knott 1977). Some X-ray-induced mutants with improved stem rust resistance were crossed with each other and two strains were selected exceeding the mother variety by 13% − 15% in yield. This positive behavior is obviously due to the resistance against strain 17/63 of the pathogen (Peixoto Gomez 1972).

Findings not yet completely understood have been obtained in Yugoslavia with regard to **leaf-rust-resitant strains**. They were found in the M_{12} to M_{16} generations of old mutant lines, which had been selected for characters other than disease resistance after X-ray and neutron irradiation. The mutational events have obviously occurred in advanced generations, probably during M_3 to M_5. One of these mutants differs in three genes from its mother variety with regard to its reaction to *Puccinia recondita tritici* (Borojević 1975, 1978, 1979).

Theoretical and practical aspects of the utilization of **stripe-rust-resistant mutants** in wheat breeding have been discussed by Konzak et al. (1977). The genetic basis of this resistance is very complicated. According to Line et al. (1974), both recessive and dominant genes become effective in this field. Moreover, they can act in a complementary way. In diallel crosses between 25 sources of resistance, at least 12 different genes were found to influence the resistance to one distinct race of *Puccinia striiformis* specifically. One out of 15 Czechoslovakian stripe-rust-resistant mutants has the prerequisites for direct utilization as a variety (Haniš et al. 1969).

Mildew-resistant wheat mutants can be obtained relatively easily. Out of 120,000 adult M_2 plants tested in Hungary, 0.6%−0.9% exhibited field resistance to *Erysiphe graminis* (Kiraly and Barabás 1974). The induction of mutations for **resistance to Septoria nodorum** is insofar of considerable importance as only little natural genetic variability for this kind of protection is available. Resistant or tolerant mutants have been obtained in Kenya, Switzerland, and Argentina (Little 1971; Brönnimann and Fossati 1974; Favret et al. 1974). The genes governing *Septoria* tolerance belong to a polymeric system with additive effects (Fossati and Brönnimann 1976).

In *Triticum durum,* mutants with a higher level of resistance to *Puccinia graminis, recondita,* and *striiformis* were isolated by Parodi and Nebreda (1977) and by Skorda (1977). Two Italian mutants, being short-stemmed and early maturing, were found to have an improved degree of resistance against bunt (*Tilletia triticoides;* Bozzini 1971).

13.1.4 Oats

In contrast to the results discussed so far, it seems to be extraordinarily difficult to induce mutations for rust resistance in *Avena sativa*. A few mutants, showing an increased tolerance against *crown rust*, were obtained in EMS treatments in the United States but they were low in yield. Recurrent EMS treatments resulted in one line with normal yield and full tolerance (Frey et al. 1976; Simons and Frey 1977). All attempts to obtain resistance against *stem rust*, however, remained without any success. In the United States, 5 million M_2 plants, following irradiation and EMS treatment, were tested 1972–1974. Out of this giant material 124 plants were selected, but no resistance was observed when they were tested further. Therefore, the presently existing methods of mutation breeding are obviously not suitable to solve this problem (McKenzie and Martens, 1974; McKenzie et al. 1976; Harder et al. 1977).

13.1.5 Maize

The use of induced mutation in maize breeding is more difficult than in the other cereals mentioned so far because of the allogamous breeding system of the species. Nevertheless, some resistant lines have been obtained (Table 22). Two Bulgarian lines show a combined resistance to *leaf blight* and *maize mosaic dwarf virus*. A mutant, completely immune to *smut*, is already utilized in maize breeding (Stoilov and Popov 1976; Stoilov and Daskaloff 1976). Some *Fusarium*-resistant lines, selected in Hungary, show a considerable increase in protein content (Bálint et al. 1977).

13.1.6 Pearl Millet

Pennisetum typhoides, likewise a cross-fertilizing species, is highly susceptible to downy mildew and ergot in India. So far, it was not yet possible to induce mutations for resistance against ergot *(Claviceps microcephala)*. However, a genetically conditioned variation for resistance against downy mildew *(Sclerospora graminicola)* is available in India and the United States (Murty 1974; Raut et al. 1974; Burton 1974). Hybrids have been released which were developed by means of mildew-resistant mutants. They show only traces to very low infection in a wide range of different ecological conditions and were found to be competitive with the standard with regard to yield (Murty 1976).

13.1.7 Sugarcane

Sugarcane is a high polyploid, highly heterozygous crop. A review on the possibilities of its improvement by means of mutations was given by Heinz (1973). Some moderately red-rot-resistant mutants are available in India and Bangladesh, one of them being developed into a released variety (Rao et al. 1966; Haq et al. 1974; Jagathesan 1976).

Table 21. Disease-resistant wheat mutants

Disease and pathogen	Authors	Mutagens	Remarks
Triticum aestivum			
Stem rust (*Puccinia graminis f.sp. tritici*)	– Edwards et al. 1969	EMS	Mutants resistant to races 111, 15B, 32
	– Peixoto Gomes 1972	X rays	
	– Ramirez Araya et al. 1972	EMS, EI, HM	3 mutants resistant to race 15B
	– Haniš 1974; Haniš et al. 1976, 1977	X rays, gamma rays, neutrons, EMS	11 resistant mutants
	– Siddiqui and Siddiqui 1974	Gamma rays, EMS	
	– Abdel-Hak and Kamel 1977	Gamma rays	
	– Parodi and Nebreda 1977	Gamma rays	
	– Skorda 1977	Gamma rays, neutrons	
Leaf rust (*Puccinia recondita tritici*)	– Haniš 1973, 1974; Haniš et al. 1977	X rays, gamma rays, neutrons, EMS	
	– Siddiqui and Siddiqui 1974	Gamma rays, EMS	
	– Borojević 1975, 1976, 1977, 1978, 1979	Gamma rays	Differs in 3 genes with regard to the reaction to *Puccinia* from the mother variety
	– Abdel-Hak and Kamel 1977	Gamma rays	
	– Parodi and Nebreda 1977	Gamma rays	
	– Djelepov 1978	Gamma rays, EMS	
Stripe rust (*Puccinia striiformis*)	– Konzak et al. 1968	EMS	
	– Haniš et al. 1969, 1976, 1977	X rays, gamma rays, neutrons, EMS	1 mutant prerequisites for direct utilization as a variety
	– Line et al. 1974	Chemical mutagens	12 different genes for resistance to one race of *Puccinia striiformis* found
	– Parodi and Nebreda 1977	Gamma rays	
	– Skorda 1977	Gamma rays, neutrons	
	– Konzak et al. 1977	NaN$_3$, EMS, MNH, ENH	
Yellow rust (*Puccinia glumarum*)	– Jech 1966b	Gamma rays	
Powdery mildew (*Erysiphe graminis*)	– Abi-Antoun 1974	Gamma rays	
	– Haniš 1973, 1974; Haniš et al. 1976, 1977	X rays, gamma rays, neutrons, EMS	1 resistant mutant released as variety
	– Djelepov 1978	Gamma rays	
	– Kiraly and Barabás 1974	Gamma rays, EMS	
	– Little 1971	Gamma rays	
Septoria nodorum	– Brönnimann and Fossati 1974; Fossati and Brönniman 1976	Chemical mutagens	3 resistant lines Some tolerant lines
	– Favret et al. 1974	Neutrons	

Cercosporella herpotrichoides		— Haniš et al. 1977	X rays, gamma rays, neutrons, EMS	1 mutant strain tolerant
Triticum durum				
Puccinia graminis		— Parodi and Nebreda 1977	Gamma rays	
		— Skorda 1977	Gamma rays	
Puccinia recondita		— Parodi and Nebreda 1977	Gamma rays	
Puccinia striiformis		— Parodi and Nebreda 1977	Gamma rays	
		— Skorda 1977	Gamma rays	
Bunt (*Tilletia triticoides*)		— Bozzini 1971		2 short-stemmed, early maturing mutants

Table 22. Disease-resistant mutants of other *Gramineae*

Species	Disease and pathogen	Authors	Mutagen	Remarks
Avena sativa	Crown rust (*Puccinia coronata avenae*)	— Frey et al. 1976; Simons and Frey 1977	EMS	Some tolerant mutants
Zea mays	Northern corn leaf blight (*Helminthosporium turcicum*)	— Popova and Popov 1971; Popova 1975; Stoilov and Daskaloff 1976; Stoilov and Popov 1976	Gamma rays, neutrons, EMS, NMU	3 lines resistant
	Helminthosporium maydis	— Cornu et al. 1977	EMS	1 resistant strain
	Fusarium sp.	— Bálint et al. 1977	Gamma rays, neutrons, EMS	4 mutants with increased resistance
	Smut (*Ustilago zeae*)	— Stoilov and Daskaloff 1976; Stoilov and Popov 1976	Gamma rays, neutrons	1 line completely immune; used in maize breeding
	Maize mosaic dwarf virus	— Stoilov and Popov 1976	Gamma rays, fast neutrons	2 mutants with combined resistance to *Helminthosporium* and the virus
Pennisetum typhoides	Downy mildew (*Sclerospora graminicola*)	— Burton 1974	Neutrons	Utilized in pearl millet breeding
		— Murty 1974, 1976	Gamma rays	
		— Raut et al. 1974	Gamma rays	
Saccharum species	Red rot (*Physalospora tucumanensis*)	— Rao et al. 1966	Gamma rays	
		— Haq et al. 1974	Gamma rays	15 mutants with different degrees of resistance
	Smut (*Ustilago scitaminea*)	— Jagathesan et al. 1974; Jagathesan 1976	X rays, gamma rays	1 mutant released
Turf Bermuda grasses	Ring and sting nematodes; root knot	— Jagathesan 1976	X rays, gamma rays	
		— Burton 1976, 1981	Gamma rays	Sterile triploid hybrids

13.1.8 Dicotyledonous Crops

In Tables 23 and 24, some species of different dicots are listed in which resistant mutants were selected after having applied different mutagens. Only a very small number of these genotypes is available in *pulses,* which play an important role as protein sources in many developing countries. This is certainly not due to the fact that it could not be possible in principle to induce resistance against diseases as in cereals. But it is

Table 23. Disease-resistant mutants in different Leguminosae and Solanaceae

Species	Disease and pathogen	Authors	Mutagens	Remarks
Phaseolus vulgaris (Haricot bean)	Bacterial diseases	– Khvostova 1967	Gamma rays	Released variety "Saparke 75"
	Golden mosaic virus	– Tulmann et al. 1980	EMS	Mutant with very few virus symptoms; lower yielding; used in cross-breeding
Lespedeza stipulacea	Root-knot nematodes *(Meloidogyne incognita)*	– Offutt and Riggs 1980	Gamma rays, neutrons	
Glycine max	Rust *(Phakipsora pachyrhizi)*	– Lu 1970	Gamma rays	
Pisum sativum	Fungal diseases	– Khvostova 1967	Gamma rays	
	Powdery mildew	– Gottschalk and Kaul 1980	X rays	1 mutant and 3 recombinants in India more tolerant than mother variety and Indian local lines
Vicia faba	Rust *(Uromyces fabae)* chocolate spot *(Botrytis fabae)*	– Abdel-Hak and Kamel 1977	Gamma rays	Mutants with lower level of the diseases
Vigna radiata	Yellow mosaic virus	– Shakoor et al. 1977; Haq and Shakoor 1980	Gamma rays	1 mutant moderately resistant; good yield potential
Cicer arietinum	Blight *(Ascochyta rabei)*	– Haq and Shakoor 1980	Gamma rays	2 mutants with varying degree of resistance
Lycopersicon esculentum	TMV Fusarium oxysporum	– Yamakawa and Nagata 1975	Gamma rays	Combined resistance to both pathogens in lines from *L. esculentum* x *peruvianum* hybrids
	Late blight *(Phytophthora infestans)*	– El-Sayed 1977	Gamma rays	5 mutants
Solanum tuberosum	Potato blight *(Phytophthora infestans)*	– Khvostova 1967	Gamma rays	
	Brown rot *(Pseudomonas solanacearum)*	– Kishore et al. 1975	Gamma rays, EMS, NMU	
	Synchytrium endobioticum	– Tarasenko 1977a	X rays	

Table 24. Disease-resistant mutants in some other dicotyledonous crops

Species	Disease and pathogen	Authors	Mutagens	Remarks
Corchorus capsularis and *C. olitorius*	*Macrophomina phaseoli*	– Singh et al. 1973	X rays	
	Stem rot	– Haq et al. 1974	Gamma rays	1 mutant fully, 4 mutants moderately resistant
	Wilt	– Khvostova 1967	Gamma rays	
Vitis vinifera	Downy mildew (*Plasmopara viticola*)	– Coutinho 1977	X rays, neutrons	
Morus species	Dogare disease (*Diaporthe nomurai*)	– Nakajima 1973	Gamma rays	2 mutants
Zinnia elegans	Leaf-curl virus	– Swarup and Raghava 1974	X rays	Resistant variety developed
Mentha piperita	Wilt (*Verticillium albo-atrum var. menthae*)	– Murray 1969, 1971	X rays, neutrons	7 strains highly resistant, early maturing, 5 strains moderately resistant
Mentha cardiaca	Wilt (*Verticillium dahilae*)	– Horner and Melouk 1977	Gamma rays	3 resistant mutants
Pirus malus	Apple powdery mildew (*Podosphaera leucotricha*)	– Lapins 1973	Gamma rays	

essentially more difficult to screen these mutants because of the larger expenditure in space. In Russia, a released *Haricot bean* variety, developed from a gamma-ray-induced mutant, is available (Khvostova 1967). A horticulturally utilized mutant strain of *Zinnia elegans,* resistant to the leaf curl virus, is available in India. The resistance is governed by two dominant genes, one of them being a repressor gene (Swarup and Rhagava 1974). The *Phytophthora* resistance of a *tomato* mutant is obviously due to the production of rishitin as a consequence of the infection with *Phytophthora infestans.* Rishitin is a natural antibiotic. Its accumulation in the plant has the effect of a resistance mechanism against late blight (El-Sayed 1977).

Intensive work in this field has been done in the United States in *Mentha piperita.* The plants of this species, which are grown as the source of peppermint oil, are heavily infected by *Verticillium albo-atrum*. After having irradiated the stolons with X rays and neutrons, about 100,000 M_1 plants were cultivated in a severely wilt-infested soil. Seven highly wilt-resistant strains were developed from an M_2 generation containing more than 6 million plants. An additional advantage of these strains is their earliness. Another five strains were found to be moderately resistant against *Verticillium.* These plants are very vigorous and show heterosis for herbage weight. With regard to oil yield and oil quality, there are no differences between the mutant strains and the initial material (Murray 1969, 1971).

13.2 Resistance Against Animal Pathogens

So far, plant diseases caused by fungi, bacteria or viruses have been discussed. A few cases are known that experimentally produced mutants show resistance or tolerance against animal pathogens of distinct crops.

Some M_2 plants of *Sorghum vulgare* show resistance against the shoot fly (Goud et al. 1970). In *castor (Ricinus communis)* – an important Indian oil plant – the capsule borer and the gall fly make plenty of damage. A dwarf mutant was found to be comparatively more resistant against these two insects than the mother variety. Plants of this genotype show normal yield but they need considerably less space, another advantage for agronomic utilization (Ankineedu et al. 1968). Some high-yielding Indian varieties of *Gossypium hirsutum* are extremely susceptible to insects, particularly to jassids *(Empoasca devastans)*. A mutant is characterized by the increase of hair density on the underside of the leaves and on other plant organs, resulting in a fair degree of tolerance to the insects (Swaminathan 1965; Raut et al. 1971). Improved resistance to brown plant hopper and gall midge pest is realized in some early ripening *rice* mutants in Indonesia (Ismachin and Mikaelsen 1976).

Some mutants were found to be protected to some extent against bird damage. A *wheat* mutant, for instance, released already in 1961 and derived from an awnless variety, is possibly less damaged by birds during ripening period because of having awns (Swaminathan 1965). Mutants protected against damage by birds are also known in *Pennisetum typhoides* (Bilquez et al. 1965) and *Sorghum subglabrescens*, whereas another *Sorghum* mutant was found to be more resistant to weevil attack (Sree Ramulu 1968, 1974a, b).

13.3 Herbicide Tolerance

A mutant of the triploid *turf Bermuda grass* was found to be herbicide tolerant (Burton 1976). This can be an important trait in specific cases. One of our *pea* mutants, a dichotomously branched type with good yielding properties, shows a high susceptibility against a specific herbicide which is not observed in the mother variety and in all other mutants of our collection. The leaves of the treated plants are strongly damaged, resulting in a decrease of the vitality of the plants.

14 Drought Resistance, Heat Tolerance, Winterhardiness

Under specific ecological conditions, **resistance or tolerance against drought** is a valuable character which can be utilized for improving the existing varieties. Some mutants showing this advantageous trait are known in different legumes. A high-yielding *fodder lupine,* released in Russia, proved to be drought resistant (Khvostova 1967). The increased grain-yielding capacity of an alkaloidless mutant strain of *Lupinus luteus* is not only due to its high vegetative biomass, but also due to an increased drought resistance (Jashovsky and Golovchenko 1967). *Vigna radiata* is preferably grown in dry areas of the western parts of India because of its tolerance to drought. An EMS-induced mutant shows a favorable combination of earliness, drought resistance, and increased yield potential (Prasad 1976). One of our *Pisum* mutants, inferior to its initial line under the normal climatic conditions of Germany, was found to show a high degree of drought resistance when grown in India (Gottschalk and Patil 1971). Finally, an Indian *rye grass* mutant should be mentioned in this context, having the capacity to withstand high temperature and drought better than the mother variety (Malik and Marie 1971).

In countries with hot climates, mutants which have been screened under moderate climatic conditions can considerably suffer from the high temperatures. Distinct recombinants of our *Pisum* collection, for instance, were found to be highly heat susceptible when grown in North India. All the plants of a specific recombinant type died in early stages of ontogenetic development, whereas the plants of four other genotypes died due to heat before completing seed ripening (Gottschalk and Kaul 1980). Therefore, 126 different *Pisum* mutants, derived from the same mother variety, were tested with regard to differences in their **tolerance against high temperatures** in our phytotron. Constant temperatures of 25°, 30°, 35°, and 40°C were given to young plantlets. In principle, all the genotypes tested were able to survive under 35°C, but none of them was able to tolerate 40°C. With regard to the time during which the plantlets died under these unfavorable conditions, clear differences between distinct genotypes were observed (Gottschalk 1981g). These findings show that a genetically conditioned variability for heat tolerance has been induced in our radiation genetic experiments which, however, cannot be discerned if the material is only grown in Germany.

An increased degree of **winterhardiness** was observed in some mutants of *Hordeum vulgare* (Enchev 1976), *Triticum aestivum* (Khvostova 1967), *Secale cereale* (Kivi et al. 1974), and *Oryza sativa*. The Japanese rice variety "Reimei," developed from a gamma-ray-induced mutant, is relatively well protected against damage from coolness because of the low position of the ear primordia (Kawai 1967). *Tifway-2 Bermuda grass,* a released turf grass variety obtained by gamma irradiation, was found to be not only more frost resistant than its mother variety, but it is also more resistant to root knot, ring and sting nematodes, and shows a better spring growth (Burton 1981). The tremendous expenditure, necessary for reaching a distinct aim of breeding, becomes evident from *coastcross 1 Bermuda grass:* 500,000 green stems with dormant buds of favorable hybrids were treated with gamma rays, giving rise to one single mutant with improved winterhardiness (Burton et al. 1980). In Finland, mutants of *Brassica campestris oleifera* have been selected, surpassing the mother variety with regard to winterhardiness (Kivi et al. 1974). An improved frost resistance is observed in an X-ray-induced *sour cherry* mutant (Zwintzscher 1967). Moreover, some *pear* mutants were found to be resistant in cold storage (Roby 1972).

15 Shattering and Shedding Resistance

Nonshattering and nonshedding are characteristic features of many crops which have been selected very early soon after the wild-growing ancestors had been transferred into the status of cultivated plants. Nevertheless, plenty of seeds get lost in cereals and in some oil plants during ripening or harvesting because of fruit shattering or shedding. Genes causing resistance to this damage may therefore be of direct agronomic interest. So far, only little information is available in this field. **Nonshattering mutants** have been selected in *soybeans* (Nalampang 1975; Rubaihayo 1976a, b), *Brassica juncea* (Nayar and George 1969), and *rice* (Majumder 1969). Mutants of *Sesamum indicum,* selected in Japan, have indehiscent capsules. This positive effect, however, cannot be utilized agronomically because of some negative features caused by the pleiotropic gene (Kobayashi 1965). The increased yield of a Russian alkaloidless mutant of *Lupinus luteus* is partly due to its **shedding resistance** (Jashovsky and Golovchenko 1967). A mutant of this type is also known in *rice,* but it shows a reduced fertility.

16 The Pleiotropic Gene Action as a Negative Factor in Mutation Breeding

One of the greatest handicaps of mutation breeding is the pleiotropic action of the mutant genes: the agronomically useful trait of the mutant is accompanied by one or several negative traits in the majority of all cases. Therefore, many mutants with characters of agronomic interest cannot be utilized for the improvement of the respective crops. A clear example of this widely realized situation is given in Fig. 18. One of our *pea* mutants, homozygous for gene *ion,* has an increased number of seeds per pod due to a considerable increase of the number of ovules per ovary. This favorable presupposition for an increase of the seed production, however, cannot be utilized in pea breeding, because the number of pods per plant is reduced. These two characteristics — the positive and the negative one — appear commonly in all generations tested so far, thus demonstrating a distinct regularity which reduces the breeding value of the strain. A similar situation is realized in many other mutants. Such a situation can be due to true pleiotropic gene action, but it can theoretically be also due to mutational events in closely linked genes. As these problems have serious consequences for the utilization of induced mutants in plant breeding, they should be discussed a little more in detail, even if these details are very specific.

16.1 The Alteration of Pleiotropic Patterns Under the Influence of Changed Genotypic Background or Environment

Normally it is not possible to eliminate the negative traits of the pleiotropic spectrum from the positive ones. On the contrary, the whole complex of diverging characters is commonly transferred from generation to generation showing a monogenic inheritance in crosses. Because of this behavior, the phenomenon is commonly interpreted in the sense of pleiotropy, although there is no possibility for evidencing this concept for methodological reasons. It is not only of theoretical but also of practical interest that the pleiotropic pattern of a mutant gene can be altered to some extent by transferring it into a specific genotypic background (Konzak 1976; *wheat;* Khvostova 1978; Sidorova 1981b, *peas*). An intensive work in this field was carried out by Gaul and coworkers using distinct traits of the *barley erectoides mutant ert 16* such as awn, culm, spike, and internode characters. They could show that the individual characters of the complex can vary independently from each other by changing the genetic background of *ert 16.* This was done by crossing the *erectoides* mutant, derived from the German variety "Haisa II," with the Abyssinian variety "Bulchi Gofa" and by studying the

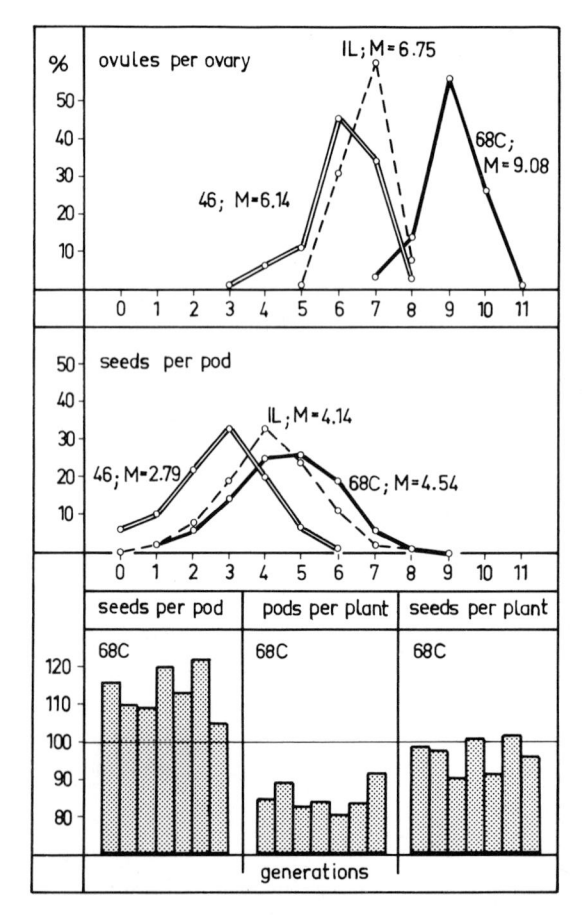

Fig. 18. Example for a pleiotropic gene action influencing yielding criteria of the *garden pea*. Gene *ion* of mutant 68C increases the number of ovules per ovary considerably *(upper part* of the graph), resulting in an increase of the number of seeds per pod *(middle part)*. However, it decreases the number of pods per plant. This effect is regularly observed in each generation *(lower part;* each *column* represents the mean value for one generation as related to the control value of the mother variety). As a consequence of the positive and negative component of the pleiotropic spectrum of gene *ion*, the number of seeds per plant is only about equal to that of the initial line

various parts of the *ert 16* pattern in the Bulchi Gofa genome. Later on, seven Swedish *erectoides* mutants were crossed with three German barley varieties and again a modification of the *erectoides* complex was observed (Hesemann and Gaul 1967; Gaul et al. 1968; Gaul and Grunewaldt 1971; Grunewaldt 1974a, b; Gaul and Lind 1976; Lind and Gaul 1976). In this way, it is possible to separate characters of agronomic interest from undesirable ones. Unfortunately, it is not possible to give any predictions with regard to a desired alteration of the pleiotropic spectrum in a specific genome. The detection of the optimal combination of a given mutant gene and a distinct genetic background is a matter of chance. A tremendous amount of work will be necessary in order to select that particular combination in which the gene expresses its most favorable pleiotropic pattern.

So far, the possibility of changing the individual constituents of a pleiotropic pattern under the influence of the genotypic background was discussed. A second possibility to reach this aim consists in cultivating the mutants under different ecological conditions. Russian *wheat* mutants, grown at climatically different areas of the country, were evaluated in this context. The respective genotypes had 10–12 diverging traits besides the main mutated character of the plants, the complex being obviously

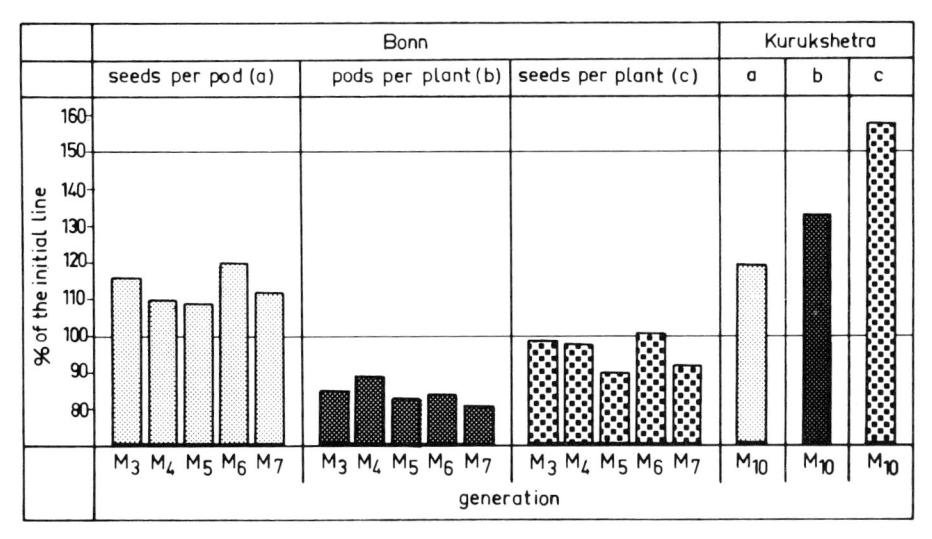

Fig. 19. Mean values of some criteria of agronomic interest of the *Pisum* mutant 68C grown in Bonn, Germany, and Kurukshetra, North India. Each *column* represents the mean value for one generation as related to the control values of the mother variety = 100%. The mutant gene increases the seed number per pod but decreases the pod number per plant under German conditions. In India, the pod number is considerably increased due to strong branching of the plants resulting in a high seed production per plant

due to a pleiotropic gene action. Some quantitative traits of the complex were found to become specifically modified under the influence of distinct ecological conditions (Khvostova 1978). Similar experiences were made in Russian *pea* mutants (Glazacheva and Sidorova 1973).

The pleiotropic pattern of one of our *pea* mutants was obviously altered by growing the mutant in North India. Gene *ion* of the *Pisum* genome increases — as already mentioned — the number of ovules per ovary, but it reduces the number of pods per plant when the mutant is cultivated in Middle Europe. In Kurukshetra (India), the positive part of the pleiotropic spectrum appears as in Germany, but the negative part does not appear. On the contrary, the number of pods per plant is considerably increased as a consequence of a high degree of branching which is not observed in the mother variety. Thus, the pleiotropic pattern of gene *ion* comprises two positive constituents distinctly visible under the influence of specific climatic conditions of North India and resulting in a clear improvement of the mutant (Fig. 19; Gottschalk and Kaul 1975).

From the examples given above, it is vividly evident that no predictions can be made about those ecological conditions in which a given mutant gene may attain a possibly favorable expression of its pleiotropic spectrum. It is again a matter of chance to find them. It would certainly be worthwhile to test small groups of genotypes, which contain mutant genes of particular interest for breeding purposes, under different climatic conditions in order to discern their ecological plasticity. Details concerning these problems are discussed in Chap. 18.

So far, only a very small number of detailed findings are available in the field of the alteration of pleiotropic spectra under the influence of the genetic background and/or of specific ecological factors. It is therefore difficult to judge to what an extent these possibilities may contribute to an improvement of the prospects of mutation breeding in the near future. A close international cooperation between those institutions in which comprehensive collections of mutants are available, would certainly help to extend our knowledge in this field.

16.2 Mutations of Closely Linked Genes

A second possibility for explaining the genetic situation of the mutants we are discussing consists in assuming that they are not homozygous for a single pleiotropic gene but for several genes, each of them being responsible for a distinct character of the "pleiotropic" pattern. In addition, we have to assume that the respective genes are very closely linked and that they have commonly mutated during the mutation treatment. Because of their close linkage, the single genes of the group are not separated from each other by crossover events. Therefore, the complex of diverging characters is transferred as a unit from generation to generation, showing a monogenic inheritance in crosses. Under these prerequisites, a pleiotropic gene action is pretended, but in reality different independently functioning genes become effective. Thus, true pleiotropic gene action and the action of simultaneously mutated neighboring genes are two phenomena which normally cannot be distinguished from each other.

We have developed a method by means of which these two principally different phenomena can be distinguished. Unfortunately, this test cannot be used on a large scale but only in a small number of specific cases. We cannot study this question in a consequent, a direct way, but we depend on distinct groups of mutants which we, however, cannot produce systematically. On the contrary, it is a matter of chance whether or not they arise in mutation treatments; some of them may be already available in the collections. Details on the method and the results can be found in a series of publications (Gottschalk 1965b, 1967, 1968a, 1976b). The principle of the tests consists in trying to split the apparently existing pleiotropic patterns of suitable mutants and to attribute the individual characters of the spectra to several independently functioning genes. For this purpose, we use mutants differing in a group of characters from the mother variety and showing a monohybrid segregation for the respective complex. In other words, they behave as if they would be homozygous for pleiotropic genes. The deciding prerequisite for the test, however, consists in the fact that we need groups of independently arisen partially similar mutants. They should be identical with regard to specific traits of their "pleiotropic" spectra, but they should differ from each other with regard to other traits of the spectra. These partially similar mutants are crossed with each other.

A clear example may demonstrate this somewhat difficult initial situation. Let us assume that two mutants with the following traits are available in a collection:

Mutations of Closely Linked Genes

(1): — narrow leaflets
      — female sterility
(2): — narrow leaflets
      — normal fertility

The following questions have to be answered:

— Are the anomalies of mutant (1) due to the action of one pleiotropic gene or of two independently functioning genes?
— Are there any genetic relations between the two mutants because of their similarity with regard to the abnormal leaflets?

Theoretically, two different kinds of F_1 hybrids can be expected after having crossed the two mutants with each other:

— Possibility I: The hybrids have normal leaflets and full fertility. In F_2, a modified dihybrid segregation occurs.
This would mean that there are no genetic relations between the two mutants. The leaf anomaly of the two mutants is caused by different genes of the genome. Moreover, it can be concluded that the anomalies of mutant (1) are possibly caused by a pleiotropic gene.

— Possibility II: The hybrids have the narrow leaflets of their parental mutants and full fertility. In F_2, all the plants have the narrow leaflets, but there is a 3:1 segregation for fertility and female sterility.
This is evidence that the anomalies of mutant (1) cannot be caused by a pleiotropic gene. On the contrary, they are caused by two different genes which could be separated from each other by hybridization. One of these two genes is also present in mutant (2). It is now necessary to test their degree of linkage.

By means of this method, more than 30 different "pleiotropic" mutants of our *Pisum* collection, belonging to five groups of partially similar mutants, were genetically analyzed. In most cases, the spectra could be splitted through hybridization and single

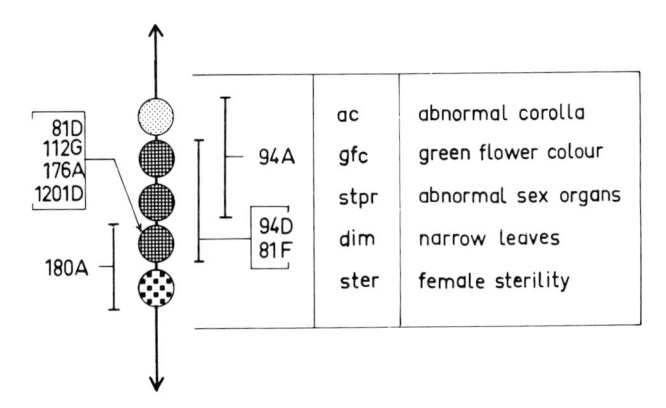

ac	abnormal corolla
gfc	green flower colour
stpr	abnormal sex organs
dim	narrow leaves
ster	female sterility

Fig. 20. Composition of the *dim* segment of the *Pisum* genome, containing five adjacent genes, and the mutants arisen by mutations of single or several genes located in the segment

traits or small groups of traits could be attributed to independent genes. In all these genotypes, the mutant genes were found to be extremely closely linked. There was only a single case in which two genes could be separated from each other by crossing over. The degree of linkage was too close so that only 2 out of 13,000 gametes of the hybrid, used for fertilization, were influenced by the crossing over.

Figure 20 shows the genetic situation of such a group of *Pisum* mutants. The genes belong to the so-called *dim*-segment, containing at least five very closely linked or even neighboring genes. The segment as a whole has obviously a somewhat higher susceptibility against mutagenic agents. After having applied X rays and neutrons, five different groups of mutants were obtained. The distribution of the anomalies of the various groups shows that it is a matter of chance how many genes of the *dim*-segment mutate in a distinct embryo and which ones of the segment mutate. In this way, partially similar mutants arise from different treated embryos as illustrated in the figure. From the five genes of the segment, only gene *dim* is of some interest for pea breeding. So far, 12 mutants, homozygous for *dim*, arose from different irradiated embryos in our radiation genetic experiments. They have narrow leaflets and stipules, small flowers, and seeds which could be suitable for canning. The seed production of these genotypes is equal to or somewhat better than that of the initial line. Relatively often, gene *Dim* mutated ± simultaneously with one or several neighboring genes of the segment, which are responsible for the realization of negative characters. These mutants are without any agronomic value. They pretend a pleiotropic situation under inclusion of the character "small seeds" which, however, is not realized.

What conclusions can be drawn from these results? It can be assumed that the situation just discussed is not a specific behavior of the species *Pisum sativum*, it is certainly also realized in other crops. But, because of the small number of findings available so far, it cannot be calculated what role it plays within the phenomenon of "pleiotropy" in the commonly used sense. There is no doubt that true pleiotropism exists and that it has very negative consequences with regard to the utilization of mutagenesis

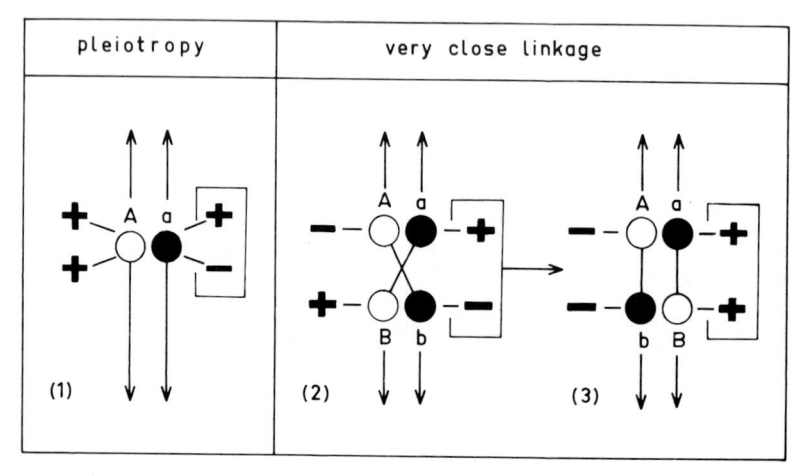

Fig. 21. Comparison of the effect of a pleiotropic gene and that of two adjacent mutant genes with regard to the utilization of a positive character in breeding (explanation in the text)

in plant breeding. But there is also no doubt that a proportion of cases of this kind is due to mutations in closely linked genes. The height of this proportion cannot be calculated because of lack of information.

The two possibilities just mentioned are very divergent with regard to their consequences for mutation breeding. If a positive trait is part of a pleiotropic complex which also contains negative traits, there is, in most cases, no chance to utilize the respective character for breeding purposes. If, however, the positive and negative characters are controlled by different genes, it is possible in principle to separate the positive trait from the negative one by a crossing-over event (Fig. 21). Such an event will be very rare because of the close linkage of the genes involved. But it is not impossible to reach this aim by working on a very large scale. This expenditure could be justified in exceptional cases, if a valuable character has appeared in mutation treatments, for instance a specific kind of resistance, accompanied by negative traits.

17 The Penetrance Behavior of Mutant Genes as a Negative Factor

Those mutant genes which are of any agronomic interest generally show a stable penetrance. Very rarely, an unstable penetrance can influence the breeding value of a mutant negatively. Such a situation is realized in a group of dichotomously branched *Pisum* mutants. The two polymeric genes *bif-1^{1201}* and *bif-2* of the *Pisum* genome are responsible for the alteration of the monopodial shoot structure of the pea plant into dichotomy. The plants show a stem bifurcation in the upper part of the shoot (Fig. 12). As a consequence of this anomaly, the number of pods per plant is increased in comparison to that of the mother variety. Number of seeds per pod, seed size, and other characters of agronomic interest are not influenced. The genes *bif-1^{1201}* and *bif-2* are identical with regard to their action on the shoot system. They obviously are not linked and become effective independently from each other without showing any kind of interaction.

Unfortunately, the genes are unstable with regard to their penetrance. Not all the plants homozygous for *bif-1^{1201}* or *bif-2* show the stem bifurcation. On the contrary, each of the two strains is composed of bifurcated and nonbifurcated plants being genetically identical. The latter ones are phenotypically normal, but their seed yield is considerably lower than that of the bifurcated plants. Very often, it is even lower than that of the initial line. This regularity appeared in each of the 15 generations tested so

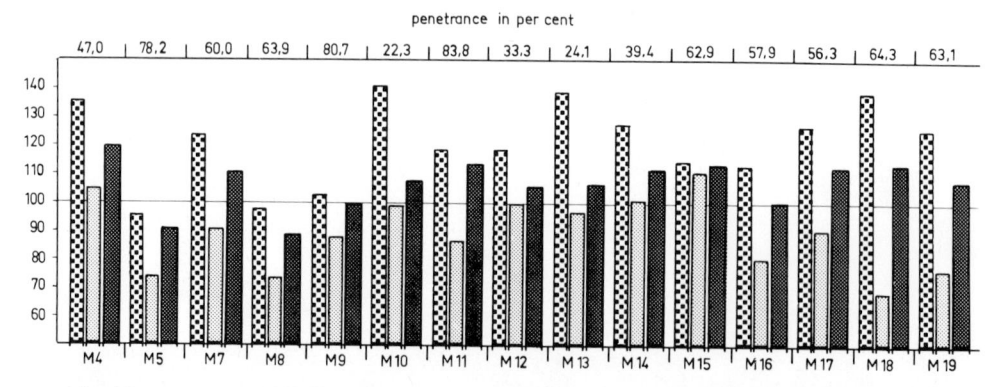

Fig. 22. Dependence of the seed production of the *Pisum* mutant 1201A on the penetrance of gene *bif-1* in 15 generations. *Left column* of each group: seed yield of the bifurcated plants; *middle column:* seed yield of the nonbifurcated plants having the same genetic constitution as the bifurcated ones. *Right column:* seed yield of the whole strain. All the mean values are related to the control values of the mother variety = 100%. The nonbifurcated plants, due to the unstable penetrance of gene *bif-1,* have lower yields influencing the yielding properties of the mutant strain negatively

far (Fig. 22). Because of these yield differences of the two modificants of the bifurcated mutants, there is a clear relation between the penetrance behavior of the genes and the seed production of the strains:

- The higher the degree of the penetrance of $bif\text{-}1^{1201}$ and $bif\text{-}2$,
- the higher the proportion of bifurcated plants in the strains,
- the higher the productivity of the strains in relation to that of the initial line.

This correlation becomes clear from Fig. 22. Thus, an increase of the degree of penetrance or even its full stabilization should lead to an enhanced seed production of the mutants.

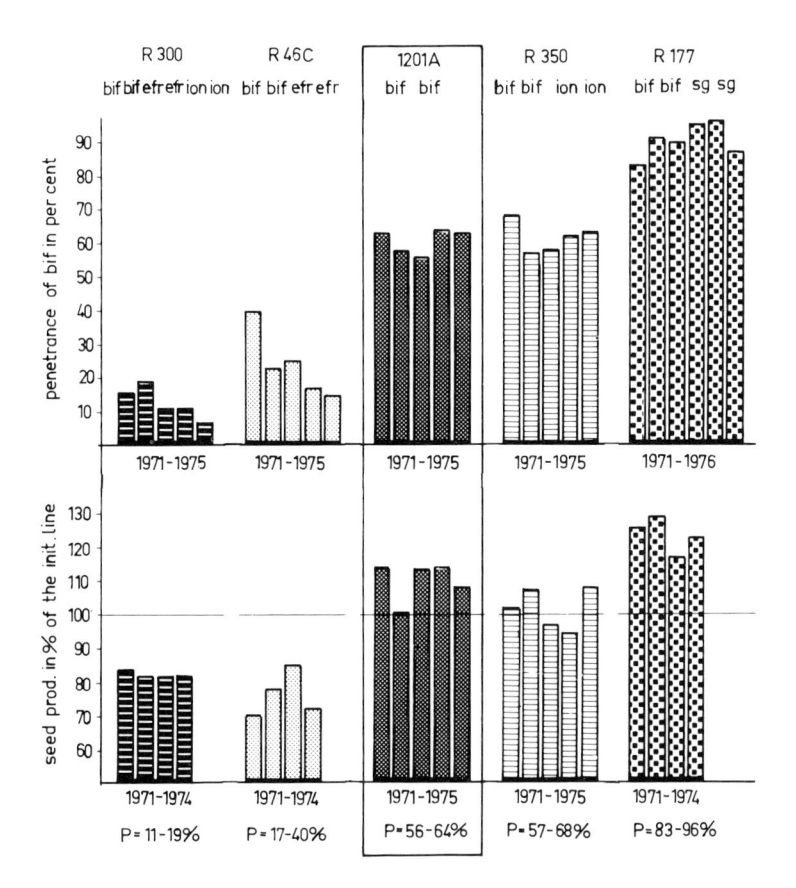

Fig. 23. Dependence of the penetrance of gene *bif-1* on the genetic background (*upper part* of the graph).

- Gene *efr* reduces the penetrance of *bif-1*.
- Gene *ion* reduces the penetrance of *bif-1* even more, but only if gene *efr* is simultaneously present in the genome (see comparison of recombinants R 300 and R 350)
- Gene *sg* increases the penetrance of *bif-1*.

The seed production of the recombinants decreases or increases, respectively, in relation to the penetrance of *bif-1* (*lower part* of the graph)

The penetrance of *bif-1*[1201] was found to depend upon both environment and genotypic background. Considering 15 generations grown in West Germany, the gene showed a penetrance ranging between 22% and 84%. These differences of the same material grown at the same location can only be due to the influence of distinct environmental factors which may vary to some extent in successive years. In Yugoslavia, the penetrance of the gene is regularly lower than in Germany; in Egypt, Ghana, and India, its penetrance is 0%. Thus, gene *bif-1*[1201] is not able to express its action under the climatic conditions of these countries, and we would not have been able to select this valuable mutant since it does not differ morphologically from the initial line. So far, we were not yet able to find an environment in which the gene shows full penetrance.

Mutant 1201A of our collection, homozygous for *bif-1*[1201], is a useful genotype in spite of the difficulties just mentioned. Its seed production as tested in many years is somewhat higher than the control values of the mother variety (Fig. 13). The mutant was crossed with some other mutants of agronomic interest and recombinants homozygous for *bif-1*[1201] and the respective other genes were selected. They were developed into pure lines. Interestingly, some genes of the *Pisum* genome were found to influence the penetrance behavior of *bif-1*[1201]. Genes *efr* for earliness, and *ion* for increased number of ovules per ovary reduce the penetrance. This reduction is accompanied by a reduction of the seed yield. Gene *sg*, however, responsible for the formation of smaller grains, increases the penetrance considerably resulting in an improvement of the seed production (Fig. 23). A similar effect was observed for two other mutant genes of our collection. The penetrance of *bif-1*[1201] is fully stabilized in the respective recombinant lines, i.e., all the plants show the stem bifurcation. Interest-

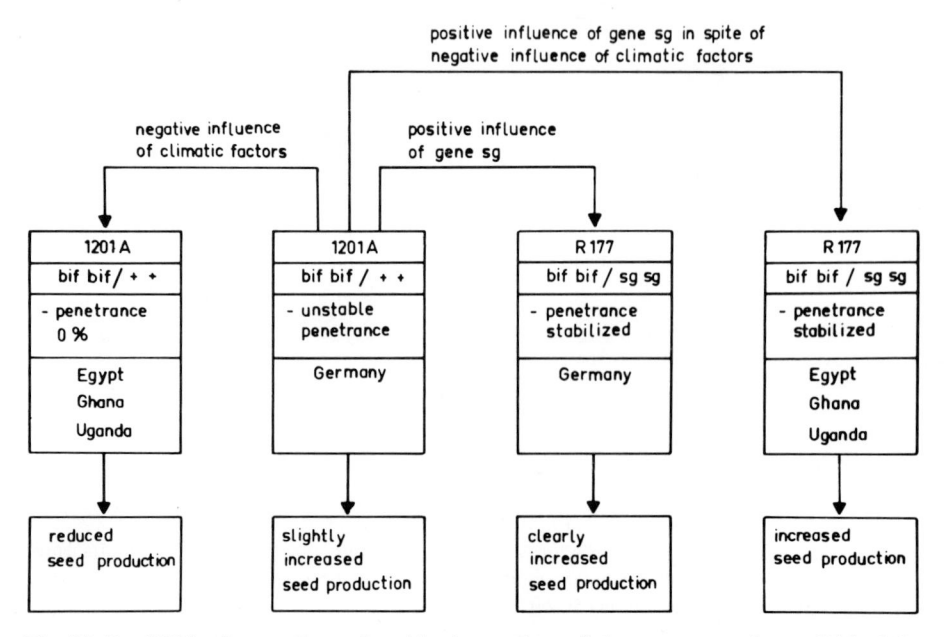

Fig. 24. Possibilities for negative and positive interactions of the penetrance of gene *bif-1* of the *Pisum* genome through environmental factors and/or other mutant genes of the genome

ingly, the plants of these strains are bifurcated also in India under those environmental conditions which normally suppress the expression of *bif-1^{1201}* completely. Thus, the genetic factors stabilizing the penetrance are stronger than the environmental factors suppressing it. It should be mentioned that a multiple allele of *bif-1^{1201}*, designated as *bif-1^{37}*, shows full penetrance. Details on the penetrance problem of these genotypes are published in a series of papers (Gottschalk 1965a, 1966, 1971a, 1972a, 1978c; Gottschalk and Chen 1969; Gottschalk and Abou-Salha 1982). The reaction of gene *bif-1^{1201}* to climatic factors and to the presence of gene *sg* is schematically presented in Fig. 24.

Similar differences in the ecological reaction were reported by Wojciechowska (1973) for a *serradella* mutant *(Ornithopus sativus)*. The plants are dwarfy and have fine leaves. These characters are, however, only expressed under field conditions. In the greenhouse, the gene is not able to manifest its action: the plants are phenotypically normal.

18 The Adaptability of Mutants to Altered Environmental Conditions

A characteristic peculiarity of many mutants lies in the fact, that their adaptation optimum to distinct environmental conditions differs markedly from that of the mother variety. So far, only little detailed information is available in this field. But already the small number of clear findings demonstrate that the utilization of mutants in plant breeding could be made more effective if the ecological reaction of prospective genotypes would be considered in an appropriate way. For intensifying these studies, a close international cooperation is necessary in order to discern specific reactions of distinct mutants and recombinants and to select specific ecotypes. There is no doubt that many mutants, existing in the various collections, could already be utilized agronomically if the whole breadth of their ecological reactions would be known. The field of gene-ecology is a very young branch of experimental mutation research which will certainly deliver interesting results in near future.

Gene-ecological problems can be studied by using two different methods. The genotypes tested can be either grown in different natural environments or they can be cultivated under the controlled conditions of a phytotron. Both these methods have advantages and disadvantages with regard to the judgment of the usefulness of the material tested; therefore, they are discussed separately.

18.1 The Reaction of Mutants to Different Natural Environments

The reaction of mutants and recombinants to differences of their natural environments can refer to the following criteria:

— Differences in the soil quality including different kinds of manuring.
— Relatively small climatic differences at the same location in successive years.
— Relative small ecological differences in different regions of the same country.
— Strong climatic differences in different regions of big countries or in different continents.

Let us discuss the findings available in the order just given.

The adaptability of distinct groups of mutants to the **soil quality** was already mentioned in Chap. 6 with regard to the capacity of *barley* and *wheat erectoides* mutants to utilize high levels of nitrogen. Additional findings in this field have been published by Scarascia Mugnozza (1966b; *Triticum durum*) and by Chakrabarti and Sen (1975; *Oryza sativa*). Under irrigated conditions, the released mutant variety "Casteldelmonte" of *durum wheat* shows a significantly increased yielding ability in South Italy (Pacucci and Scarascia Mugnozza 1974). It would certainly be worthwhile to test a high number of mutants, preferably those of different cereals, for tolerance against high salinity of soils. In this way, genotypes could be selected which are possibly suited for cultivation in salt regions being not yet utilized agronomically.

Very often, mutants grown at the same locality in **subsequent years** show a completely different behavior with regard to their seed yield as related to that of the initial line. These differences are obviously due to a varying response of the respective mutants and the mother variety to climatic differences. Specific ecotypes can already be discerned in this way provided that they are tested during a relatively large number of generations. This becomes clear in Fig. 25 for ten *Pisum* mutants tested over a period of 6 to 18 generations in West Germany (Gottschalk 1978b). With regard to their ecological reaction, these genotypes can be subdivided into two different groups. One group shows the same reaction to the climatic conditions as the mother variety. The

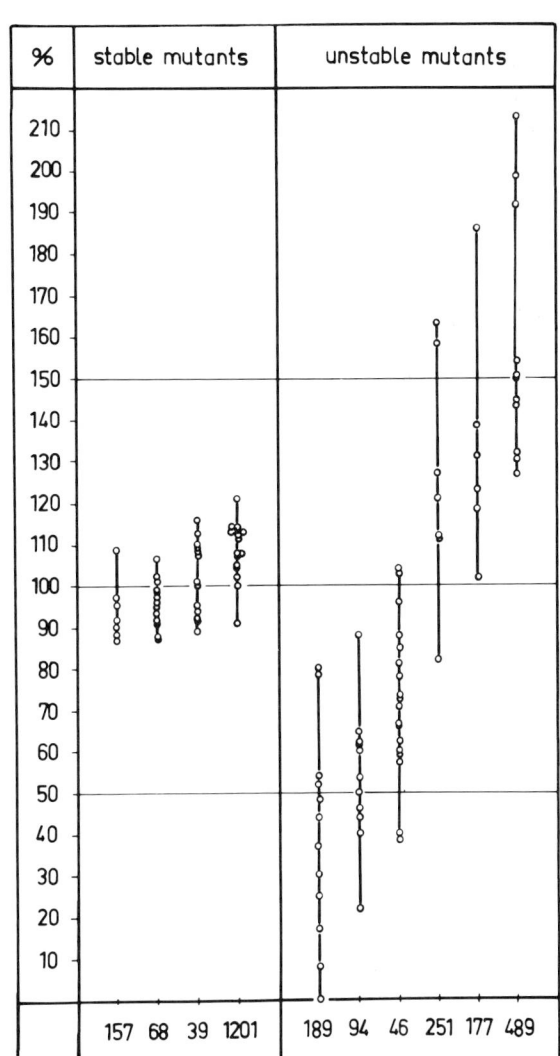

Fig. 25. The seed production of ten mutants and recombinants of *Pisum sativum* grown at Bonn over a large number of generations. Each *dot* represents the mean value for the character "number of seeds per plant" for one year as related to the control value of the mother variety = 100%. *Left group:* Genotypes having an adaption optimum to the environmental conditions similar to that of the initial line; *right group:* Genotypes having adaption optima different from that of the initial line

genotypes of the second group, however, are characterized by having strongly divergent relative mean values for the seed production in different years. This is obviously a consequence of the response to distinct climatic differences which are not observed in the initial line. The early ripening recombinant 46C of our collection, for instance, was tested over 18 generations. The plants reached only in three generations values comparable to those of the mother variety. In all the other generations, their grain yield was essentially lower, in two years only about 40% of the control values. This recombinant was found to be adapted to hot and dry conditions. It would certainly be superior to the initial line in such a climate and is just being tested in India.

Some *soybean* mutants represent likewise new ecotypes better adapted to years with unfavorable climatic conditions (Zacharias 1967). Similar results were obtained in a late-maturing mutant of *oats* which responds to seasonal and other environmental conditions in a way different from the mother variety (Ahokas 1976).

A group of *durum wheat* mutants was tested together with their mother variety at **different localities** in Italy. A number of mutant lines were found to be superior with regard to yield, culm length, lodging resistance, and some other characters (Scarascia Mugnozza et al. 1972; Pacucci and Scarascia Mugnozza 1974).

Of particular interest are those experiments in which mutants are tested in **different areas of big countries** with sharp differences in their ecological conditions. In this field, especially the findings obtained in Russia and India have to be discussed. *Pisum* mutants were grown at various localities of the U.S.S.R. differing from each other with regard to soil quality, altitude, and meteorological conditions. A number of quantitative characters, such as plant height, date of flowering, and various yielding criteria, were evaluated. The yield of a compact mutant was as follows (Sidorova and Khvostova 1972):

—	Omsk Buryat ASSR Krasnoyarsk	level of the initial line
—	Kirovsk Tomsk Bashkir ASSR	much lower than that of the initial line
—	Moscow	9.0% higher
	Kalinin	22.5% higher than that of the initial line

Similar results with different pea mutants were published by Sidorova and Uzhintzeva (1969); Sidorova et al. (1972); Glazacheva and Sidorova (1973); Sidorova and Bobodzhanov (1977). Also some Russian *soybean* and *wheat* mutants were found to have better adaptability to distinct environmental conditions (Josan and Nicolae 1974; Kalashnik 1972; Khvostova 1978).

An Indian *rice* mutant with short culms and improved grain quality was tested at ten different locations of the country with regard to its yield (kg/ha). The mean values obtained varied between 91% and 129% of the corresponding control values of the mother variety (Fig. 7; Reddy 1975).

Some mutants of *cereals* and *legumes* were tested in **different countries** or even **continents**. Very interesting results were obtained in *Triticum durum* grown at up to 28 different localities in Italy, Greece, Turkey, Cyprus, Israel, Lebanon, Jordan, Iran,

Iraq, Pakistan, India, Egypt, Syria, Tunisia, and Morocco over a period of several years (Tessi et al. 1968; Bogyo et al. 1969; Pacucci and Scarascia Mugnozza 1974). The mutants were not only compared with their mother varieties but also with local varieties of the respective countries. At each locality, at least one mutant was found to outyield the best local variety. Important characters, such as straw length, lodging resistance, flowering and ripening time, grain volume weight, were clearly improved in some of the mutants. The various mutants showed, however, different relative yielding capacities in the different countries demonstrating the differences in their adaptability to the various ecological conditions. Some early ripening, short-stemmed, lodging resistant mutants proved to be especially adapted to semiarid conditions. Moreover, other environmental factors, such as temperature, fertility level of the soils, insect and disease conditions, were found to influence the yielding capacity of the genotypes tested in a specific way. The Swedish *barley* variety "Mari," developed from an early ripening mutant, was compared with the variety "Edda" which because of its pronounced earliness is cultivated in North Sweden.

— In Angermanland (North Sweden, 63°N), Edda was found to be 8 days earlier than Mari.
— In Svalöf (South Sweden, 55°N), Edda is 2 days earlier.
— In Tenerife (Canary Islands, 28°N), Edda is 16 days later than Mari (Åkerberg 1966).

These differences are obviously due to the adaptability of "Mari" to high temperatures, particularly during the last period of the generative phase of the plants.

Some agronomically interesting *Pisum* mutants and recombinants of our collection are being tested in Egypt, Ghana, Uganda, Brazil, and at six climatically different localities of India. Clear differences in both quantitative and qualitative characters were observed due to a specific reaction of distinct genes to the altered climatic conditions. The morphology, fertility, flowering and ripening time, yield, penetrance behavior, composition of pleiotropic spectra, and many other criteria were evaluated. Details are found in a series of publications (Gottschalk 1967a, 1977a, 1978b, 1980a, 1981b, c; Gottschalk and Patil 1971; Gottschalk and Kumar 1972; Gottschalk and Imam 1972; Gottschalk and Kaul 1975, 1980, Gottschalk and Wolff 1977; Gottschalk et al. 1978; Müller and Gottschalk 1978; Gottschalk and Müller 1979). One of our early flowering recombinants shows a high plasticity with regard to the number of days to flowering as related to the corresponding behavior of the initial line. In Germany, it is 7–10 days earlier; in Kurukshetra (North India) 16, in Shillong (North East India) 19, in Jabalpur (Central India) 23 days earlier. Even stronger differences were observed in some other genotypes of our collection (Table 25). Similar results on a narrower scale were obtained in an early ripening *pea* mutant cultivated at three different locations in Russia (Sidorova and Khvostova 1972).

Of particular interest is the reaction of some fasciated mutants. They are the highest yielding genotypes of our collection when grown under the moderate climatic conditions of Germany (Figs. 9, 26). Some mutants of this group do not flower at all in Egypt, India, and Brazil. In Ghana, flowering of the strongly fasciated mutant 489C was considerably delayed and its seed production was very poor. The linearly fasciated mutant 251A flowered richly in North India but it was completely seed sterile. In South Australia, finally, all the plants of this mutant died prematurely, whereas the

Table 25. The flowering time of four *Pisum* mutants and four recombinants under different climatic conditions as related to that of the mother variety "Dippes gelbe Viktoria"

Locality and country	Mutants				Recombinants			
	68C	251A	489C	1201A	R 46C	R 300	R 350	R 177
Bonn, Germany	± 0	+ 4	+ 8	± 0	− 10	− 9	± 0	± 0
Cairo, Egypt	−	−	NF	−	− 7	− 11	−	−
Cape Coast, Ghana	± 0	−	+ 20	± 0	−	−	−	−
Kampala, Uganda	+ 7	+ 20	+ 55	+ 7	− 7	− 7	+ 7	+ 7
Piracicaba, Brazil	−	−	−	−	− 15	−	−	−
Varanasi, North India	− 3	−	NF	± 0	− 13	− 13	± 0	−
Kurukshetra, North India	± 0	+ 3	NF	+ 6	− 16	− 11	± 0	± 0
Bombay, West India	± 0	−	NF	−	− 10	−	−	−
Udaipur, West India	± 0	+ 7	+ 30–40	± 0	− 19	−	−	+ 6
Jabalpur, Central India	− 7	± 0	NF	−	− 23	−	−	−
Shillong, N.E. India	− 5	−	NF	± 0	− 19	− 15	− 8	−

− 10 = 10 days earlier than Dippes; + 14 = 14 days later, ± 0 = equal to Dippes; NF = not flowering; − = not tested at that locality

- 68C: increased ovule number per ovary.
- 251A: stem linearly fasciated.
- 489C: stem strongly fasciated.
- 1201A: stem dichotomously branched; unstable penetrance.

- R 46C: early flowering, stem dichotomously branched.
- R 300: early flowering, stem dichotomously branched, increased ovule number per ovary.
- R 350: increased ovule number per ovary, stem dichotomously branched.
- R 177: stem dichotomously branched; penetrance stabilized.

other genotypes and the mother variety showed normal ontogenetic development. This unexpected behavior can only be understood as a negative reaction of that particular mutant to a specific ecological factor of this region not yet known in detail. Thus, these genotypes, highly productive in Middle Europe, are completely useless for pea breeding in the countries just mentioned and in others having similar climates. The early flowering recombinant R 46C mentioned above is not of interest for German pea breeding because of its low and unreliable yield. In Bombay, however, it outyielded the initial line considerably, due to its tolerance against fungal and viral diseases. In the western parts of India, this genotype was found to be highly drought resistant. A few years ago, when there was a complete lack of the monsoon rains in this area, it was the only genotype surviving and producing some seed, whereas all the other genotypes cultivated, inclusive the Indian local cultivars, died from drought. Other mutants differed from their behavior in Germany in dependence on specific climatic conditions

Fig. 26. The yielding properties of four *Pisum* mutants grown in Bonn, Germany, and at two Indian localities. Each *column* represents the mean value for the character "number of seeds per plant" for one generation. The mutants show specific reactions to the diverging climatic conditions resulting in drastic alterations of the yielding potentialities. The fasciated mutant 489C, high yielding in Germany, is not able to produce any seeds in India

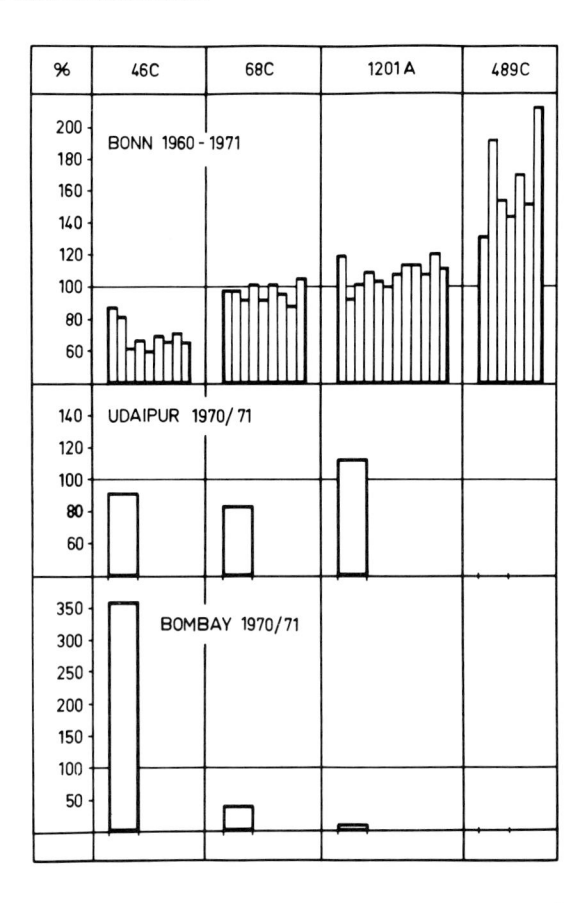

with regard to number and length of internodes, degree of branching, flowering and ripening behavior, general fertility, and seed production among others.

Some *Pisum* genes of agronomic interest were found to be completely unable to express their action under distinct climatic conditions. Three specific genes of the *Pisum* genome showed the following behavior in climatically different countries:

- *Gene bif-1^{1201} Causing Dichotomous Stem Bifurcation (Mutants 1201A, 239CH)*. Under German conditions, this gene shows an unstable penetrance which reduces the breeding value of the two mutants to some extent. In Yugoslavia, its penetrance was regularly lower than in Germany. In Egypt, Ghana, Brazil, and India, gene *bif-1^{1201}* is not able to express its action; all the plants homozygous for this gene are nonbifurcated and phenotypically identical with the mother variety. The penetrance of *bif-1^{1201}* could be stabilized by gene *sg* for reduced grain size. All the plants of this recombinant were bifurcated when grown in India. Thus, the stabilizing effect of gene *sg* is stronger than the action of that environmental factor which suppresses the action of *bif-1^{1201}* (Gottschalk 1978c).

- *A Gene for a Very Weak Degree of Stem Fasciation*. This gene belongs to the group of 18 genes present in the strongly fasciated mutant 489C. It is hypostatic and can only display its action if the respective epistatic genes have been eliminated. All the plants of two weakly fasciated recombinants, having this genotypic constitution, were found to be nonfasciated in the northern and western regions of India. Under the ecological conditions of these areas, the gene is not effective (Gottschalk et al. 1978 Gottschalk and Kaul 1980).

– *Gene Ion for an Increased Number of Ovules per Carpel (Mutant 68C).* Under the moderate climatic conditions of Middle Europe, gene *ion* results regularly in an increase of the number of seeds per pod. In Udaipur (western India), however, the mean values for the trait "number of seeds per pod" of mutant 68C and its mother variety were equal, and in Ghana the mutant had even a lower mean than the initial line. The pleiotropic pattern of gene *ion*, however, can be positively influenced by environmental factors. The reaction of that gene to the climates of different countries is schematically given in Fig. 27.

The examples just discussed are insofar of interest as they demonstrate that distinct mutant genes can only be utilized agronomically under specific ecological conditions, while they are useless under altered conditions. It would not have been possible to select the genotypes mentioned in Egypt, Brazil, or India in mutagenic treatments because the actions of the mutant genes do not become discernible.

In the examples just given, the nonfunctionability of mutant genes in specific climates is a negative effect because the mutants in question are useful under those ecological conditions under which they have been selected. But also the opposite situation is realized. Gene *dgl* of the *Pisum* genome causes complete degeneration of the leaves during ontogenetic development, resulting in a very low seed production. In Egypt and India, however, this gene cannot express its action. Plants homozygous for

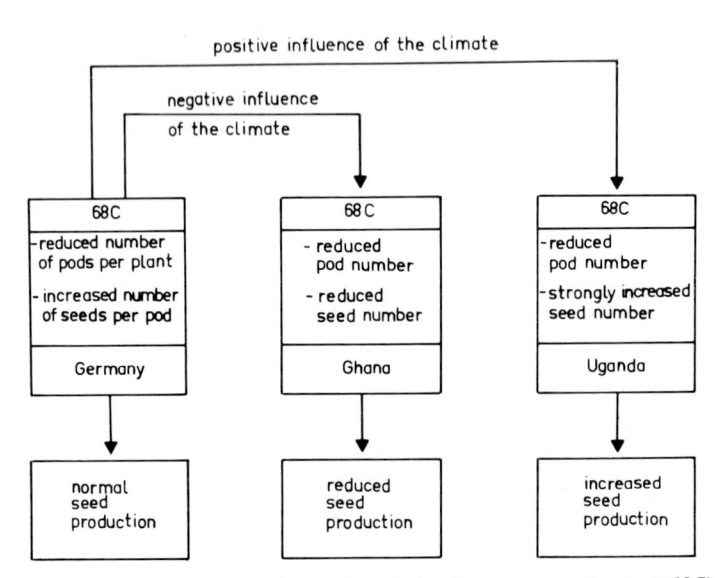

Fig. 27. The expression of gene *ion* of the *Pisum* genome (mutant 68C) under the influence of different climatic conditions

dgl are phenotypically normal; they show full physiological activity and good yield (Gottschalk 1979b). Similar reactions are certainly possible for genes controlling characters of agronomic interest but having a negative selection value under specific ecological conditions which can be positively altered in a different environment.

It was already mentioned in Sect. 16.1 that even the composition of the pleiotropic spectra can be altered under the influence of specific climatic factors. Findings in this field are available in *Pisum sativum* (Glazacheva and Sidorova 1973; Sidorova and Bobodzhanov 1977; Gottschalk and Kaul 1975), and *Triticum aestivum* (Khvostova 1978).

18.2 The Reaction of Mutants Under Controlled Phytotron Conditions

The breadth of the adaptational capacities of a given genotype to different ecological factors cannot be discerned under field conditions. The controlled conditions of a phytotron are needed for analyzing the specific influence of a distinct environmental factor or a distinct combination of several factors. Impressive results in this field of mutation research have been obtained in different *barley* mutants by Gustafsson's group utilizing the Stockholm phytotron (Dormling et al. 1966; Gustafsson 1969; Dormling and Gustafsson 1969; Gustafsson and Dormling 1971, 1972; Gustafsson et al. 1973a–c, 1974, 1975; Gustafsson and Lundqvist 1976). By means of this method, the most suitable environmental conditions for an optimal vegetative development and seed production of specific genotypes can be determined.

Of particular interest in this field is the reaction of some *erectoides, eceriferum*, and some *early flowering mutants* to different photo- and thermoperiods. The mutant variety "Mari," developed from Bonus barley, ripens in South Sweden 7–10 days earlier than Bonus and is extremely lodging resistant. It shows a somewhat reduced grain yield, but it gives very high yield in other parts of the country and in distinct regions of Norway and Finland. Moreover, it was used for developing some new varieties of improved breeding value (see Chap. 5). The differences between mother variety and the mutants are very pronounced under phytotron conditions. "Mari" and "Mona," for instance, are adapted to widely different climatic constellations and proved to be very productive under optimal conditions in both vegetative and reproductive characters. Most of the mutants studied show a specific reaction to the combination of various photo- and thermoperiods, the reaction being not only different from that of the mother variety but also from each other. They were found to be much more flexible with regard to their ecological demands as compared to the initial line; they have a wider range of adaptability. Moreover, two allelic early ripening mutants gave a pronounced overdominance under short-day conditions, while the degree of heterosis was less pronounced under long-day conditions. Heterosis in all photoperiods used with the exception of permanent light was also observed for two allelic *erectoides* mutants.

We have made similar phytotron studies with our mutants and recombinants of *Pisum sativum*. So far, more than 100 genotypes were studied, each of them showing its specific reaction to the various conditions different from the behavior of the mother variety and the reactions of all the other genotypes tested. These findings demonstrate the broad variability of our material with regard to the adaptability to specific ecological conditions, which is an important prerequisite for specific kinds of selections. I shall give a brief survey considering some genes controlling characters of agronomic interest; details are found in a series of publications (Gottschalk 1981b,c,e, 1982a–c). Many fasciated and early flowering genotypes were tested under controlled long- and short-day conditions, all the other environmental factors were equal in these treatments. It was shown that the nonflowering of some of our fasciated mutants under short-day conditions, observed in Egypt, Brazil, India, is due to the action of gene *fis* controlling the photoperiodic behavior of the plants. This gene suppresses the initiation of flower formation. Genotypes, homozygous for *fis*, need long-day for flowering. Their flowering behavior, however, does not exclusively depend on the photoperiod, but to some extent also on the thermoperiod. This becomes clear from the divergent reactions of the fasciated mutants under long-day phytotron and field conditions.

Gene *efr* for earliness was found especially susceptible to reacting to both, alterations of the photoperiod and changes of the genotypic background. This becomes clear from the behavior of more than 50 different recombinant types homozygous for *efr* and for specific other mutant genes or gene groups. Some of the additional mutant genes influence *efr* positively, resulting in an increased degree of earliness. Many of the mutant genes studied, however, have the opposite effect, causing a considerable delay of flowering, obviously due to certain negative interactions between *efr* and distinct other genes (Fig. 10). The respective recombinants are genetically early as a consequence of the presence of gene *efr* in their genomes, but in reality they are extremely late because of the negative interactions just mentioned. Four different *efr* recombinants were unable to produce normal flowers under the short-day conditions offered to them. Tiny floral buds were produced at very low nodes of the stem in accordance with the behavior of recombinant R 46C, the donor of gene *efr*. These buds, however, did not undergo any further development, they remained in that tiny stage. This reaction is not related to the nonflowering of the fasciated mutants in short-day due to gene *fis*. On the contrary, the initiation of flower formation takes place in all the plants of the four *efr* recombinants just mentioned, but a specific gene, not yet exactly identified, hinders further development of the floral buds into functioning flowers. This gene, however, becomes only effective under short- but not under long-day conditions. This is an interesting cooperation of a specific gene with a specific photoperiod, resulting in the suppression of the normal action of a useful mutant gene. Furthermore, this example demonstrates the close interrelations between genes and environment for realizing distinct characters.

The phytotron can be used not only for studying problems of basic research, but can also be utilized for preselecting genotypes adapted to specific ecological conditions. It was just mentioned that our fasciated *Pisum* mutants do not flower in short-day; therefore, they cannot be grown in developing countries with short-day climate. These high-yielding genotypes with good protein production, however, would be a very valuable material in countries in which the protein gap has to be closed by means of seed proteins. Therefore, we have tested some of our fasciated genotypes under short-day phytotron conditions. Recombinant R 729B of our collection derives from the cross of the fasciated mutant 489 C × Blixt's *cochleata* 5137. The plants are very tall and strongly fasciated. In contrast to the parental mutant 489C, they flowered richly, their begin of flowering being equal to that of the mother variety. This behavior is obviously due to the elimination of gene *fis* for long-day requirement; moreover, a hypostatic gene for relatively early flowering, deriving from 489C, becomes effective in the recombinant because the respective epistatic gene has likewise been eliminated. Seed and dry matter production of these plants were essentially better than in the mother variety, demonstrating thereby a high physiological activity under the phytotron conditions offered to them (Gottschalk 1982c). Because of the long internodes, the recombinant would not be suited for field cultivation in its present genotypic constitution. But we know now that these plants are adapted to short-day climates with relatively high temperatures. In further crosses, the gene for long internodes has to be replaced by a gene for shorter ones maintaining the favorable flowering behavior and yield potential of the recombinant. In other phytotron trials, genotypes have been selected producing more seeds and dry matter under the unfavorable short-day than under the more favorable long-day conditions. Furthermore, a preselection of *Pisum* mutants was carried out, showing a certain degree of tolerance to temperatures over 30°C.

In all these trials, only preselections of prospective genotypes can be made, whereas the definite efficiency of the selected mutants or recombinants can only be studied in a country having that specific climate in close cooperation with a geneticist of that country. Many mutants, having existed for years or even decades, could already be utilized for breeding purposes if their ecological adaptability would be known. The phytotron can provide plenty of information in this field.

The findings presented in this chapter demonstrate the relativeness of the selection value of mutant genes and the questionableness of terms such as "micromutants" and "useful" or "useless mutants." A genotype, selected in a country with moderate climate, may be without any agronomic interest because of its reduced yield. The same mutant can be of high interest in a hot and dry country because of its drought resistance. This favorable peculiarity, however, cannot be discerned under those ecological conditions under which the selection was carried out. Thus, a mutant is useless in Middle Europe and very useful in Central India. In Germany, it would have been eliminated in a mutation breeding program, whereas it is used in India for developing a new commercial variety.

19 The Alteration of Morphological and Physiological Seed Characters

The nature of the seed of a crop plays not only a role with regard to the seed yield of the plant but also comprises an important qualitative trait. This holds particularly true in respect to distinct storage substances such as proteins, carbohydrates, alkaloids, or medicinal substances among others. These problems are discussed in detail in Chap. 20. In the present chapter, the alteration of some morphological and physiological seed characters under the influence of mutant genes is reviewed. Some *barley* genes altering kernel shape have been allocated to distinct chromosomes of the genome by crossing the mutants with translocation tester lines (Häuser and Fischbeck 1976).

19.1 Seed Size

Mutants with larger grains are mainly known in *Oryza sativa* (Marie 1967, 1970). The grains of an Indian gamma-ray-induced rice mutant are about one and a half time as long as normal ones, causing thereby an enormous increase of the thousand grain weight. The gene is pleiotropic and influences plant height, tillering, panicle length, number of grains per panicle, and time of maturity partly positively, partly negatively. The main advantage of this mutant lies in its having an improved lodging resistance in comparison to the mother variety. Therefore, the mutant strain "Jhona 349" is widely cultivated in Haryana (North India) although it is not yet released officially (Bhan and Kaul 1974). The Indian *wheat* variety "Arjun" has high yield and a high degree of resistance against fungal diseases but its grains tend to mottle, reducing the market value. After treatment with nitrosomethyl urea, three mutants with nonmottled grains were selected in an M_2 generation of about 30,000 plants. The seeds of these genotypes are considerably larger than those of Arjun. The yielding properties of the mutants and their resistance behavior are not negatively influenced by the mutated genes (Sawhney et al. 1982).

In *legumes,* a large-kerneled *peanut* mutant with improved yielding properties was released in India as a new variety already in 1973. Moreover, the mutant was used for crossing and some lines with even higher yield were selected (Patil 1975, 1980). We have obtained two large-grained *pea* mutants in our radiation genetic trials but both are inferior to the mother variety with regard to seed production because of the reduced number of grains per pod (Gottschalk 1966). Large-grained *Pisum* mutants are also available in Russia (Sidorova et al. 1969). Similar types of economic value have been selected in *Lupinus mutabilis* (Pakendorf 1974), *Linum usitatissimum* (Pospisil 1974) and *Ricinus communis* (Kulkarni 1969).

The **reduction of the grain size** is realized in some *rice* mutants. The Indian strain "Rageni" shows a considerable increase in grain weight per plant in spite of the reduced kernel size. The mutant has some other useful characters, such as earliness, lodging resistance, good grain quality, increased protein production, possibly also an increased amount of some essential amino acids (lysine, methionine). The strain is adapted to late sowing (Khan 1973). Also other Indian rice mutants of this category are high-yielding, one of them having a desirable combination of short grains and earliness (Kaul 1978; Ghosh Hajra and Halder 1981). In *Pisum,* more than ten small-grained mutants have been selected in our X-ray and neutron trials, most of them being homozygous for the same recessive gene. With regard to the character "number of seeds per plant," they are either equivalent or somewhat better than the mother variety. Because of the strongly reduced grain size, however, the mean values for the character "grain weight per plant" are considerably lower. These types could be of some interest for canning. A gene for reduced grain size is also present in our fasciated pea mutants, likewise influencing their breeding value negatively (Gottschalk 1966, 1977a).

19.2 Seed Shape

Alterations of the grain shape toward roundness due to the action of mutant genes was found in *rice* (Reddy and Reddy 1973a, b; Mallick 1978). A very desired character in rice breeding is fine, slender grains, somewhat longer but narrower than the normal ones. They have been isolated by Siddiq and Swaminathan (1968); Swaminathan et al. (1970); Ram (1974) among others. Mutants, in which this character is combined with other useful traits, seem to be promising for rice breeding. This holds true for the following combinations:

— Fine grains,
  earliness,
  improved milling and cooking quality (Reddy and Reddy 1973a, b).
— Fine grains,
  increased number of panicles,
  high yield (Basu and Basu 1970).
— Fine grains,
  earliness (Kaul 1978).
— Fine grains,
  earliness (18 days earlier),
  improved grain quality,
  normal yield (Mallick 1978).
— Fine or long slender grains,
  improved grain quality (Ismail and Ahmad 1979).
— Fine grains,
  better agronomic performance
  (a recombinant selected after having crossed two fine-grained mutants with each other; Kaul 1981).

19.3 Seed Color

In some cases, alterations of the seed coat color are of direct importance for the improvement of a crop. Some high-yielding Mexican *wheat* varieties, for instance, were not accepted for commercial cultivation in India because of their red grain color. By means of EMS and NMU, mutations of amber grain color were induced. The respective strains do not differ from the red-colored varieties with regard to yield (Bansal et al. 1972). A similar situation is valid in *grain sorghum*. Pure white pearly grains of *Sorghum subglabrescens*, obtained in mutation treatments in India, have a higher market value than those with red wash and are therefore preferred in cultivation (Sree Ramulu 1968, 1974a, b). *Phaseolus vulgaris* is a basic food crop in many South and Middle American countries. In some areas, beans of specific colors are preferred. Mutants with white, yellow, or brown seed coats increase the marketability of black-colored varieties in Costa Rica and represent an improvement (Moh 1971, 1972). A similar situation exists in *Lens culinaris*. The increase of the genetic variability for seed coat color after application of gamma rays and NMU may therefore be of practical interest (Sharma and Sharma 1979b).

19.4 Physiological Seed Characters

The grain quality can be improved in different crops in a different way as shown in the following examples:

- *Triticum aestivum:*   Mutants with good "chapati"-making quality in India (Khan 1973).
  Mutants with improved baking quality in Germany (Wagner 1969).
- *Oryza sativa:*   Mutants with better milling quality (Marie 1967, 1970; Reddy and Reddy 1973b).
  Mutants with better cooking quality (Swaminathan et al. 1970; Reddy and Reddy 1973a; Reddy 1977; Samoto 1975).
- *Sorghum subglabrescens:*   Mutants, the grain of which pop better and freer (Sree Ramulu 1968).

Some more mutants with superior grain quality have been isolated in *barley* (Trofimovskaya and Zhukovsky 1967), *rice* (Reddy and Reddy 1972), *durum wheat* (Mosconi 1967), and *bread wheat* (Orlyuk 1973). Mutants with soft, medium hard, and very hard grains of a hard grain wheat variety belong also to this category (Kumar 1978).

The economic value of some mutants is due to the alteration of certain physiological characters of the seeds. The Swedish *barley* variety "Kristina" arose by crossing "Mari" (developed from an X-ray-induced mutant) with the Norwegian variety "Dormen." It has high yield, good lodging resistance, and excellent malting properties but short dormancy period. This lowers the breeding value of Kristina to some extent because the seeds tend to germinate on the culm, particularly in years with rainfall during harvest period. Some mutants of Kristina are now available showing a reduction of this tendency, resulting in a prolongation of dormancy (Gustafsson et al. 1972). The opposite situation prevails in *black gram (Phaseolus mungo).* The seeds of this species have hard coats which prevent the uptake of water, giving a pronounced dormancy. After X-ray and EMS treatment, two types of nondormant mutants were selected. This represents an improvement of the species in this respect (Appa Rao et al. 1975). A similar case is known in *Melilotus albus* (Scheibe and Micke 1967). Of normal sweet clover seeds 90%–95% have an impermeable seed coat, causing a considerable delay of germination. The mutant has only 10% seeds of this category at most. The trait is controlled by a single recessive gene.

In *rice,* finally, easy threshing types have been obtained in India besides mutants with fine grains, compact panicles, reduced plant height, and altered ripening times. These *indica* mutants in *japonica* type are insofar of direct practical value as the Indian population does not like the stickiness of the *japonica* grains (Siddiq and Swaminathan 1968).

On a very large scale, the possibilities for selecting **sprouting-resistant** *wheat* mutants were tested in Finland using 5000 M_3 families of gamma-ray-treated material. A genetically conditioned variation of this trait was observed and some slowly sprouting pedigrees were selected (Kivi and Ramm-Schmidt 1969).

20 The Alteration of Seed Storage Substances

The seed is a storage organ providing nutritional substances to the germinating embryo. Two main groups of compounds are found in every seed: the carbohydrates, in general comprising the greatest part of the seed, and the proteins. Moreover, the lipids and a considerable number of further substances are present. The amount and composition of these substances are genetically controlled. Most information is available in seed proteins, therefore this part is discussed more in detail.

20.1 Seed Proteins

Protein mutants cannot be selected by visible traits; on the contrary, complicated biochemical methods are necessary for discerning alterations of the protein situation. Theoretically, all the M_2 plants of a mutation program should have to be analyzed biochemically in order to isolate mutants of this category. This is often impossible because respective quick methods are not yet available. It was already mentioned in the introduction that a great number of lines or of M_2 plants have been tested, in exceptional cases resulting in the selection of a small number of protein mutants. The usual procedure, however, consists in analyzing the seed proteins of mutants which show any morphologically discernible alterations. This holds particularly true for those genotypes which are of any agronomic interest, especially for mutants with satisfying yield potentials. A small proportion of these genotypes shows alterations in seed proteins which can be caused by two basically different mutational events:

– The altered protein situation can be part of a pleiotropic pattern of that mutant gene which controls the morphologically discernible trait.
– The altered protein situation can be due to the action of an independent "protein gene" existing besides that gene which gave rise to the selection of the mutant. In this case, the mutant is homozygous for at least two mutated genes.

Undoubtedly, many protein mutants, existing in our collections, are not discerned because of these difficulties. Others are eliminated in the M_2 and M_3 generations as a consequence of the methodological insufficiencies just mentioned.

Before going into detail, some general considerations on seed proteins are given. Moreover, the influence of internal and external factors on production and composition of these substances is discussed in order to get a basis for understanding the action of mutant genes effective in this field and for judging their economic importance.

20.1.1 The Characterization of Seed Proteins

Cereals contribute 55% of protein to human diet, *legumes* only 15%. The seed protein content differs considerably in these two groups, ranging between 10% and 20% in cereals, whereas more than a quarter of legume seed substances can be proteins. Especially high percentages are found in the *soybean* with values up to 40%. The proteins are composed of several subfractions with different chemical behavior due to different amino acid patterns and different molecular weights.

Cereal proteins are composed of four main subfractions as follows:

— the water-soluble albumins,
— the salt-soluble globulins,
— the alcohol-soluble prolamins,
— the glutelins, soluble in diluted alkali.

These subfractions can be further subdivided into a considerable number of smaller components. The *prolamins* are the typical storage proteins of *maize, barley, wheat,* and *sorghum*. They are characterized by a high proportion of arginine, glutamine, and proline, whereas their content of lysine and of the sulfur-containing amino acids is extremely low. Their proportion on total seed protein varies considerably in different cereals. The amount of prolamins is negatively correlated with lysine content. In low-prolamin species total lysine content is comparatively high, while it is low in high-prolamin species (Table 26). This is insofar of interest as lysine is the limiting amino acid in cereals, and thus a measure for the nutritional value of these seed proteins. In contrast to the cereals just mentioned, the prolamin content of *rice* and *oats* is very low. In the case of *rice,* the *glutelins* are the main storage fractions, in the case of *oats* the *globulins*. The lysine content of these two protein groups is higher than that of the prolamins, therefore the nutritional quality of rice and oats seed protein is higher. Detailed information on cereal seed proteins is given by Danielsson (1949b); Mossé (1966, 1968); Paulis et al. (1975b); Miflin and Shewry (1979); Bewley and Black (1979); Nwasike et al. (1979); Wilson et al. (1981).

Table 26. Seed protein content and solubility fractions (% protein) of cereal grains (after Johnson and Lay 1974; and Kakade 1974)

Cereal	Protein content % dr.m	Albumin (H$_2$O soluble)	Globulin (salt soluble)	Prolamin (alcohol soluble)	Glutelin (alkali soluble)	Lysine (% protein)
Rice	8–10	trace	2–8	1–5	85–90	4.4
Rye	9–14	5–10	5–10	30–50 (secalin)	30–40	4.3
Oats	8–14	1	80	10–15 (avenin)	5	4.0
Barley	10–16	3–4	10–20	35–45 (hordein)	35–45	3.6
Wheat	10–15	3–5	6–10	40–50 (gliadin)	30–40	3.4
Corn	7–13	trace	5–6	50–55 (zein)	30–45	2.8
Sorghum	9–13	trace	trace	60 (kafirin)	Considerable	2.2

The *legume seed proteins* are composed of only two main subfractions:

- The water-soluble albumins (mostly enzymatically active proteins), which can be subdivided into a great number of further subfractions.
- The salt soluble globulins consisting of two major fractions (in *Pisum sativum*, for instance, of vicilin and legumin). These proteins, too, can be subdivided further.

The globulins are typical storage proteins, characterized by a high amount of amides (glutamine and asparagine) and an extremely low percentage of S-containing amino acids. In *Pisum sativum,* the ratio of globulin to albumin is 1:1.6; 60% of seed proteins are globulins. By ultracentrifugal separation, the globulin fraction can be subdivided into several (possibly 19) fractions, some of which are obviously under control of separate genes. Further detailed information has been given by Danielsson (1949a, b, 1952); Derbyshire et al. (1976); Sutcliffe and Pate (1977); Boulter (1979); Casey (1979); Bewley and Black (1979); Müller and Werner (1979); Thomson et al. (1979); Khan et al. (1980).

Seed proteins in cereals and legumes are mostly deposited in *protein bodies.* The composition and size of these organelles depends highly on the species and on the position of the protein bodies within the seed. 70%–80% of their dry matter are storage proteins. Among the other substances especially phytic acid is of importance as phosphor storage; in addition certain enzymes are present. In *monocotyledonous crops,* protein bodies of different structure are found within the seeds. The bodies of the aleurone layer do not contain prolamins and their protein content was found to be lower. The bodies of the starchy endosperm, however, contain predominantly prolamins. In *dicotyledonous crops,* the protein bodies seem to be uniform within the seeds, containing in the case of *peas* legumin and vicilin (Pernollet 1978).

20.1.2 Factors Influencing Protein Content and Composition

The amount and compostion of seed proteins are widely influenced by both environmental and endogenous factors. They can cover the action of protein genes; therefore, it is necessary to discuss their influence.

20.1.2.1 Environmental Factors

The values for the protein content of the seed flour vary considerably within the same genotype, ranging between 8% and 18% in *wheat* (Johnson et al. 1969) or 5% and 12% in *rice* (Coffman and Juliano 1979). The variation range of the seed protein percentage in *pea plants* of a given genotypic constitution is presented in Fig. 28 using the variety "Dippes Gelbe Viktoria," the initial line of our radiation genetic experiments, grown in the same year in different plots within the same field. The distribution curve ranged between 19% and 25% having a maximum at 22%. Each value is based on a mixed seed sample from about 30 plants, thus representing a mean value. Because of this, the variation range is comparatively small. In two subsequent generations, extensive quantitative investigations on total seed protein were carried out (Quednau and Wolff 1978a, b). The variety was grown in 65 replications. From five single plants of each

Fig. 28. Variation in seed protein content of 38 replications of the variety "Dippes Gelbe Viktoria" of *Pisum sativum* used as initial line for our radiation genetic experiments

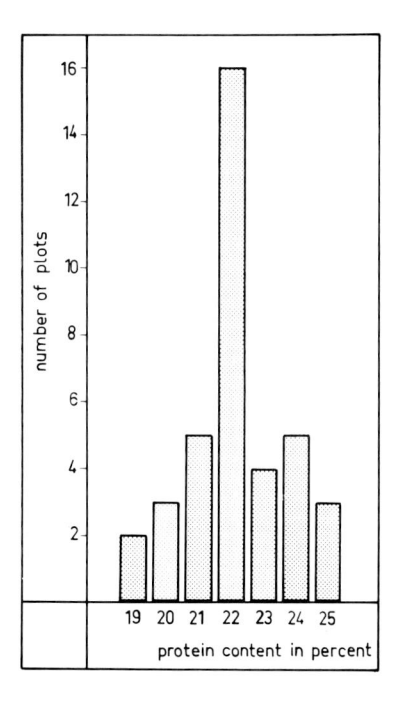

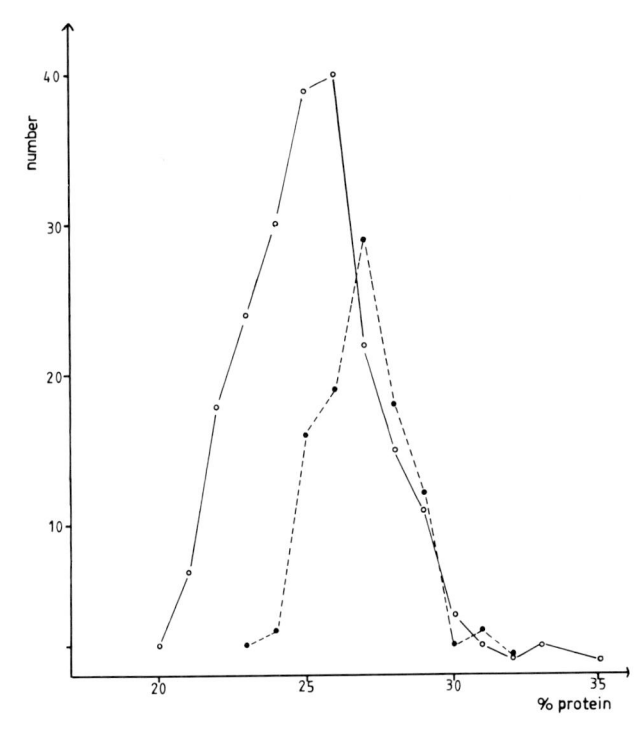

Fig. 29. Distribution of seed protein values in two subsequent generations. Material: *Pisum sativum*, variety "Dippes Gelbe Viktoria". 1978, 218 single values ○——○; 1976, 88 single values ●----●

replication, the seeds of one pod were separately analyzed. As expected, the variation range of the protein values was essentially greater, the values ranging between 20% and 35% in the first, and between 23% and 32% in the second year (Fig. 29; mean values 25.5% and 27.3% respectively). These differences within the same genotype can only be due to differences in distinct nongenetic factors influencing seed protein inlet. Of particular interest in this respect are the higher values, demonstrating that pea plants are potentially capable to produce very high amounts of seed protein under certain circumstances. They are normally not discernible when the determinations are carried out on the basis of material derived from several plants. Thus, optimizing the growing conditions of a crop might be one of the possibilities for improving the protein content of its seed flour.

It is not yet clear which factors are responsible for the differences in the protein content in the two subsequent years, because the *Pisum* material was not grown under controlled phytotron and soil conditions. A possible factor might be the *temperature*, which was considerably higher in the second year as compared to the first one. This assumption is supported by findings obtained in *soybeans* (Schuster and Marquardt 1973). Also phytotron experiments in *barley* have shown that the protein content is increased when the plants are grown at higher temperatures (Gustafsson 1969), whereas low temperatures reduce protein accumulation in higher plants in general (Zeller 1957). Also other climatic and cultivation factors are known to influence protein inlet into seeds. Photoperiod and day length are reflected in the protein content of *barley* (Gustafsson 1969). High solar radiation and dry location was found to have a negative effect on protein content in *rice*, while low plant density and improved water management resulted in increased protein percentages (Coffman and Juliano 1979). Investigations of Lawrence (1976) in *wheat*, on the other hand, showed that an increased water supply has no effect on seed proteins.

Certain findings reveal that the reactions of different genotypes to climatic factors and conditions of cultivation are not uniform with regard to the protein content of their seed meal. They depend obviously on the specific genotypic constitution of the material analyzed. This becomes clear from different reactions of a group of *Pisum* genotypes comparatively grown in West Germany and India (for details see Sect. 20.1.4.2). Moreover, it was shown that the density of plant cultivation is differently responsed by pea genotypes (Hiemke 1973). A similar behavior was observed in *wheat* varieties with regard to air temperature (Lawrence 1976).

An important factor seems to be *fertilization*, especially N-fertilization. In the case of *legumes*, it is of minor influence on seed protein content because of the legume – *Rhizobium* interactions (Richards and Soper 1978; Röbbelen 1979). On the other hand, Slinkard (1972) found an increased protein content in the *garden pea* after nitrogen application. In *cereals*, the positive effect of N-fertilization on seed proteins has been repeatedly evidenced in that way that the amount of fertilizers is to some extent correlated with the degree of improvement. Findings are available in the following crops:

- maize (Decau and Pollacsek 1970; Rendig and Broadhent 1978; Zink 1979);
- barley (Andersen and Køie 1975; Andersen 1977; Ewertson 1977; Pomeranz et al. 1977; Chery 1979; Winkler and Schön 1980);

- wheat (Swaminathan et al. 1969; Nadeem et al. 1978; Nardi and Brunori 1980; Hahne 1981);
- rice (Swaminathan et al. 1969; Harn et al. 1973; Allen and Terman 1978).

Similar effects were found in *cotton* seeds (Elmore et al. 1978). Moreover, the consequences of fertilization in general depend on time of application. In *barley,* late application after anthesis results in increasing protein content, apparently at the expense of starch (Andersen 1977; Chery 1979). The various protein fractions, however, are influenced in a different way. Especially the prolamin fraction is increased (Andersen and Køie 1975; Ewertson 1977; Winkler and Schön 1980). Within this fraction, a special subfraction is affected predominantly (Køie et al. 1976b). At the same time, an increase of the amount of amides is observed (Chery 1979; Röbbelen 1979; Winkler and Schön 1980). Similar findings are reported for *maize* (Rendig and Broadhent 1978). The prolamins are of low nutritional value because of their low lysine content. Thus, the nutritional value of the respective cereal protein is reduced by late N-fertilization. The application of N-fertilizers at an earlier time, for example at sowing, influences the protein fractions more homogeneously, increasing also the albumin and globulin fractions with better nutritional values. Thus, seeds of plants which have been manured during early ontogenetic stages have a better quality than those which have been manured after anthesis (Køie and Kreis 1977; Elwinger 1978; Winkler and Schön 1980). The increase of protein content by N-fertilization − similar as at higher temperatures − results in a stabilization of the protein values; the variation coefficient is lowered (Ulonska et al. 1973, 1975; Ulonska and Baumer 1976). The reduction of the variation range improves the possibilities to select protein mutants, particularly those which deviate only slightly from the control material.

The question whether virus or fungal infection influence the seed protein content has not yet been studied thoroughly enough; the findings available are contradictory. In *Pisum* varieties, virus infection resulted in increased protein values (Slinkard 1974), whereas no differences between healthy and sick plants were found in our mutant pea material (Müller 1976). Crown rust infection of *oats* was found to have no effect on protein percentage (Simons et al. 1979).

These findings show that a greater number of external factors can influence the seed protein content. They interfere with genes controlling this trait. Therefore, the material should be grown with several replications, in different locations as well as in different years with the aim to calculate variation ranges of the respective quantitative characters.

20.1.2.2 Endogenous Factors

The seed protein content is not only influenced by environmental but also by endogenous factors, such as culm length, leaf area, time of maturation, seed size, number of seeds per plant and others. Certain correlations between these factors and the protein production seem to be generally valid for the crops studied; on the other hand, many exceptions were found demonstrating that generalizations in this field should be done with caution.

Characters of Plant Structure. In *peas,* the protein content varies according to the *position of the seed along the stem,* but no strong gradient seems to exist (Hiemke 1973).

The deviation in seed protein content from pods of the same nodes are as big as those from different nodes, but seeds of lower nodes were found to have a greater kernel weight than those of the upper ones. Thus, a gradient seems to exist in such a way that the absolute amount of protein per seed is the greater, the longer the time of seed development is. Also in *wheat*, different seed sizes along the ear are found. Smaller seeds are present in the upper as well as in the lower part, while bigger kernels are present especially in the middle parts. Measurements of protein percentage in the different seed-size classes revealed a negative correlation between seed size and protein content (Hahne 1981).

Seeds grown on *different tillers* of the same plant can considerably vary in their protein content, in *wheat* by up to 2.5% (Bhatia et al. 1970). Similar findings are available in *barley* (Favret et al. 1970). The differences between the spikes are greater than the differences within the same spike. A prolonged period of development or reduced seed set are made responsible for an increased protein content. In *rice*, the differences seem to be not so profound (Mikaelsen and Kartoprawiro 1973).

The *culm length* has obviously also some influence on the protein content of the seeds. A negative correlation was observed in *bread wheat* (Nagl 1978a; Johnson et al. 1979) and *rice* (Tanaka and Takagi 1970; Harn et al. 1973; Hillerislambers et al. 1973; Harn and Won 1978). In *barley*, however, the seeds of long culms were found to have higher protein content than those of shorter ones (Toft Viuf 1969).

Yielding Characters. In *cereals* and *pulses*, a negative correlation exists between seed production and protein content as well as between seed size and protein content. The relatively high amount of protein in smaller seeds may be due to the fact that a lesser amount of starch is deposited, thus reducing their size. The correlations between number of seeds per plant and protein content become understandable by assuming that a distinct capacity of protein accumulation is combined with a low number of developing seeds, this resulting in a relatively high protein amount in each single seed. The basic reason for this kind of increased protein content would thus be the reduced seed number. The negative correlations just mentioned were found in many crops, for instance in:

– barley	(Ibrahim et al. 1966; Favret et al. 1969; Scholz 1971, 1972, 1975, 1976a);
– bread wheat	(Dumanović et al. 1970, 1973; Kumar 1976; Denić 1978; Singhal et al. 1978; Bhagwat et al. 1979; Johnson et al. 1979);
– rice	(Hillerislambers et al. 1973; Kushihuchi et al. 1974; Monyo and Sugiyama 1978; Bhagwat et al. 1979; Chutima 1979; Kaul 1980a);
– beans	(Gridley and Evans 1979);
– soybeans	(Hartwig 1979);
– broad beans	(Lafiandra et al. 1979; Sjödin et al. 1981a);
– peas	(Kaul and Matta 1976; Blixt 1979).

Exceptions from these correlations are reported.

Investigations in *peas* contribute some more details to this topic, demonstrating that generalizations are not justified. After analyzing small and large seeds of the same genotype, no significant deviations in protein content were found (Fig. 30). When

Endogenous Factors

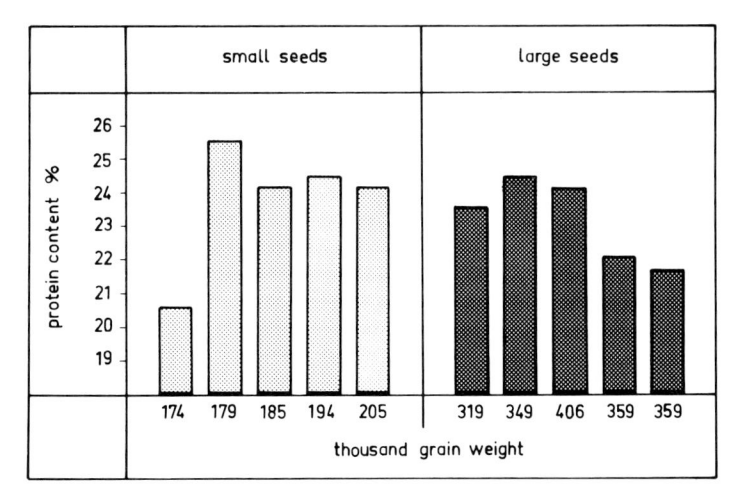

Fig. 30. Seed protein content in small and large seeds. Material: *Pisum sativum*, variety "Dippes Gelbe Viktoria"

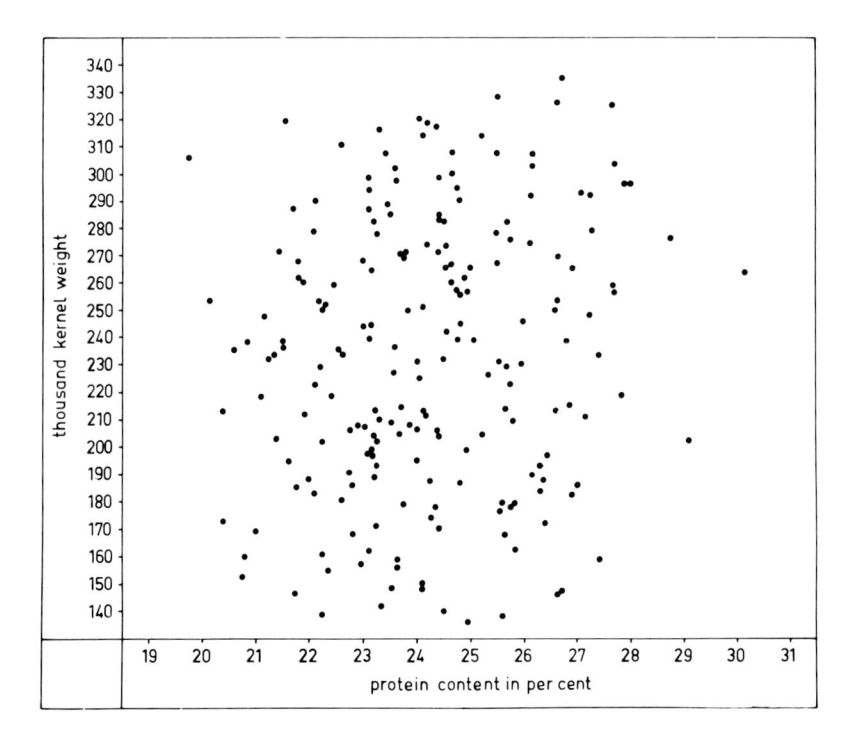

Fig. 31. Relations between seed size and protein content of the seed meal in 19 *Pisum* mutants, 73 recombinants and the mother variety

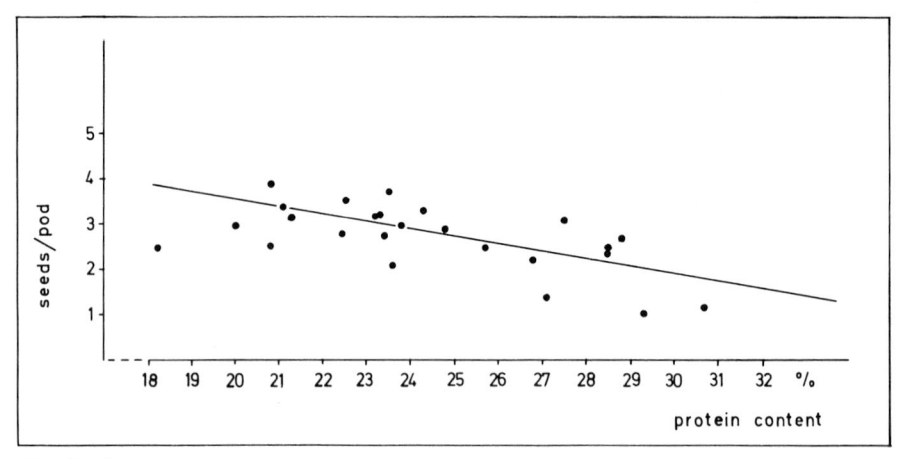

Fig. 32. Correlation between seed protein content and number of seeds per pod in 25 different *Pisum* mutants and recombinants

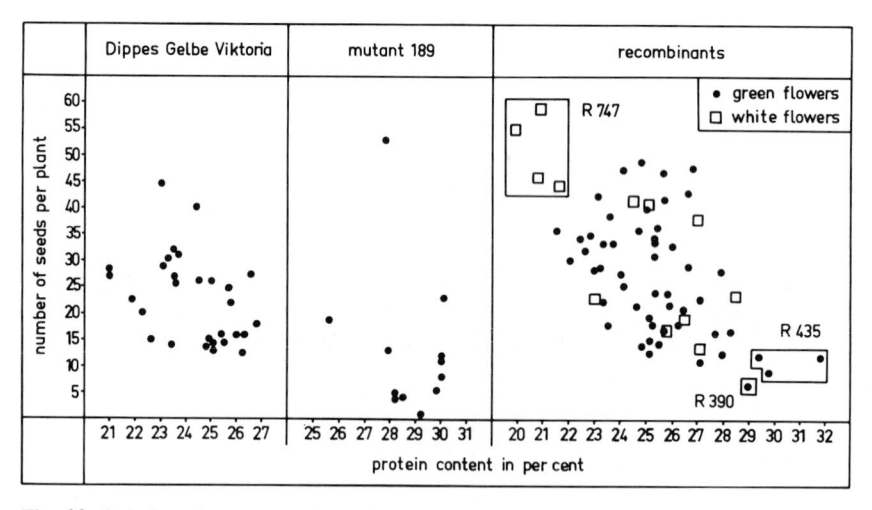

Fig. 33. Relations between seed production and protein content of the seed meal in the *Pisum* variety "Dippes Gelbe Viktoria", mutant 189 and in 24 recombinants obtained from crosses between mutants 489C × 189. The *dots* represent mean values either for different generations or for different replications grown in the same generation, respectively

thousand kernel weight is plotted against protein content, no correlations were found either (Fig. 31). Thus, seed size has apparently no influence on protein content in peas. But a negative correlation seems to exist between number of seeds per pod and protein content (Fig. 32). With regard to the seed yield, the correlations mentioned above are obviously not generally valid; on the contrary, the genetic constitution of the material analyzed may be of importance in this concern. The left-hand part of Fig. 33 shows that no correlation exists between number of seeds per plant and protein content of the seed flour within the two genotypes studied, the X-ray-induced mutant 189 of our

Endogenous Factors

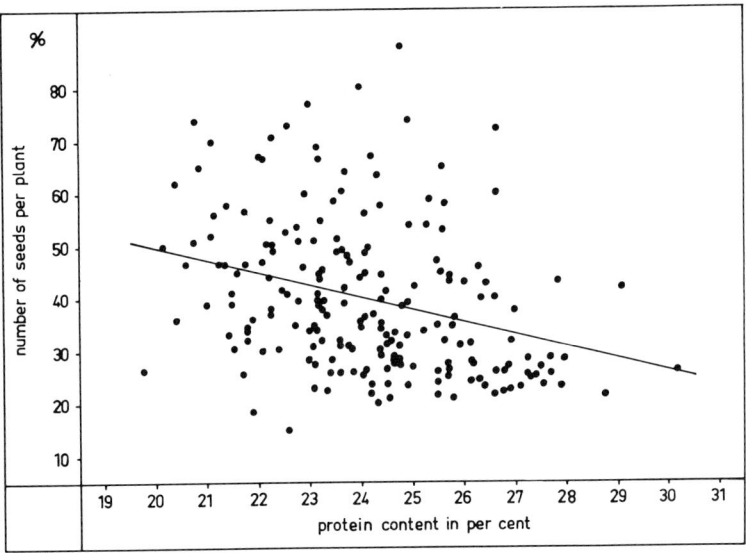

Fig. 34. Relations between seed production per plant and protein content of the seed meal in the mutant material considered in Fig. 31

collection and its mother variety "Dippes Gelbe Viktoria." This holds obviously also true for a group of *Pisum* recombinants (right-hand part of the figure): a negative correlation is indicated but the statistical evaluation of the values reveals that it is not significant. If a greater number of *Pisum* genotypes is considered, a slight but statistically significant negative correlation between seed production per plant and protein content of the seed meal seems to exist, but numerous exceptions are observed (Fig. 34; for further details see Wenzel 1981; Gottschalk 1983; Gottschalk and Wolff 1983).

A further group of factors which might influence seed protein inlet are those involved in nitrogen metabolism. If NO_3^- is taken up by the plant as nitrogen source, it has to be reduced to NH_4^+. NH_4^+ is introduced into α-ketoacids, forming amino acids which are the basis for protein synthesis. The availability of the respective compounds and of the enzymes catalyzing these reactions could play a role. The results available are not uniform and no clear relations were found which could be generalized. Genetically conditioned differences in the uptake of soil nitrogen are reported for a *wheat* variety with high seed protein percentage (Johnson et al. 1979), moreover for *maize* (Beauchamp et al. 1976; Chevalier and Schrader 1977). Different amounts of certain enzymes are proved too, particularly with regard to the amount and reactivity of nitrate reductase (*barley:* Kleinhofs et al. 1978; Kleinhofs and Kuo 1980; Young et al. 1980; *wheat:* Rao and Croy 1972; Rao et al. 1977; Johnson et al. 1979; *pea:* Feenstra and Jacobsen 1980). Furthermore, the rate of seed protein synthesis and the duration of this process seems to be reflected in the protein amount (*wheat:* Johnson et al. 1969; Brunori et al. 1977; Donovan et al. 1977; Bhagwat et al. 1979; *rice:* Perez et al. 1973; Bhagwat et al. 1979).

In *legumes,* the root nodules should be expected to have an influence on seed protein production. In different *pea* varieties, an extraordinarily great variability with regard to the number of nodules per plant was found. There is a positive correlation between number of nodules and total plant production, but the protein content is obviously not involved (Richter 1974). In pea mutants, a correlation between number of root nodules and seed production per plant was not found (Müller 1978b).

20.1.3 Seed Protein Content of Different Varieties of the Same Species

Before considering the influence of mutated genes on seed protein content, it may be of interest to discuss the potentials of protein production within a given species. This can be done by determining the variation range of the protein content in variety collections existing all over the world. As the varieties differ from each other in a more or less great number of genes, it can be assumed that some of them influence the protein content too; thus, a certain variation range should become discernible.

Table 27. Ranges of variability of the seed protein content in varieties or lines of different crops

Material tested		Range of protein content (%)	References
Cereals			
Zea mays			
114 inbred lines		9.8–16.3	Davis et al. 1970
117 inbred lines	Flint	10.4–17.9	Pollmer et al. 1977
	Dent	10.8–17.2	
Hordeum vulgare			
1000 varieties from WBC		8.8–16.9	Hagberg and Karlsson 1969
650 spring varieties		9.4–18.1	Toft Viuf 1969
5828 from Gatersleben collection		10.0–22.9	Müntz et al. 1979
	Mainly	12.0–18.0	
Triticum ssp.			
12613 varieties from WTC		6.0–23.0	Johnson et al. 1973
8007 from Gatersleben collection		10.0–24.3	Müntz et al. 1979
	Mainly	13.0–19.0	
Oryza sativa			
7000 genetic stocks		5.0–18.0	Tong et al. 1970
4445 varieties		5.4–12.5	Kaul 1973
Avena sativa			
3300 lines		7.8–21.9	Frey 1973
289 varieties		12.4–24.4	Robbins et al. 1971
Legumes			
Pisum sativum			
1450 varieties		14.0–36.0	Gottschalk et al. 1975a
1400 varieties		20.4–30.9	Slinkard 1972
Vicia faba			
33 varieties and lines		22.9–38.5	Griffiths and Lawes 1978
Vigna radiata			
432 strains		16.2–29.3	Sandhu et al. 1979

WBC = World Barley Collection
WTC = World Triticum Collection

Table 27 gives variation ranges of seed protein contents of several crops. The mean values within the same species vary extremely. In general, the lowest value is about half as high as the highest one. In some cases, for instance in *Pisum*, the two opposite values even differ by a factor of three from each other. These findings indicate that an improvement of the protein production could already be reached by utilizing "high protein strains" directly or by incorporating them into crossbreeding programs.

In *maize*, selection over a period of 70 generations resulted in a strain with 23.5% protein (IHP strain) beginning from 10.9% in the initial line. This increase, however, is combined with reduced seed size and reduced seed production (cited after Frey 1977). The variety "Atlas 66" of *bread wheat* with 18.3% protein is one of the best wheat varieties. Its high protein content is caused by an extended period of synthesis rate (Johnson et al. 1969; Brunori et al. 1977). The high protein content seems to be controlled by several genes, deriving probably from "Frondoso". Atlas 66 has been intensively used in crossbreeding and increments of the respective protein values by 15%–25% were obtained.

Table 28. Ranges of seed protein content of mutants of different crops

Material tested	Range of protein content (%)	References
Cereals		
Zea mays		
37 mutants	8.0–14.4 Control value: 9.8	Bálint et al. 1970
Hordeum vulgare		
92 mutants	9.3–14.9 Control values: 9.7–10.3	Doll 1972
Triticum ssp.		
78 mutants	13.1–19.1 Control value: 14.7	Narahari et al. 1976
420 mutants	11.1–14.3 Control value: 13.0	Parodi and Nebreda 1979
Oryza sativa		
545 mutant lines	4.2–16.5 Control value: 6.5	Tanaka and Takagi 1970
1041 mutants	4.3–13.6	Harn et al. 1973
718 mutant liens	4.0–16.0	Tanaka 1973
236 mutants	6.3–22.0 Control value: 8.5	Narahari et al. 1976
Avena sativa		
1580 mutant lines	16.05–22.75 Control values: 15.95–20.9	Jalani and Frey 1979
Legumes		
Pisum sativum		
138 mutants	16.0–33.0 Control value: 24.0	Gottschalk et al. 1975a

20.1.4 Alteration of Seed Proteins Through Mutant Genes

Since the rediscovery of the famous *opaque* and *floury* mutants of *Zea mays* in the early 1960's, it is generally accepted that seed protein content and composition can be altered under the influence of mutated genes. Genes that are involved in the accumulation of carbohydrates are known since the 1940's. These findings mark the beginning of intensive investigations on the genetically conditioned improvement of seed composition. First research work was done using the world collections and well-known genotypes, such as *hiproly barley* (Hagberg and Karlsson 1969) or *hl-sorghum* (Singh and Axtell 1973), were selected. In the following years, comprehensive mutation programs with high expenditure in material and work were started. In all the genotypes selected, the improved protein situation is combined with reduced yield. But, as the yield of these mutants is not too much reduced, they are of some interest for plant breeding.

Table 28 presents ranges of protein values found in lines deriving from mutationally treated material. The variation range of the original lines was considerably increased by mutation (Tanaka 1969; Bhatia et al. 1970; Dumanović et al. 1970; Chu et al. 1973; Harn et al. 1973; Krausse et al. 1974; Adam 1975; Parodi 1975; Parodi et al. 1976; Ismachin 1976). But it is obivous that the highest values found in the mutants hardly surpass the natural variability of the varieties or strains of the respective species. The literature does not allow any predictions whether further improvement is possible by means of induced mutations when a variety with a high protein content is chosen as initial line. It is probable that the natural variation range comprises already all the protein values which can be tolerated by the respective crop.

Seed protein content is generally considered to be a complex character of a crop controlled by many genes located on several chromosomes (Frey 1977; Konzak et al. 1978; Coffman and Juliano 1979; Soave et al. 1979, 1981, and others). From electrophoretic studies in *wheat*, Dhaliwal (1977) estimates that even a minimum of 60–70 and a maximum of 360–420 genes are involved in seed protein control.

Specific genes belonging to this group are known in *maize*. Table 29 gives a list of mutant maize genes influencing seed storage substances, most of them being recessive. Further details are given later. Also in *barley* (Balaravi et al. 1976; Jensen 1979a, b, 1981a; Table 35) and in *sorghum* (Singh and Axtell 1973; Axtell and Lafayette 1976) protein genes are known. In some cases, information on their basic effects is available, though the metabolic details are not yet clear. The mutant genes influence in almost all the cases studied the amount of the prolamins synthesized during kernel development combined with deviations in amount and composition of carbohydrates. Apparently, mutual relations between the various biochemical pathways exist. In the case of the *opaque 2* and *floury 2* genes of *maize*, it is assumed that they have the function of regulatory genes controlling several structural genes. Similar conditions are discussed for the *lys 3a* gene of *barley*. In *legumes*, specific regulatory genes of this group are not yet known.

20.1.4.1 Protein Mutants in Cereals

Comprehensive mutation programs have been performed in the self-fertilizing cereals *barley, wheat,* and *rice* during the past decades, resulting in voluminous collections of experimentally produced mutants, many of which were investigated with regard to their seed proteins. In *maize*, it being a cross-fertilizing crop, mutation breeding is in the very beginning, but many spontaneously arisen mutants with altered seed proteins are available. Plenty of information was obtained from the biochemical and genetic analysis of these genotypes, which has to be considered for judging the experimentally produced protein mutants of other cereals. Therefore, the mutants of *Zea mays* with quantitative and qualitative alterations of their seed proteins will be discussed in detail.

Maize. Zea mays is the best-studied crop as far as mutant genes are concerned, causing deviations in seed protein content and composition; some details are given in Table 29. It seems to be the rule that genes affecting seed proteins are also influencing seed carbohydrates. Further mutants of this big group, not included in the table, are known in maize, but their genetic situation is not yet completely clear and most of them are not yet designated.

A good deal of literature concerns *opaque 2 (o 2)* (Mertz et al. 1964) and *floury 2 (fl 2)* mutants (Nelson et al. 1965), which had been refound by Nelson and co-workers in the early 1960's. Both these mutants were the first ones in which seed protein was found to be changed under the influence of mutated genes. The findings obtained in these two genotypes are the theoretical basis for comprehensive national and international research programs, therefore, we shall discuss them more in detail. The *fl 2* gene reacts semidominantly, whereas the *o 2* gene is recessive (Nelson 1967; Jones 1978). *Opaque 2* is located on chromosome 7, *fl 2* on chromosome 4 (DiFonzo et al. 1980). Both these genes are responsible for the synthesis of endosperm proteins with higher amounts of lysine and tryptophane combined with a shift in the amounts of other amino acids (Table 30). The reason for these deviations are altered proportions of seed protein subfractions. In *floury 2,* an additional increase of the amounts of methionine (Nelson et al. 1965) and of not extractable seed proteins is observed (4.3% in normal maize, 19.0% in *fl 2;* Jimenez 1966; Mossé 1966).

Obviously, the *opaque 2* gene can be obtained relatively easily by mutation. Several *o 2* mutants have been selected, arisen independently from each other after application of different mutagens (Nelson 1967; Bálint et al. 1980; Prešev et al. 1973; Stoilov and Popova 1974/1975).

Opaque Maize. In normal maize, 47% of endosperm protein belong to prolamin fraction called zein, in the case of maize representing the main storage proteins of this crop. The glutelins comprise 35%. Under the influence of the *o 2* gene, these proportions are changed: 50% of endosperm proteins are glutelins and only 22% are prolamins. In addition, the amounts of the albumin and globulin fractions are increased. Because of the varying lysine amounts in the different fractions, this shifting is mainly responsible for the increased lysine content (Nelson 1967). By using a different method, Sodek and Wilson (1971) found a stronger reduction of the prolamin fraction in *o 2* (Table 31). Because of the drastic effect on prolamins detailed investigations on these proteins have been carried out. Although the biochemical data for the prolamin subfractions are not yet completely clear, a general statement on certain relations between prolamin polypeptides and the genes controlling their production seems to be possible.

Table 29. Mutant genes in maize which are of economic interest

Mutant gene	Characteristics	References
amylose extender *ae* recessive	– Starch content reduced – Amylose content increased – System of branching enzymes changed – Prolamin fraction reduced – Lysine content increased – Bigger protein bodies – Multilayer aleurone	Kramer and Whistler 1949; Zuber et al. 1958; Creech 1968 Black et al. 1966; Boyer and Preiss 1978 Misra 1978 Baenziger and Glover 1977 Nelson 1970
brittle *bt1* recessive	– Starch content drastically reduced – Amount of free sugars increased – Prolamin fraction reduced – Lysine content increased – No protein bodies	Creech 1968 Cameron and Teas 1954 Misra 1978 Misra et al. 1975 Baenziger and Glover 1977
bt2 recessive	– Starch content drastically reduced – Reduced activity of ADPG-pyrophosphorylase – Content of free sugars increased – Prolamin fraction drastically reduced – Lysine content increased – No protein bodies	Creech 1968 Tsai and Nelson 1966 Cameron and Teas 1954 Misra 1978 Misra et al. 1975 Baenziger and Glover 1977
dull *du* recessive on chromosome 10	– Starch content reduced – Amylose content increased – Prolamin fraction reduced – Small protein bodies	Zuber et al. 1958 Kramer and Whistler 1949 Misra 1978 Baenziger and Glover 1977
floury *fl2* semidominant on chromosome 4	– Starch content reduced – Prolamin fraction reduced – Increased content of lysine, tryptophane and methionine – Both main prolamin subfractions are reduced to the same extend – No protein bodies – Greater embryo	 Nelson et al. 1965 DiFonzo et al. 1980 Misra et al. 1976 Baenziger and Glover 1977 Nelson 1969b
fl3 semidominant	– Lysine content increased	Ma and Nelson 1975
opaque *o2* recessive on chromosome 7	– Starch content reduced – Prolamin fraction drastically reduced – Increased content of lysine, and tryptophane – Main prolamin subfractions with greater MW suppressed – No protein bodies – Amount of free amino acids increased – Amount of trypsin inhibitor increased – RNase activity increased – Greater embryo	 Nelson 1967 Misra et al. 1972, 1975, 1976 DiFonzo et al. 1980 Baenziger and Glover 1977 Sodek and Wilson 1971 Halim et al. 1973; Reed and Penner 1978 Nelson 1967

Table 29 (continued)

Mutant gene	Characteristics	References
o7 recessive on chromosome 10	– Starch content reduced – Prolamin fraction drastically reduced – Gene controls minor subfractions of the prolamins – Lysine content increased	Misra 1978 Misra et al. 1972 DiFonzo et al. 1980 Salamini et al. 1979
shrunken sh1 recessive on chromosome 9	– Starch content reduced – Reduced activity of sucrose synthetase – Prolamine fraction reduced – Lysine content increased – No protein bodies	Creech 1968 Chourey and Nelson 1979; Chourey 1981 Misra 1978 Misra et al. 1975 Baenziger and Glover 1977
sh2	– Starch content drastically reduced – Higher amount of sucrose – Reduced activity of ADPG-pyrophosphorylase – Prolamin fraction drastically reduced – Lysine content increased – No protein bodies	Creech 1968 Laugnan 1953 Tsai and Nelson 1966 Misra 1978 Misra et al. 1975 Baenziger and Glover 1977
sh4	– Starch content reduced – Prolamin fraction drastically reduced – Salt soluble protein fraction increased – Lysine content increased – Production of pyridoxal phosphate lower	Nelson 1980 Misra 1978 Misra et al. 1975 Burr and Nelson 1973
sugary su1 recessive on chromosome 4	– Starch content drastically reduced – High amount of phytoglycogen – Amount of free sugars increased – System of branching enzymes changed; special branching enzymes – Glutelin fraction increased – Prolamin fraction reduced – Increased content of lysine and methionine – No protein bodies	Creech 1968 Black et al. 1966 Culpepper and Magoon 1924 Black et al. 1966; Boyer and Preiss 1978 Paulis et al. 1978 Misra 1978 Misra et al. 1975 Baenziger and Glover 1977
su2 recessive	– Starch content reduced – Amylose content increased – Prolamin fraction reduced – Small protein bodies	Zuber et al. 1958 Kramer and Whistler 1949 Misra 1978 Baenziger and Glover 1977
waxy wx recessive on chromosome 9	– Only amylopectin (quantity of starch unchanged) – Reduced activity of granule bound glucosyltransferase – Prolamin fraction reduced – Smaller protein bodies	Wheatherwax 1922; Sprague et al. 1943; Eriksson 1969 Nelson and Rines 1962 Misra 1978 Baenziger and Glover 1977
De*30 dominant	– Starch content reduced – Prolamin fraction reduced – Increased content of lysine and tryptophane	Salamini et al. 1979

ADPG = adenosine diphosphate glucose

Table 30. Amino acid distribution in total seed protein (1) and endosperm protein (2) of normal, *opaque 2* and *floury 2* maize (Nelson 1969b)

Amino acid	normal 1	normal 2	o 2 1	o 2 2	fl 2 1	fl 2 2
Lysine	3.0	1.6	5.0	3.7	4.8	3.3
Tryptophane	0.7	0.3	1.3	0.7	–	0.8
Histidine	2.6	2.9	3.5	3.2	2.9	2.2
Arginine	4.9	3.4	7.2	5.2	6.3	4.5
Aspartic acid	9.2	7.0	8.8	10.8	10.5	8.1
Threonine	4.1	3.5	3.8	3.7	4.1	3.3
Serine	5.6	5.6	4.7	4.8	5.2	4.8
Glutamic acid	22.6	26.0	17.2	19.8	18.5	19.1
Proline	9.6	8.6	8.4	8.6	8.8	8.3
Glycine	4.7	3.0	5.1	4.7	4.7	3.7
Alanine	9.2	10.1	6.7	7.2	8.0	8.0
Cystine	1.7	1.8	2.0	0.9	1.6	1.8
Valine	5.7	5.4	5.2	5.3	5.7	5.2
Methionine	1.3	2.0	2.2	1.8	2.7	3.2
Isoleucine	4.2	4.5	3.4	3.9	4.0	4.0
Leucine	14.6	18.8	9.3	11.6	12.0	13.3
Tyrosine	5.2	5.3	4.2	3.9	4.6	4.5
Phenylalanine	5.8	6.5	4.4	4.9	5.2	5.1
% Protein	9.0	12.7	10.5	11.1	17.0	13.6

The separation of the prolamin fraction by isoelectric focusing (IEF) results in a great number of polypeptides, the patterns of which seem to be genotype-specific. They belong to four size classes. It is assumed that the respective structural genes lie on the short arms of chromosomes 4 and 7 and on the long arms of chromosomes 4 and 10. The *o 2* gene particularly causes a reduction of the fraction with the highest molecular weight. It is assumed to control the prolamin structural genes. Further details have been published by Misra et al. (1972, 1975, 1976); Gianazzi et al. (1976); Salamini et al. (1979); Thomson and Doll (1979); DiFonzo et al. (1980); Soave et al. (1979, 1981).

The maize proteins derive from three different parts of the seeds as follows:
— about 80% from the endosperm,
— about 15% from the embryo,
— about 5% from the seed coat.

Embryo proteins have higher nutritional values than endosperm proteins, as their amino acid composition is more favorable. The lysine content of embryo protein is 6.9%, while it is only 1.6% in endosperm protein. In *opaque 2* maize, the embryo is bigger than in normal maize, contributing to the increased lysine content of the mutant (Nelson 1967; Bjarnason and Pollmer 1972). Moreover, the amount of free amino acids is increased in the *o 2* genotypes (Sodek and Wilson 1971; Table 31).

Total seed protein of *o 2* maize tends to be reduced as compared to the control material, if expressed as percent of endosperm (11.1% instead of 12.7%). Related to

Table 31. Results of protein fractionation from maize endosperm of R 802+, R 802 o 2 and R 802 fl 2 (after Sodek and Wilson 1971)

Fraction			R 802+	R 802 o 2	R 802 fl 2
Albumins		mg N/g	0.16	0.42	0.27
		%	0.90	3.3	1.3
Globulins		mg N/g	0.27	0.6	0.54
		%	1.5	4.7	2.7
Prolamins	Zein 1	mg N/g	6.66	0.37	7.47
		%	36.9	2.9	36.6
	Zein 2	mg N/g	3.34	1.73	3.12
		%	18.5	13.6	15.3
Glutelins		mg N/g	4.13	4.72	5.6
		%	22.9	37.4	27.4
Free amino acids		mg N/g	0.8	2.5	0.44
		%	4.4	19.7	2.2

% = total protein

Table 32. Normal vs *opaque* 2 corn comparison (after Nelson 1967)

Hybrid		Yield [a]	wt/10 kernels (g)	% Protein in endosperm	in whole kernels	mg prot./ 10 kernels
R 801 x R 802	+	119	3.42	11.6	11.7	400.1
R 801 x R 802	o 2	115	2.57	9.3	11.4	293.0
R 803 x R 109B	+	143	3.67	9.9	11.0	403.7
R 803 x R 109B	o 2	143	3.07	8.8	10.8	331.6
R 801 x o7N	+	150	3.90	8.7	10.3	401.7
R 801 x o7N	o 2	130	3.02	8.3	10.3	311.1
R 801 x R 75	+	125	3.48	9.7	11.3	393.2
R 801 x R 75	o 2	77	3.08	8.3	10.6	326.5

[a] Bushels per acre

the whole seed, however, it is similar to the control or somewhat higher (10.5% instead of 9.0%; Nelson 1969b and Table 32). These differences are due to the increased embryo size of the mutant. In absolute amounts, the protein content of *o 2* is strongly reduced, contributing to the smaller grain size. The main reason for this, however, is the fact that the *opaque* kernels are incompletely filled with starch. They lose translucence because they contain air spaces after drying. Combined with this, a soft seed texture is found making the seeds more susceptible to rots and insect attacks (Vasal et al. 1979). Besides seed size, yield can be negatively influenced by the *o 2* gene (Table 32).

The values found in the literature vary to some extent because the effect of the *o 2* gene is influenced by the genotypic background (Tables 33, 34). Protein percentages range between 10.3% and 11.4%. The loss in seed weight is especially variable, ranging between 3.7% and 31.1%; losses in yield vary in the same order (Table 32). This variability gave rise to optimistic expectations with regard to *opaque* genotypes with satisfying yielding capacities. Therefore, intensive crossing experiments were carried out with the aim to find an appropriate background for the *o 2* gene. Results have been given by Alexander et al. (1969); Pollacsek (1970); Salamini et al. (1970); Annapurna and Reddy (1971); Pollacsek et al. (1972); Zima et al. (1972); Paez (1973); Rakha et al. (1974); Vasal (1975); Ottaviano and Camussi (1975); Bjarnason et al. (1976, 1977); Sgarbieri et al. (1977); Loesch et al. (1978); Misra (1978); Tsai et al. (1978); Klein et al. (1980). All these attempts failed to attain the goal desired. Though protein and lysine content were satisfying in most of the genotypes tested, some showed a reduced protein content, and all a reduction in seed yield. In the papers mentioned, the influence of the genotypic background on some further characters, expressed by the *o 2* gene, is discussed.

The *opaque 2* gene influences also seed development, stopping dry matter accumulation earlier than in normal maize. Detailed findings on seed development, activity of distinct enzymes, incorporation of specific amino acids, production of subfractions, and on related problems of basic research were published by Dalby and Davies (1967); Denić (1970); Sodek and Wilson (1970); Wilson (1973); Lodha et al. (1974); Reed and Penner (1978); Tsai et al. (1978); Baenziger and Glover (1979); Burr (1979); Denić et al. (1979); Mehta et al. (1979); Nelson (1980). Also seed germination and development of the seedlings are altered under the influence of *o 2* (Jolivet et al. 1970; Kadam et al. 1973; Chibber et al. 1977; Jones and Tsai 1977; Loesch et al. 1978). Certain findings indicate that *o 2* plants can be grown more densely than normal maize (Singh and Singh 1977). In this way, their reduced yield could be compensated to some extent.

Floury Maize. Deviations in the protein subfractions are not only found in *opaque 2* but also in the *floury 2* mutant of *Zea mays*. In normal maize, 47% of the seed proteins are prolamins; this value is reduced to 22% in *fl 2*. The glutelins, albumins, and globulins are compensatingly increased (Nelson 1967). Sodek and Wilson (1971) found smaller differences between the two genotypes after having used other methods (Table 31). The *fl 2* gene affects the two main prolamin subfractions to the same extent; in addition, a new component appears. The genes for the two subfractions are located on chromosome 4 near to the *fl 2* gene. A possible regulatory function of the *fl 2* gene is discussed (Salamini et al. 1979; Soave et al. 1979; Thomson and Doll 1979; DiFonzo et al. 1980).

In the seeds of the mutant, the proportion of embryo/endosperm tissue is altered, the embryo being bigger than in normal maize (Nelson 1969b). Because of this, protein values are extremely diverging, depending on the reference value. If we relate the protein content to the endosperm, a mean value of 13.6% is obtained for *fl 2*, whereas the control value is 12.7%. Related to the whole seed, however, the two values are 17.0% or 9.0%, respectively. Also in absolute amounts, the protein content of the *floury 2* seeds is increased (Table 34).

Table 33. Average kernel weight in mg for normal maize and o 2 sibs (after Nelson 1967)

Line	Normal	o 2	% loss
R 181 B	161	155	3.7
07	210	200	4.8
A 545	230	215	6.6
(Wf9 x W6 40$_2$)	276	249	9.8
B37	272	241	11.7
H89	253	195	22.9
A239	301	218	27.6
A632	257	175	31.1

Table 34. Comparison for normal maize and fl 2 versions of 3 inbreds (after Nelson 1969b)

Inbred		% total protein	mg protein in seed	mg protein in embryo
W64A	+	10.9	26.3	6.3
W64A	fl 2	15.7	29.3	10.7
Oh43	+	9.7	23.3	5.6
Oh43	fl 2	14.8	33.4	8.5
B14	+	11.5	31.2	7.5
B14	fl 2	14.5	36.9	8.3

Other Maize Mutants with Altered Seed Proteins. Data on some other spontaneously arisen maize mutants, in which the seed proteins are influenced by distinct genes, are summarized in Table 29. Most of these genotypes show an increased lysine content *(ae, bt 1, bt 2, fl 3, o 7, sh 1, sh 2, sh 4, su 1* and *De* 30).* This increase is — similarly as in *o 2* and *fl 2* — due to the reduction of the prolamin fraction. In the case of *sh 4,* the increase of lysine additionally is due to an increased amount of salt-soluble proteins, in the case of *su 1* an additional amount of glutelins being the reason for the increased methionine content (Paulis et al. 1978). An interesting finding is the increased number of aleurone layers observed in the *ae* genotype, which seems to be responsible for the relatively high amount of seed proteins (12.5%) with an increased proportion of nonprolamin proteins. The aleurone layer is that place of the kernel where preferentially nonprolamins are deposited (Duvick 1961). Further details are given in Table 29. The dry matter accumulation in *su 2, o 2, wx,* and the two recombinants *wx/o 2* and *su 2/o 2,* shows that a shorter grain-filling period is responsible for the reduced seed size of the respective genotypes, which also show an increased water content during seed development. The disappearance of this water during the ripening period causes the wrinkled surface of the seeds (Baenziger and Glover 1979). The seeds of mutants *ox 7749, ox 7455* and *ox 7537* have an increased lysine content without reaching the high values of the *opaque 2* genotype. They do not show the strong reduction of the prolamin fraction chracteristic for *o 2* (Nelson 1979).

The performance of mutation programs in this crop is more difficult than in other cereals because maize is a cross-pollinating species. They have been started in Hungary and Bulgaria, and some genotypes with increased protein content have been found

(Bálint et al. 1970; Stoilov and Popova 1974/75; Popova 1975; Stoilov and Hristova 1978). One of them proved to be highly resistant to *Fusarium* (Bálint et al. 1977). Three other maize mutants are known in which the amount of free proline in the embryo is diminished. In two of them, also the proline-rich zein fraction is reduced, whereas the albumin and globulin fractions as well as the free amino acids are more abundant than in the control material (Dierks-Ventling 1982).

Protein Bodies in Different Maize Mutants. Seed proteins, especially the storage proteins, are deposited in protein bodies (Pernollet 1978). In maize, prolamins are present preferably in protein bodies of the starchy endosperm, while in those of the aleurone layer mostly other proteins are deposited (Duvick 1961). The protein bodies of the *opaque 2* mutant have a diameter of only about 1/20 of those of normal maize (Wolf et al. 1967) or they are completely absent (Baenziger and Glover 1977). Their presence or absence depends obviously on the genetic background. In the genotypes *fl 2, su 1, sh 1, sh 2, bt 1,* and *bt 2,* no protein bodies were detected at all. In a further group of genotypes *(su 2, wx,* and *du),* the protein bodies were found to be smaller than normal. This holds true especially for those of the starchy endosperm. In all the cases in which they are changed, also deviations in the amount of the prolamins are observed. Thus, it was assumed that these facts are basically related with each other. On the other hand, it was shown that prolamins are not necessarily deposited in these cell organelles; they are also found in seeds not having any visible protein bodies, for example in the *bt 1* or *fl 2* genotypes. From normal maize, too, it is known that more protein is present than the number of protein bodies allows.

Barley. The most intensively studied barley protein mutant is *1508* obtained in Risø, Denmark (see below). Besides, a considerable number of further mutants of this group have been selected (Favret et al. 1969; Bansal 1970, 1971; Scholz 1971; Doll 1972, 1973, 1975; Gaul et al. 1973; Mikaelsen 1973; Nettevich and Sergeev 1973; Ulonska et al. 1973, 1975; Krausse et al. 1974; Lehmann et al. 1975; Walther 1975; Walther et al. 1975; DiFonzo and Stanca 1977; Walther and Seibold 1978, 1979). According to Scholz (1971, 1972, 1975, 1976a), the negative correlation between protein content and some yielding criteria can hardly be broken. An interesting example for this correlation are some desynaptic mutants having a very low seed production due to their meiotic anomalies. The protein content of their seeds is higher than that of the control material (Ahokas 1977). Investigations of other authors, however, show that this correlation obviously does not exist in mutants having only a small increase in protein content. There are even examples for positive relations between the traits just mentioned (Ulonska et al. 1975; Hadjichristodoulou and Della 1978; Walther and Seibold 1979).

Hiproly Barley. Before discussing the experimentally produced mutants, some findings on hiproly barley should be mentioned. Hiproly is a rather primitive stock cultivated in Ethiopia having small seeds and reduced yield selected from 2500 varieties of World Barley Collection in 1968 (CI 3947; Hagberg and Karlsson 1969). Is is characterized by both a high content of total protein (17.0%) and a high lysine content (4.1%). The high lysine character is simply inherited as a recessive trait which is not necessarily connected with high protein content (Hagberg et al. 1970, 1979; Munck et al. 1970,

Table 35. Protein mutants in barley

Mutant	Characteristics	References
notch 1 notch 2	– Protein content increased – Lysine content increased – Amount of free amino acids increased	Bansal 1970; Balaravi et al. 1976
C 61 C 63 C 64	– Protein content increased – Seed size increased, except of C 63	Favret et al. 1969
Risø mutants all genotypes with increased lysine content, but with shrunken seeds, except of mutants 3, 7 and 440; protein content and shrunkeness controlled by the same gene.		Doll 1975; Karlsson 1975, 1977; Jensen 1979a, b, 1981a; Køie and Doll 1979
3		
7	– Hordein B more reduced than C	
8	– *lys 4d*, dominant, on chromosome 5 – Hordein B and C proportionally decreased	
9		
13	– Allelic to 29 and 86 – On chromosome 6, linked to genes: *sex 1f, xn* and *Amy 1* – Hordein B more reduced than C	
16	– On chromosome 1 ? – Hordein B more reduced than C	
17	– On chromosome 1, linked to gene *n* – hordein B more reduced than C	
18	– *lys 3b*, on chromosome 7	
19	– *lys 3c*, on chromosome 7	
56	– On chromosome 5, near to *hor 2* locus – Hordein B decreased, hordein C increased	
440		
527	– On chromosome 6, linked to genes *xn* and *Amy 1* – Hordein B more reduced than hordein C	
1508	– Allelic to 18 and 19 – *lys 3a*, on chromosom 7, linked to gene *sex 3c* – Hordein B and C proportionally decreased	
hiproly	– *lys*, recessive, on chromosome 7 – Independent of *19, 56* and *1508* – Special water and salt soluble seed proteins increased	

Mutants 29, 56 and 86 from Carlsberg II, others from Bomi barley

1979). The gene responsible for this improvement was formerly designated as *hl*, later on as *lys 1* and nowadays as *lys*. It is proposed to lie on chromosome 7 between genes *r* (smooth awn) and *s* (short rachilla hair) (Fig. 35; Karlsson 1975; Hagberg 1978). The *lys* gene was the first gene of the barley genome which was found to influence the seed proteins.

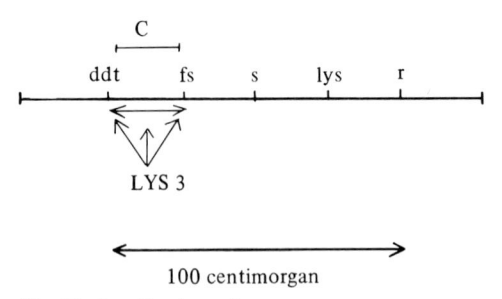

Fig. 35. Localization of some genes on chromosome 7 of barley (after Robertson 1971; Karlsson 1975, 1977; Hagberg 1978; Jensen 1979b). *C* centromer region; *ddt* DDT resistance; *fs* fragile stem; *s* short rachilla hairs; *lys* high-lysine gene from hiproly barley; *r* smooth awn; *LYS 3* lys 3 locus with allels lys 3a, b, c

The reduced seed size of hiproly is due to a reduced dry-matter incorporation into the ripening seeds; especially the starch deposition is reduced. By application of crossed immuno electrophoresis, four salt soluble proteins were found to be strongly increased. In normal barley, they account for 7% of grain lysine, in hiproly however for 17%. Fifty per cent of the lysine increase of this genotype is due to this deviation. In addition, special proteins of the albumin fraction are increased, contributing furtheron to the high lysine content. Details on the biochemical differences between hiproly and normal barley have been published by Munck et al. (1970, 1979); Hagberg (1978); Hejgaard and Boisen (1980); Soendsen and Martin (1980); Jonassen (1980a, b).

By incorporation of the *lys* gene into the genomes of several high-yielding bread barley varieties, high lysine strains with improvements in seed production and seed size as compared to the original hiproly with only 30% of the yield of a normal standard variety were achieved. After several backcrosses, yield rose up to 90% of the control values of the standard varieties. The seeds of a part of this material are still smaller but not shrunken (Hagberg 1978).

The Risø Mutants. In Risø, Denmark, a group of protein mutants of barley is available originating from the varieties Carlsberg II and Bomi after application of EMS and EI. Especially mutant 1508 has been studied intensively. In 1970, investigations on 15,000 M_2 plants were carried out resulting in 20 genotypes with increased lysine content (Tables 35, 36; Doll 1972, 1973, 1975, 1977; Ingversen et al. 1973a; Doll et al. 1974; Doll and Køie 1975). Some of these mutants have in addition a higher protein content, but the whole group shows reduced yield. Mutant 1508 was found to have the highest lysine content of the group (43% more than Bomi), furthermore a 36% increase of tryptophane and a relatively high grain yield (Table 36; Ingversen et al. 1973b). The gene responsible for these alterations lies on chromosome 7 near to the centromere region. It is designated as *lys 3a* (Fig. 35), being allelic to *lys 3b* and *lys 3c* of mutants 18 and 19 (Karlsson 1977; Jensen 1979b) and nonallelic to gene *lys* of hiproly (Muench et al. 1976). In addition, the gene was found to be closely associated to *sex 3c*, expressing shrunken endosperm (Ullrich and Eslick 1978b).

Gene *lys 3a* does not only influence the composition of the seed proteins but also the carbohydrates and the grain weight. The seeds of mutant 1508 are incompletely filled. Dry-matter incorporation in ripening grains is inferior in all high-lysine genotypes

Table 36. Survey of some high-lysine barley mutants selected in the varieties Bomi and Carlsberg II (after Doll 1975)

Variety or mutant	Lysine content g/16 g N	Protein content 6.25 n N%	Seed size mg	Grain yield g/m^2
Bomi	3.8	8.8	49.2	611
Mutant 7	4.3	8.4	49.2	544
Mutant 8	4.2	9.7	41.5	486
Mutant 9	4.0	10.7	31.9	328
Mutant 13	4.7	10.2	39.1	443
Mutant 16	4.1	11.0	44.2	475
Mutant 17	4.2	12.1	43.7	320
Mutant 440	4.4	10.4	35.6	412
Mutant 527	4.3	9.7	39.4	424
Mutant 1508	5.2	9.5	43.5	472
Carlsberg II	4.0	8.7	48.2	556
Mutant 29	4.3	9.5	42.6	473
Mutant 56	4.7	9.8	42.5	431
Mutant 86	4.2	10.3	37.9	441

Table 37. Distribution of nitrogen in the seed protein fractions of the Risø mutant *1508* and Bomi barley (after Tallberg 1981a)

Protein fraction	*1508*	Bomi
Albumin/globulin	43	29
Prolamin	17	49
Glutelin	22	17
Insoluble rest	18	5

as compared to corresponding nonmutated material. The amount of starch per grain is reduced. The reduced starch content partly might be due to the higher α-amylase activity which is probably responsible for a premature breakdown of starch (Munck 1975; Doll and Køie 1978; Hagberg 1978). The reduction in yield is mainly due to the reduced seed size while the number of seeds per unit area is more or less unchanged (Doll and Køie 1978; Oram and Doll 1981).

The increased lysine content of mutant 1508 is due to changes in the proportions of endosperm proteins; in addition, the amounts of free amino acids, especially of free lysine, are increased. The hordein fraction is reduced to one-third, whereas a strong increase of the albumin/globulin fractions is observed (Table 37; Ingversen et al. 1973b; Brandt 1975, 1976; Køie et al. 1976a; Munck 1976; Tallberg 1977, 1981a, b). The seed development of mutant 1508 shows characteristic deviations from normal barley, particularly with regard to the timing of hordein synthesis. The various seed protein fractions and subfractions of the mutant have been intensively studied biochemically; details were published by Brandt (1976); Brandt and Ingversen (1976); Sozinov et al. (1976); Køie et al. (1976b); Køie and Doll (1979); Doll and Brown (1979); Shewry et al. (1979, 1980).

The barley seed proteins are located in protein bodies which predominantly consist of homogenous spheres accompanied by a granular component. The homogenous component represents obviously a storage organelle with a higher concentration of prolamin; the granular component is associated with glutelins. In mutant 1508, the granular component is the most prominent one (Ingversen 1975). These findings are in concordance with the reduced amounts of prolamins found in the mutant's seeds.

Some characters of other protein mutants available in Risø are summarized in Tables 35 and 36. Nearly all these genotypes are characterized by an increased lysine and protein content but, with one exception, seed size and seed yield are reduced. In mutant 7, seed size is unchanged and total protein content shows likewise no deviations. Prolamin content is reduced, whereas the amount of the other fractions is more or less the same. The quantity of not extractable nitrogen compounds is increased (Tallberg 1981b). In mutant 56, the increased lysine content is mainly due to a considerable reduction of a specific subfraction of the prolamins. These deviations are probably due to a mutation at or near the *Hor 2* locus (Køie and Doll 1979; Doll 1980). Some of the mutant genes of this group are already localized (Jensen 1979a,b, 1981; Table 30).

Mutant 1508 was crossed to a hiproly backcross (hiproly × Mona) and double recessives of the constitution *lys 3a/lys* were selected (Hagberg et al. 1979; Tallberg 1981). They produced 15% more lysine than mutant 1508, and 70% more than Bomi barley. This increase is due to further reduction of the prolamin fraction, whereas the amounts of glutelins and water-soluble proteins are increased. Thus, an additive effect of the two lysine genes becomes evident in the 1508/hiproly double recessives. Their seeds are still shrunken and seed size is reduced. Double recessive plants from crosses between hiproly and the protein mutant 7 show a slightly increased lysine content due to an increase of the glutelins and the water-soluble fraction. The seeds are smaller but they are well filled without shrunkeness (Hagberg et al. 1979; Tallberg 1981b). An almost complete suppression of the prolamin synthesis during seed development is observed in 527/1508 double recessives. In plants of the constitution 29/1508, the prolamin content is only slightly less than that of 1508. The lysine content of both these genotypes is similar to that of mutant 1508 (Kreis and Doll 1980).

Further Protein Mutants of Barley. In the course of a mutation program carried out in Argentina, four mutants of special interest were selected (Favret et al. 1969). Their seed characters are given in Table 38. In three of them, the increased protein content is combined with an increased seed weight. Mutant 64 of this group seems to be the most promising one: Seed weight is increased by nearly 40% and protein content by almost 60% resulting in a protein production per seed which exceeds that of the control material by 116%. The relation of protein to carbohydrates is changed from 1:8 in the mother variety to 1:6 in the mutant, indicating a more balanced content of nutritionally valuable substances. Crossing experiments with this mutant revealed that seed size and mg protein per seed are controlled by the same gene. Further investigations showed that the increased protein content is due to a longer period of deposition, while the rate of synthesis is the same as in the initial line (Favret et al. 1970).

The *notch mutants* obtained in India (*notch 1* and *notch 2;* Bansal 1970) have 40% more seed proteins with 20% more lysine than the control material. In both the mutants, the albumin and globulin fractions are increased (20% of the seed proteins in

Table 38. Seed characters of barley mutants and their mother lines (after Favret et al. 1969)

Line or mutant	Seed weight (mg)	Protein per seed (mg)	Protein content (%)
M.C.20	40.7	4.05	9.8
Mutant C.61	49.1	7.77	15.2
Mutant C.62	55.9	5.73	10.5
Mutant C.63	28.9	4.31	14.3
Mutant C.64	56.9	8.75	15.5

the control, 29% and 28% in the mutants, respectively; Balaravi et al. 1976; Singh and Sastry 1977a). The prolamin fraction is somewhat reduced, containing a specific subfraction which might be responsible for a better digestibility of these proteins (Singh and Sastry 1977b).

In Russia, a barley mutant with increased protein content and good yielding properties was isolated by Nettevich and Sergeev (1973). So far, only M_2 and M_3 values are available. If the findings are confirmed in later generations, the mutants would be of interest for practical use. In addition, mutants with increased lysine content were found (Nigmatullin and Muminshoeva 1977; DiFonzo and Stanca 1977).

A group of 36 barley mutants was investigated by Mikaelsen (1973). The *erectoides* mutant H-14 exceeds the initial line by more than 10% in crude protein content. This mutant was crossed with mutant H-5, being similar to H-14 in morphology. The hybrids had 40% more protein than H-14 obviously due to heterosis.

Finally, it should be mentioned that some spontaneously arisen barley mutants were found to have an increased protein and lysine content (Ullrich and Eslick 1978a). In four further genotypes, quantitative differences in the water-soluble endosperm proteins in relation to the control material were found by means of electrophoresis. The specificities of the banding patterns are correlated with the nutritional value of the respective genotypes (Kreft et al. 1975).

Wheat. In the frame of comprehensive mutation programs in *Triticum* species, investigations on the protein situation of mutants have been carried out. Some of the favorable protein chracters were found to be heritable, but they are often correlated with reduced yield and kernel size. In some cases, the negative correlation between seed size and protein content is due to increased starch deposition in bigger seeds. Details were published by Banerjee and Swaminathan (1966); Popović et al. (1969); Bhatia et al. (1970, 1978); Dumanović et al. (1970, 1973); Golovchenko et al. (1970); Siddiqui and Doll (1973); Adam (1975); Parodi (1975); Parodi et al. (1976); Narahari et al. (1976); Denić (1978); Hermelin and Adam (1978); Nadeem et al. (1978); Parodi and Nebreda (1978); Rajni et al. (1978); Singhal et al. (1978); Autran and Branlard (1979); Bhagwat et al. (1979).

In spite of the negative correlations just mentioned, mutants are available having an increased protein production per hectar as compared to the control material (Golovchenko et al. 1970; Khan 1973; Parodi and Nebreda 1978, 1979). Besides, mutants were found showing both a moderate protein improvement and a higher yield (Nagl 1973). A Chilean mutant of this group, induced by gamma irradiation, has been approved for registration.

Investigations on protein quality revealed that some mutant genes become effective also in this respect. Denić (1978) found two mutants with increased amounts of free amino acids. Also with regard to the amino acid composition, deviations were observed related to differences in the protein subfractions. Similar findings were obtained by Bhatia et al. (1970); Dumanović et al. (1970); Golovchenko et al. (1970); Nagl (1976); Sašek et al. (1977); Favret et al. (1979). Attempts to localize genes controlling seed protein production were made by Lawrence and Shepherd (1981a, b).

Rice. A great number of Japanese rice mutants, selected according to visible traits, were analyzed for their protein content. The content of a group of native rice varieties ranges between 5% and 12%, that of several commercial varieties between 5.5% and 10%. This range was considerably enlarged by irradiation. In the mutants studied, a range between 4.2% and 16.3% was found as compared to 6.5% of the initial line. If the seed protein content of the mother variety is defined as 100%, the mutants range between 72% and 194%. Some genotypes were found to have better seed production in addition to good protein content of the seed meal. Details on this giant amount of work have been published by Tanaka (1969–1978); Tanaka and Tamura (1968); Tanaka and Takagi (1970); Tanaka and Hiraiwa (1978).

Protein mutants of rice, some of them showing a suitable yielding capacity, were furthermore selected in:

— India   (Siddiq et al. 1970; Vilawan and Siddiq 1973; Narahari et al. 1976; Kaul 1978; Ghosh Hajra and Halder 1981; Ghosh Hajra et al. 1982);
— Bangladesh   (Shaikh et al. 1976);
— Burma   (Haq et al. 1973);
— Indonesia   (Ismachin et al. 1978);
— Korea   (Harn et a. 1973, 1975; Harn 1976; Harn and Won 1978);
— Japan   (Kataoka 1974);
— Tanzania   (Monyo and Sugiyama 1976, 1978; Monyo et al. 1979);
— Hungary   (Sajo 1978).

Also in this material, an increased variation range for the trait "protein content of the seed flour" becomes discernible. Moreover, the favorable protein situation is often correlated with negative traits such as reduced grain yield (Kataoka 1974; Kushihuchi et al. 1974; Narahari et al. 1976; Harn and Won 1978; Monyo and Sugiyama 1978; Monyo et al. 1979) or reduced kernel size (Harn et al. 1973; Hillerislambers et al. 1973). Most of the authors, however, report exceptions from these correlations. According to preliminary findings obtained by Bollich and Webb (1968), at least three genes are involved in the genetic control of seed protein content in rice. It can, however, be assumed that the true number will be considerably higher.

Some of the rice mutants were analyzed with regard to qualitative alterations of their seed proteins, and deviations in the amino acid composition were found (Tanaka and Tamura 1968; Tanaka 1969; Siddiq et al. 1970; Monyo et al. 1979). Increase in the amounts of methionine, phenylalanine, arginine, threonine, lysine, and corresponding decrease in other amino acids were observed. These findings indicate a shifting in protein subfractions. In some Korean mutants of this group an increase of the glutelin fraction was observed combined with a lower content of basic amino acids (Harn et al. 1973).

In rice varieties, the protein distribution within the kernel can be different (Chu et al. 1973). As rice is dehulled and polished, it would be of importance that less protein is deposited in the outer layers. Mutants with a changed protein distribution are available (Tanaka and Tamura 1968; Sorala and Reddy 1979; Ghosh Hajra and Halder 1981). The structure of the starch layer is different in some genotypes; in others the protein bodies, normally placed more in the peripherical zones of the grain, are found more in the middle zone. In some high protein mutants of this category, a better utilization of fertilizers has been observed representing an additional advantage to the increased protein content with regard to practical use (Harn et al. 1973). One of the Japanese mutants shows a prolonged capacity of the leaves to absorb low molecular nitrogen cmpounds which may contribute to the improvement of the protein content of the seeds (Tanaka 1978).

Sorghum. Approximately 9000 lines of *Sorghum* World Collection were analyzed for floury endosperm. Sixty-two lines with this character were found, two of them (IS 11167, IS 11758) having an increased lysine content (3.34% and 3.13% as compared to 2.09% in normal kernels). In both these varieties, relatively high levels of proteins are found (15.7% and 17.2%; 12.0% in a comparable commercial line). The high lysine content in both the genotypes is assumed to be controlled by two recessive genes which are possibly allelic. The kernels are partially dented resulting in a somewhat reduced seed weight. The seed yield of these *hl* varieties is considerably reduced; nevertheless, they are cultivated in Ethiopia, from where they derive. These sorghums are agriculturally used because of their good taste; they are prepared together with high-yielding varieties. The difference in lysine content in both the genotypes is due — similarly to maize and barley — to shiftings in protein subfractions. The lysine-poor prolamins are reduced from 55% of total seed protein to 39%, whereas the high-lysine fractions (albumins, globulins, and glutelins) are increased (Singh and Axtell 1973; Axtell et al. 1974, 1979).

Following mutational treatment, 23,000 M_3 heads were investigated and 33 mutants were isolated showing at least 50% increase in lysine. Mutant P 721 of this group seems to be very promising. It is assumed that a single gene, expressing partly dominance, is responsible for this trait. Protein content of this mutant is increased from 12.9% in normal sorghum up to 13.9%, lysine content from 2.09% to 3.09%. Also in this case, the increase in lysine originates from alterations in the proportions of specific subfractions. More salt-soluble and alcohol-insoluble reduced glutelins are found, both these fractions being comparably lysine rich. On the other hand, the alcohol-soluble proteins, called kafirins, are drastically reduced (Axtell et al. 1974; Axtell and Lafayette 1976; Guiragossian et al. 1978; Paulis and Wall 1979).

In mutant P 721, dry-matter accumulation ceases 7 days earlier than in the control, resulting in a 11%–14% reduction in seed size. After having crossed the P 721 mutant (an opaque form) with vitreous lines, genotypes with increased yield were obtained (Axtell et al. 1974). A great disadvantage of sorghum seeds is their high content of specific polyphenols called tannins (for details see Chap. 21).

Other Cereals. In contrast to the great number of findings on protein mutants of maize, wheat, rice, and sorghum, only little information is available on other cereals. Protein mutants of *oats* show reduced yield (Frey and Chandhanamutta 1975; Frey 1977; Jalani and Frey 1979). Some data on the avenins, the prolamins of *Avena,* are available (Kim et al. 1979). The inheritance of the groats proteins has been studied by Sraon et al. (1975).

A more intensive mutation program has been carried out in *pearl millet* (Rabson et al. 1978). After application of different mutagens, 10,200, 5100 and 1470 samples of three inbred lines were investigated with regard to high DBC values. A computer program was developed for handling the analytical data obtained. The selected material was grown in a second season and was reanalyzed. So far, there are no indications on the existence of mutants comparable to *opaque 2* in *maize* or *1508* in *barley,* but several of the genotypes selected have an increased protein content. Also mutants with increased lysine content were isolated (Rabson et al. 1979).

Eragrostis tef is a crop exclusively utilized in Ethiopia. A mutation program has been startet already 10 years ago but no detailed results have been published so far (Berhe 1973). In *Secale cereale,* the seed protein content is negatively correlated with seed size (Ruebenbauer and Kaleta 1980). Effective mutation genetic experiments have not been performed in this cross-fertilizing crop.

20.1.4.2 Protein Mutants in Legumes

Fifteen per cent of nutritionally used proteins derive from legumes. In general, they are consumed together with other nutritions, especially with cereals. In the industrial countries, legumes play only a minor role; in many other countries, however, they are basic foodstuffs. Investigations on the nutritional efficiency have shown that the percentage of utilizable protein (u.p.) in meals being composed solely of cereals is comparably small (100% maize: u.p. = 2.93%). The addition of beans in a ratio of 26% (meal composed of 74% maize and 26% beans) raises the value up to 6.26% in rat assays (Bressoni 1972). This improvement is especially evident in maize. In oats or rice, the improvement is less important because the percentage of utilizable protein is by far higher than in maize (u.p. in rice 4.10%, in oats 8.22%).

The nutritional efficiency of a crop with regard to its seed proteins highly depends on the composition of these substances. While in cereals, besides the S-containing amino acids, lysine and tryptophane are limited, in legumes lysine is available in sufficient amounts, but the sulfur-containing amino acids are deficient. The improvement of the u.p. by the maize/bean meal just mentioned can be entirely attributed to the improved and more balanced amino acid uptake. Therefore, an important aim of legume breeding consists in improving the quality of the seed proteins by increasing the amount of the limiting amino acids. An improvement of the total protein content is less important, as 20%–40% of legume seeds are proteins.

Influenced by the results obtained in cereals, mutation programs in many pulses were started in the early 1970's. Particularly intensive work was done in *Glycine, Phaseolus,* and *Pisum;* therefore, these crops are discussed in more detail. Moreover, single programs are being carried out in:

- *Vigna radiata* (Ojomo 1976; Ojomo and Omueti 1978; Bhagwat et al. 1979; Sandhu et al. 1979; Singh and Chaturvedi 1981b);
- *Lens culinaris* (Shaikh et al. 1978);
- *Cicer arietinum* (Shaikh et al. 1978; Abo Hegazi 1980);
- *Vicia faba* (Abo Hegazi et al. 1973; Nagl 1978b; Abo Hegazi 1979; Hussein and Abdalla 1979).

In these crops, the investigations are still in the state of selecting mutants, and in most of the publications mentioned only M_2 and M_3 findings are available. Varieties of *Vicia faba* are investigated with the aim to find strains with improved nutritional value. It seems possible to select genotypes with high protein content and reduced quantities of antinutritional factors such as hemagglutinins, trypsin inhibitors or tannins (Sjödin et al. 1981b).

Soybeans. The soybean *(Glycine max)* is of special agronomic interest because of its extraordinarily high protein content ($\sim 40\%$ of the seed meal) combined with a high amount of fat (18%). It is used in human nutrition as forage and for green manuring. Mutation programs are running:

- in the Philippines (Santos et al. 1970);
- in Japan (Hiraiwa et al. 1976; Hiraiwa and Tanaka 1978);
- in Thailand (Smutkupt and Gymantasire 1975);
- in Jamaica (Panton et al. 1973, Panton 1975);
- in Uganda (Rubaihayo and Leakey 1973; Rubaihayo 1967a, b, 1978);
- in Russia (Bazavluk and Enken 1976).

In some cases, an increase of the protein content was observed, but final results are still lacking. In three mutants studied by Lowry et al. (1974), differences in the protein composition were found by means of electrophoresis.

Beans. Beans (*Phaseolus* species) play an important role in human nutrition. Especially in South America, but also in some African countries, meals are composed up to 50% of beans. The relatively high protein content of their seeds (about 25%) makes this crop a valuable foodstuff. A disadvantage is the content of toxic compounds such as hemagglutinin, moreover, a trypsin inhibitor and a specific globulin which interferes with several proteases (Jaffé 1972). All these antinutritional substances can be destroyed by heat if the exposure is long enough. Thus, the usefulness of cooked beans is not seriously affected. This might be the reason for the fact that hardly a research program exists the aim of which consists in reducing the amounts of these substances by means of mutant genes.

Mutation programs with the aim to select protein mutants were started:

- in India (Dahiya 1973; Rao et al. 1975);
- in Jamaica (Panton et al. 1973);
- in Brazil (Crocomo et al. 1978, 1979;
- in Egypt (Hussein and Abdalla 1978);
- in Uganda (Rubaihayo 1975, 1976a, 1978).

In the Brazilian material selected, the overall protein content was found to be higher than in the control population. More than 36% of the samples analyzed were about 1% higher than the mean. Correlations between protein content and other agronomic characters were not found. One protein mutant was analyzed with regard to its soluble proteins without finding clear differences (Crocomo et al. 1978, 1979).

Peas. The garden pea *(Pisum sativum)* is of minor importance for world nutrition. Especially in European countries, the seeds are used as vegetable; furthermore, whole plants are used as forage. Since Gregor Mendel's days, *Pisum* is used as a model plant for studying various general and specific problems of genetics. Many details on morphology, physiology, biochemistry, and genetics of the species are available (cf. Sutcliffe and Pate 1977 and the volumes of *Pisum Newsletter 1969–1982*). Comprehensive collections of spontaneously arisen and experimentally produced mutants exist in Landskrona (Sweden), Wageningen (The Netherlands), Bonn (Federal Republic of Germany), Gatersleben (German Democratic Republic), Norwich (England), Poznan (Poland), Saskatoon (Canada), and Delhi (India). Mutation programs were furthermore started in Bulgaria (Vassileva 1976) and in India (Jain 1976; Kaul and Matta 1976; Kaul 1980b).

Detailed investigations on protein content and protein composition of many mutants and recombinants are being carried out in Bonn since 1970 (Gottschalk and Müller 1970, 1974, 1979; Müller and Gottschalk 1973, 1978; Dümke 1973; Gottschalk 1975b, 1978f, 1983; Gottschalk et al. 1975a, b, 1976; Wolff 1975, 1979; Müller 1976, 1978a, 1979; Hussein 1976, 1979; Gottschalk and Hasenberg 1977; Gottschalk and

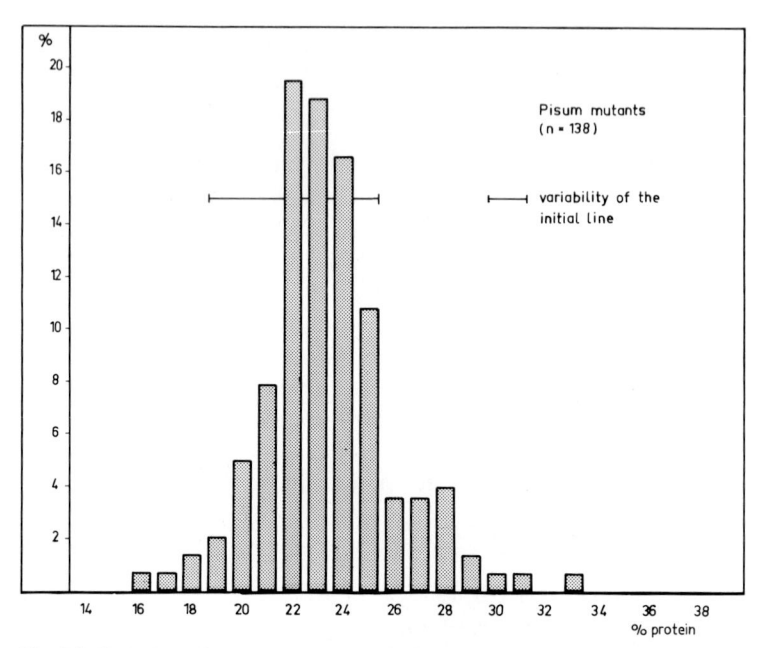

Fig. 36. Variation of the seed protein content of 138 radiation induced *Pisum* mutants as compared to the variability of their mother variety

Wolff 1977, 1983; Quednau and Wolff 1978a, b; Müller and Werner 1979; Gaul 1981; Hartmann 1981).

The seed protein content of 138 mutants of our *Pisum* collection, most of them being homozygous for a single gene, is given in Fig. 36. In 1974, the mean value of the control material was 23.5%, single values ranging between 19% and 25%. The corresponding means of the mutants studied ranged between 16% and 34%. Investigations in the following years principally confirmed these values. The high protein content of some mutants seems to be merely a consequence of their low seed production and not the effect of the respective mutant genes. Mutant 189 of our collection, homozygous for three or four mutant genes, contains gene *ipc*, which is obviously responsible for the strongly increased protein content of the seed flour (Gottschalk and Wolff 1983; Fig. 33). Basic biochemical reasons for the changed protein contents are not yet known, but gene *ipc* seems to interfere with the accumulation of total seed protein. Investigations on the protein subfractions and the amino acid pattern indicate that the protein amount as a whole is increased or reduced in the respective mutants without any drastic deviations in protein composition (Wolff, unpublished).

After having found the broad variation range of seed protein values in *Pisum* mutants, a group of recombinants was analyzed in order to study whether there are any interactions between distinct mutant genes with regard to this trait. Figure 37 shows that this is obviously the case in some genotypes. The values of recombinants R 600, R 176X, and R 605 are considerably lower than the control values of their parental mutants which are similar to the values of the mother variety. Thus, the protein content of the seed meal is negatively influenced if we combine specific mutant genes with each other. So far, we did not find combinations which influence the protein content positively. Further details on the seed proteins of *Pisum* recombinants have been given by Gottschalk and Müller (1970, 1974, 1979); Müller and Gottschalk (1973, 1978); Hussein (1976, 1979); Müller and Werner (1979); Hartmann (1981).

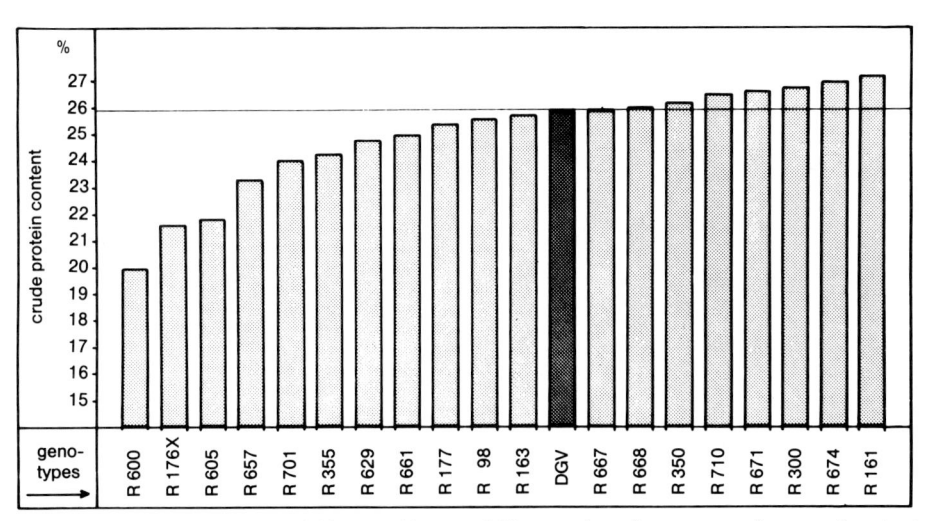

Fig. 37. Seed protein content of 19 recombinants of *Pisum sativum* homozygous for several mutant genes as compared to the control value of the mother variety

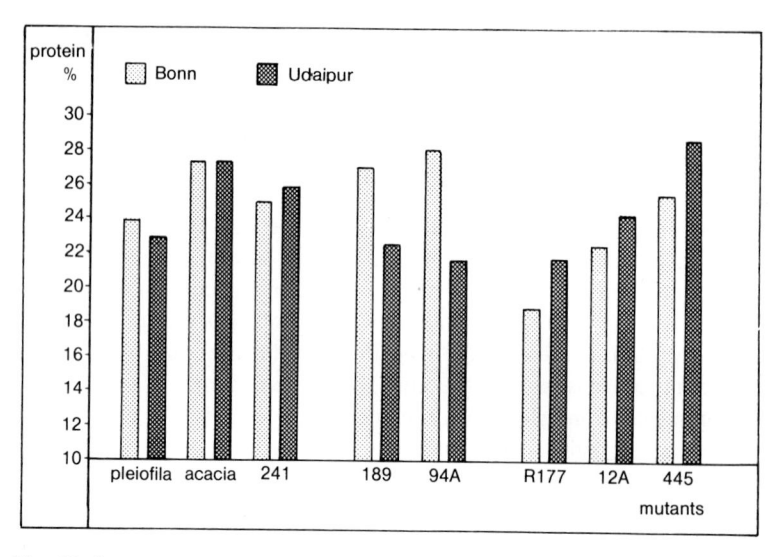

Fig. 38. Seed protein content of six *Pisum* mutants and two recombinants comparatively grown in Bonn, West Germany, and Udaipur, India

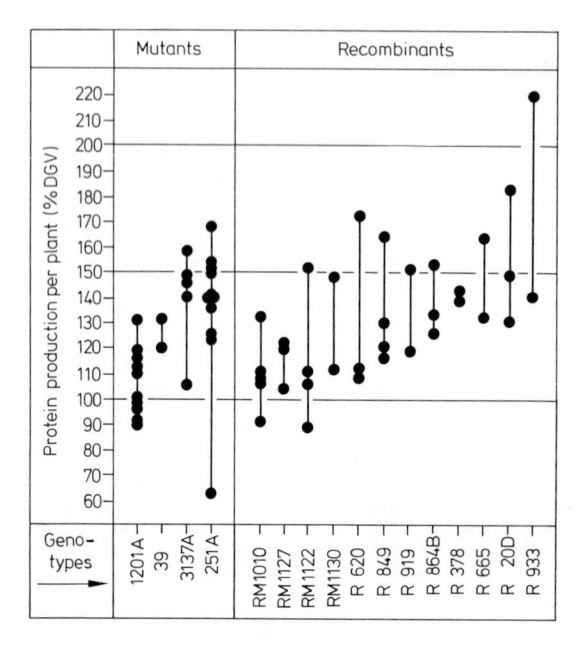

Fig. 39. The protein production per plant of four high-yielding *Pisum* mutants and 12 recombinants. Each *dot* gives the mean value for one generation as related to the control value of the mother variety "Dippes Gelbe Viktoria" = 100%

Certain findings indicate an influence of climatic factors on seed protein content. Some of our mutants and recombinants were grown together with the mother variety in different regions of India. In Fig. 38, protein values of genotypes grown in Bonn (Germany) and Udaipur (Western India) are given. The genotypes studied do not show a uniform response to the strongly divergent climatic conditions of the two countries. In some genotypes, no (*pleiofila, acacia*, 241) or only slight deviations (R 177, 12A, 445), as compared to the control values of the mother variety, were observed. In others, however, clear differences were found (189, 94A). More details have been published by Hiemke (1973); Gottschalk et al. (1975a, 1976); Müller (1978a); Müller and Gottschalk 1978). Our findings indicate that the seed protein production depends under specific climatic conditions to some extent on the genotypic constitution of the mutant material studied.

An increased protein content of the seed meal is only in those cases of direct agronomic interest in which other yielding characters are not negatively influenced by the mutant gene. According to the experiences obtained in many crops, this favorable situation is only rarely realized. This does not mean, however, that mutants and recombinants are not able to exceed their mother varieties with regard to the seed protein production per plant or per unit area. This kind of improvement is possible in all those cases in which an increased seed production per plant is not combined with reduced seed size and reduced protein content of the seed flour.

Some examples of this category of *Pisum* mutants and recombinants are given in Fig. 39. For judging the protein capacity of strains or varieties, it is necessary to consider that all the traits just mentioned are highly influenced by both mutant genes and environmental factors. Therefore, yielding analyses of several years are needed before the usefulness of the genotypes tested can be estimated. This becomes especially clear from the values obtained for the high-yielding fasciated mutant 251A of our collection. Protein content of the seed meal and seed size are somewhat reduced, but the number of seeds per plant is strongly increased. In this way, mean values for the trait "seed protein production per plant" were obtained, lying 23% to 67% higher than the control values of the mother variety, considering ten subsequent generations. In the 11th generation, however, the mutant produced 35% less protein than the initial line due to both a relatively low number of seeds and an abnormally strong reduction of the seed size caused by the extreme weather conditions of that year. The high protein production of the genotypes considered in Fig. 39 is exclusively due to the increased seed production, whereas the protein content of the seed flour has not been altered significantly. Further details on the protein production of a great number of *Pisum* genotypes have been published by Gottschalk (1983).

So far, only quantitative differences of the *Pisum* seed proteins were discussed. With regard to the protein quality, i.e., the amino acid composition, no mutants with significant differences to the initial line were found so far. Figure 40 gives the amino acid distribution of the mother variety and of two mutants. Similar distribution patterns were found in some further 50 mutants and recombinants of our collection (Wolff, unpublished). Theoretically, qualitative alterations of the *Pisum* seed proteins under the influence of mutant genes should be expected. In the seeds of different pea varieties, Jermyn et al. (1976) found a genetically conditioned variability, for instance in the methionine content. The number of genes controlling these traits is obviously

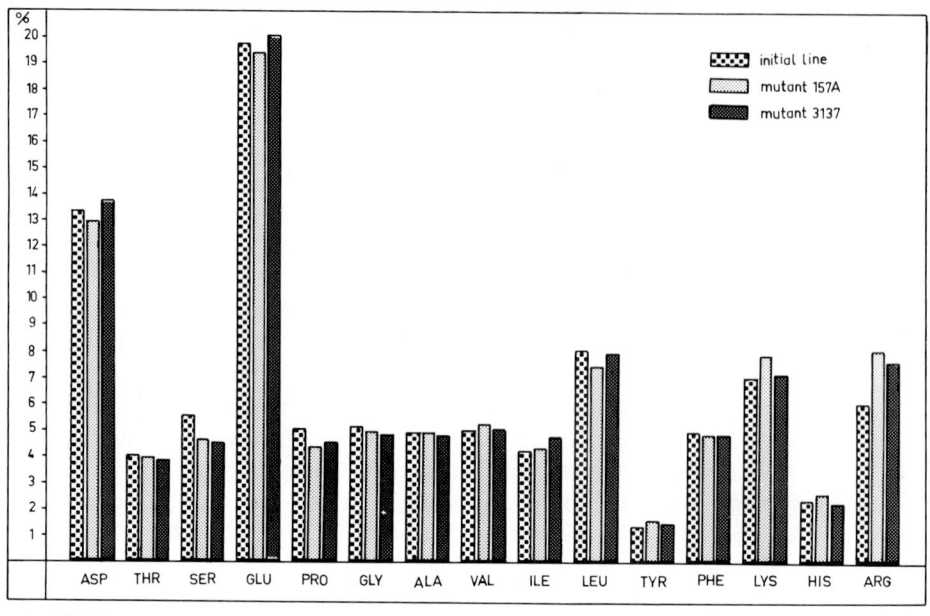

Fig. 40. The amino acid composition of the seed proteins of three *Pisum* genotypes

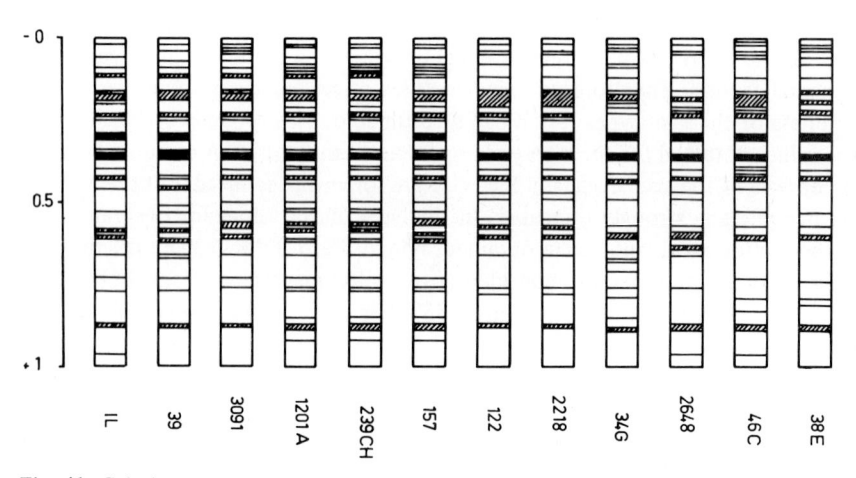

Fig. 41. Gel electrophoretic banding patterns of seed albumins of 11 *Pisum* mutants and their mother variety

very small, thus, it is difficult to obtain mutants homozygous for one of them. The situation in pea seems to be similar to the situation in barley in this respect.

The *Pisum* seed proteins can be subdivided into two main fractions; the albumins comprising about 40% and the globulins comprising 60% of the proteins. Both these groups are composed of heterogeneous subunits. The albumins can be separated into a great number of subfractions by means of normal PAA electrophoresis (Fig. 41). If SDS electrophoresis is used, their number is still greater. During seed development, the number of subfractions increases; in addition, the distribution pattern changes. The action of mutated genes becomes discernible in slight deviations from the pattern of the mother variety, but an overall similarity of the pattern remains visible. The influence of mutant genes is also discernible when specific isoenzymes are investigated (for details see Chen 1970; Gottschalk and Müller 1974; Gottschalk et al. 1976; Müller 1978a, b, 1980; Müller and Werner 1979; Wolff 1980a, b; Gaul 1981). The amount of albumins and the pattern of their subfractions are relatively stable against environmental influences. This is not the case with regard to the globulins. Their amount is highly correlated with total crude protein content (Gerhards 1981; Wenzel 1981). Electrophoretic separation of this fraction results in several subfractions (Müller and Gottschalk 1973; Gottschalk and Müller 1974, 1979; Gottschalk et al. 1976; Müller 1978a, b, 1980; Gerhards 1981). The *Pisum* globulins seem to be composed of 19 polypeptides (Müller and Werner 1979). Albumins and globulins are subsequently synthesized during seed development, albumins first, globulins later. During the first steps of this process, high molecular subunits are found; later on, the typical bandings of vicilin are observed followed by legumin. This pattern is not seriously changed under the influence of the mutant genes studied so far (Gaul 1981).

20.2 Seed Carbohydrates

Investigations on deviations of seed carbohydrates in mutants are not as numerous as those on proteins. This may partly be due to certain methodological difficulties with regard to the quantitative determination of amylose and amylopectin (Detering 1978; *Pisum*). Nevertheless, the problems related to the genetic control of carbohydrate production and their alteration through mutant genes are — at least with regard to basic research, to some extent also for practical purposes — as important as the corresponding problems of seed proteins. Therefore, intensive research work in this field seems to be worthwhile.

Seed carbohydrates are composed of free sugars and starch. The basic elements of starch are glucose molecules which are connected by 1–4 and 1–6 α-glucosidic linkages. If glucose is only combined by 1–4 linkages, unbranched glucose chains arise called amylose. The amylose content of the starch is about 27% in *maize*, 37% in seeds of *smooth peas,* and 68% in *wrinkled peas*. The second component of starch is amylopectin, in which the glucose molecules are connected with each other by 1–4 as well as by 1–6 linkages resulting in a branched chain. In some genotypes (for instance in *su 1 maize*), a third starch component called phytoglycogen is found, this being extremely branched.

An increase of the total carbohydrate content is not of special interest as these substances generally are available in sufficient amounts. More than 50% of seed dry matter consist of carbohydrates, especially of starch. Of importance, however, is the proportion of amylose and amylopectin, because it influences the usefulness of crops and pulses for cooking, consumption and storage (Kellenbarger et al. 1951; Schneider 1951; Creech and McArdle 1966; Madhusudana and Siddiq 1976; Richter 1976). During heating, amylose tends to coagulate giving relatively hard kernels and less availability of this polysaccharide. In addition, a high amount of amylose gives a soft seed texture, which is the reason for inferior storage qualities. On the other hand, industry is interested in high amylose quantities because it is required for the production of solid films, tight fibers, and rigid gels. The amount of free sugars is responsible for the flavor of the seeds.

Particularly from *maize* and *peas,* mutants with altered carbohydrates are known. Most of the information available in this field has been obtained from spontaneous mutants. Our knowledge could certainly be increased considerably, if the collections of experimentally produced mutants would be analyzed systematically in this respect.

20.2.1 Maize

In maize, genes affecting the carbohydrates are known since long (Wheatherwax 1922; Culpepper and Magoon 1924; Sprague et al. 1943; Andrew et al. 1944; Cameron 1947; Kramer and Whistler 1949, Laughnan 1953; Cameron and Teas 1954; Zuber et al. 1958). In Table 39, some genes are summarized influencing carbohydrates. All these mutants have reduced seed size because of decreased starch content. Starch content is normal if distinct genes are dominant. If they are replaced by their recessive alleles, changes in starch quantity as well as quality arise. Additional details are given in the table.

Genes Influencing Free Sugars. Table 39 shows that especially in *sh 2* and *su 1* genotypes the amount of free sugars is very high (5.9% and 6.3%, respectively, as compared to 1.4% in normal maize). This increase is mainly due to higher quantities of sucrose. Also some other genotypes have an increased sugar content but to a lesser extent (Creech 1965; Creech and McArdle 1966). This holds also true for *o 2* genotypes (Barbosa 1971) and for *bt 1* and *bt 2* (Cameron and Teas 1954; Okuno and Glover 1980). Increased sugar content in general is the basis of a high taste index.

Some of the genes involved were used in extensive crossing experiments in order to study their interactions. Two or three genes were combined with each other (Creech 1965; Creech and McArdle 1966; Tsai and Glover 1974). These experiments reveal

Table 39. Genes influencing seed carbohydrates in maize (after Creech and McArdle 1966, and Hess 1968)

Genotype	Dry wt. g/100 kernel	Total sugar % d.wt.	WSP[a] % d.wt.	Starch % d.wt.	Amylose % starch	Amylopectin % starch	Taste index
Normal	22.3	1.40	4.00	76.1	27.0	73.0	1.3
ae	13.1	1.85	5.60	49.1	50–60	40–50	6.0
du	20.9	1.92	4.60	63.7	38.0	62.0	6.3
sh2	11.3	5.90	4.40	26.0	–	–	7.7
su1	10.8	6.33	19.20	32.9	29.0	71.0	5.3
su2	20.9	1.70	4.30	61.9	40.0	60.0	1.3
wx	18.7	1.45	7.43	68.7	0.0	∼100.0	2.7

[a] WSP = Water-soluble polysaccharides

that the action of genes *sh 2* and *su 1* with respect to the production of free sugars is negatively influenced by gene *du* (*sh 2/du:* 3.47%; *su 1/du:* 4.13%). Gene *su 2*, on the other hand, intensifies the action of *su 1* (*su 1/su 2:* 9.34%), whereas the effect of *sh 2* is reduced in the presence of *su 2* (*su 2/sh 2:* 3.79%). Gene *ae* is epistatic on *su 1* (*ae/su 1:* 1.36%). Further interactions are discussed by Creech (1965) and by Creech and McArdle (1966).

Genes *su 1* and *sh 2* could be useful for breeding maize with high sugar content. Their expression varies in different genetic backgrounds. In the sweet corn variety "Golden Cross Bantam", *su 1* is less effective than in normal dent backgrounds. It is assumed that modifier genes of "Golden Cross Bantam" are responsible for the unexpected action of the *su 1* gene (Creech and McArdle 1966). Also gene *sh 2* was incorporated into commercial varieties (Laughnan 1953). The sugar content of these genotypes was found to be satisfying, but their usefulness was reduced by a negative effect on seed texture and by poor germination ability.

Genes Influencing Polysaccharides. All the genes summarized in Table 39 cause a reduction of the starch content; their combination has no additional effect. Of special interest is the composition of the starch. As industry is interested in high quantities of amylose, deviations toward a higher proportion of this component are desired. In normal maize kernels, 73% of starch is amylopectin and only 27% is amylose. As Table 39 shows, genes *du* and *su 2* are responsible for the increase of the amylose proportion. A strong increase of this component is furthermore found in *ae* genotypes where more than 50% of starch is unbranched. This gene is used for developing amylose-rich lines. The main problem is to overcome the reduced amount of total starch which is responsible for the reduced seed size.

From a theoretical point of view, gene *wx* is of interest because plants homozygous for this gene do not produce amylose at all. In analogy to this situation, mutants can be expected producing exclusively amylose at the expense of amylopectin. Such a mutant has not yet been found.

The highly branched phytoglycogen is found in the fraction of water-soluble polysaccharides. Under the influence of gene *su 1*, a very high amount of this substance is produced (Table 39). If gene *bt 1* is simultaneously present, the synthesis is reduced (Tsai and Glover 1974). The genes *du*, *su 2*, and *wx* have a slightly negative effect on the action of *su 1*, while *ae* suppresses it completely (Black et al. 1966; Creech and McArdle, 1966). The additional presence of genes *du* or *wx* in the *su 1/ae* genotype restores the capacity to produce phytoglycogen. Genes *du* and *wx* alone do not produce this substance; in combination, however, they produce small amounts. Because of these relations, it seems uncertain that the *su 1* gene controls the production of phytoglycogen from amylose (Black et al. 1966). It is assumed that phytoglycogen is a normal intermediate in starch synthesis which in normal kernels is degraded and thus cannot be found in ripe seeds (Erlander 1960).

Genetic Control of Maize Starch Synthesis. In spite of the importance of carbohydrates for nutrition, the pathways of starch synthesis are not yet fully understood. Our knowledge on certain biochemical relations, however, has been widened considerably by comparing the situation in various mutants with that of the nonmutated control material.

Gene *sh 1* of the maize genome becomes effective during the early steps of starch synthesis. In the *sh 1* genotype, reduced amounts of sucrose synthetase are found (Chourey and Nelson 1979; Chourey 1981). It is assumed that two genes are responsible for the synthesis of this enzyme because two fractions were found by electrophoretic means. It regulates the formation of UDPG — the main glucose donor for starch synthesis — from sucrose and UDP. In genotypes containing genes *sh 2* and *bt 2*, ADPG-pyrophosphorylase is lacking, this being responsible for the transfer of glucose-1 phosphate to ATP from ADPG (Tsai and Nelson 1966). Waxy genotypes *(wx)* are deficient for a starch granule bound transglucosidase responsible for the 1−4 α-glucosidic linkages between the glucose molecules; therefore, they are not able to produce amylose (Nelson and Rines 1962; Nelson and Tsai 1964). Genes *ae* and *su 1* are involved obviously in the synthesis of branched polysaccharides (Black et al. 1966; Boyer and Preiss 1978). Also in *o 2* genotypes, deviations in starch synthesizing enzymes were observed (Mehta et al. 1979).

Starch synthesis is closely connected with the formation of starch granules. Their size increases during development and the proportion of amylose within the granules increases. Within the seed, the size of the granules varies according to the physiological age of the organelle. In ripe seeds of various genotypes, starch granule size is different. The biggest granules were found in normal and *wx* maize, whereas they are very small in *ae/su 1* seeds (Boyer et al. 1976). Investigations on the influence of mutant genes revealed a similar development of the granules during the first 12 days after anthesis in all the genotypes studied. Afterward, it varies according to the genotypic constitution (Brown et al. 1971).

20.2.2 Barley and Other Cereals

The protein mutant 1508 of *Hordeum vulgare*, intensively discussed in Sect. 20.1.4.1., has also been analyzed with regard to its carbohydrates. The reduced seed size of this genotype is due to a lower level of starch. The mutant has 20% less starch and an increased amount of free sugars as compared to its mother variety, caused by a slower synthesis during seed development. A deviation in amylose/amylopectin ratio is not observed. The higher amount of free sugars may be due either to a block in starch synthesis or a premature degradation of starch conditioned by the increased amount of α-amylase (Munck 1975; Køie et al. 1976a, Køie and Kreis 1977; Kreis 1978; Køie and Doll 1979, Kreis and Doll 1980; Table 40). A similar reduction in starch and increase of free sugars is realized in mutants 527 and 29 of the Risø collection. Recombinants of the constitution 1508/527 and 1508/29 show an additive effect of the two genes

Table 40. Characteristics of normal and *1508* barley (according to Doll and Køie 1978)

Characteristics	Normal lines	High-lysine lines having the *1508* gene
Grain yield g/m^2	357	287
Starch g/m^2	189	123
Sugar g/m^2	10	14
Protein g/m^2	44	39

involved, resulting in further reduction of the amount of starch (68% or 43%, respectively, less than the control material; Kreis and Doll 1980).

In some *rice* mutants, certain differences in the structure of the starch layer were found (Tanaka and Tamura 1968). Other mutants deviate from their initial lines with regard to the proportion of amylose (Haq et al. 1973; Shaikh et al. 1976; Reddy and Sarala 1979). The *hl* variety of *sorghum* shows an increased content of free sugars (3.46% instead of 1.34% of the control material). Total starch content is only slightly reduced indicating that the *hl* gene is not seriously involved in starch synthesis (Axtell et al. 1974).

20.2.3 Peas

The differences between smooth and wrinkled peas were found to be due to two alleles of the *R* locus, the dominant allele being responsible for smooth, the recessive one for wrinkled seeds (Gregory 1903; Darbishire 1908; Kappert 1915; Table 41). The starch grains of the smooth kernels are oval, those of wrinkled ones are roundish with radial splits. These differences are connected with different amounts of amylose in the starch (Schneider 1951). Moreover, the sugar content of wrinkled peas is considerably higher than that of the smooth ones (Kappert 1915); therefore, they are preferred as vegetable. A negative correlation between amylose and total starch content exists (Kellenbarger et al. 1951) which, however, can be broken. Therefore, it is assumed that the carbohydrate production of the pea is controlled by two gene pairs: $R_a r_a$ and $R_b r_b$. The genotypic constitution of the old smooth peas is $R_a R_a R_b R_b$, that of the old

Table 41. Characteristics of smooth and wrinkled peas (according to Kellenbarger et al. 1951; Kooistra 1962; Richter 1976)

Characteristics	Smooth peas *(RR)*	Wrinkled peas *(rr)*
Starch grains	Oval, simple	Roundish, with radial splits
Starch content	High ($\sim 46\%$)	Low ($\sim 34\%$)
Content of amylose (% starch)	Low ($\sim 38\%$)	High ($\sim 69\%$)
Content of free sugars	Low ($\sim 5\%$)	High ($\sim 10\%$)
Water absorption during development	Low	High
Water loss during maturation	Low	High
Germination	Normal	Reduced

Table 42. Characteristics of some pea varieties (After Kooistra 1962)

Variety	Seed shape	Starch grains	Genotypes	Sugar % d. mt.	Starch % dr. mt.	Amylose % starch
Alaska (old type of smooth peas)	Smooth	Simple	$R_a R_a R_b R_b$	6.2	54.5	37.2
Kelvedon Wonder (old type of wrinkled peas)	Wrinkled	Composed	$r_a r_a R_b R_b$	12.1	30.4	68.1
Cennia	Wrinkled	Simple	$R_a R_a r_b r_b$	8.6	33.1	24.2
Cennia × N 2067	Wrinkled	Composed	$r_a r_a r_b r_b$	10.5	29.2	43.6

wrinkled ones is $r_a r_a R_b R_b$. Later on, new types with wrinkled seeds and simple starch grains $(R_a R_a r_b r_b)$ as well as wrinkled seeds with composed starch grains $(r_a r_a r_b r_b)$ were found (Kooistra 1962). The factor a in combination with R reduces amylose content drastically while b in combination with R increases it. Only when both starch genes are dominant, is starch content high. Carbohydrate values of these four genotypes in different pea varieties are given in Table 42. The new genotype unites the positive characters of the two parental forms: a good flavor combined with convenient cooking and storage properties.

In pulses, the proportion between amylose and amylopectin, as well as between these two starch components and sugar, can easily be changed through mutation (Pape et al. 1969). Therefore, a considerable genetically conditioned variability in amylose content (Whistler and Johnson 1964) and in other seed substances (Richter 1976) exists, which ban be utilized agronomically. Some mutants and recombinants of our *Pisum* collection were analyzed in this concern. According to the preliminary findings available, specific mutant genes of the genome influence the quantity of both, single free sugars and the total amount of free sugars, furthermore the starch content per dry matter and the amylose content of the starch (Thierfeldt 1972; Gottschalk and Hasenberg 1977; Detering 1978; Steinbrück 1977). The investigations in this field are being continued.

20.3 Seed Lipids

Lipids are deposited in oil bodies as energy rich storage substances of the seeds. They are produced from sugars and are easily degraded to sugars by lipase. Thus, they are closely related to the carbohydrates with regard to their metabolism. Lipids are a mixture of various triglycerids. Some examples may show that their amounts vary considerably in seeds of different crops:

- maize   5%
- rice   8%
- peas   1%
- peanuts 56%

There is a general tendency that seeds with a large amount of lipids are high in protein and low in carbohydrates (*soybean:* 18% fat, 40% protein, 17% carbohydrates). Carbohydrates and fats are obviously exchangeable as storage substances. Within the same crop, the amount of lipids varies depending on the genotypic constitution of the material tested:

- oats: ranging between 3.1% and 11.6% (Youngs 1978);
- maize: in American high oil lines oil content was increased from 4.7% to 17.0% by selection (Weber 1978).

Oil is predominantly deposited in the germ (Fugino 1978; Morrison 1978; Youngs 1978; Rooney 1978). The high values of the American *maize* lines just mentioned may partly be due to the bigger embryos in their seeds. A similar situation is observed in *sorghum:* 21.1% of the lipids are found in the germ, whereas only 0.6% are deposited in the endosperm (Rooney 1978). According to Morrison (1978), the distribution of lipids within *wheat* seeds is as follows:

- 0.8% – 2.2% in endosperm,
- 10.0% – 16.3% in embryo axis,
- 12.6% – 32.1% in scutellum,
- 6.0% – 9.9% in aleurone layer.

The deposition of fats in the seeds as well as the composition of the fatty acids seems to depend on environmental factors (Zeller 1957; Schuster and Marquardt 1973; Youngs and Forsberg 1979; Arnholdt and Schuster 1981).

There is only little information on mutationally caused differences in seed oil. In seeds of *o 2* and *fl 2* genotypes of *Zea,* oil content is higher than in normal maize (Nelson 1967; Arnold et al. 1977; Mortinello et al. 1978). Under the influence of the 1508 gene of *barley,* an increased fat content was found in the seeds too (1508:4.5%, Bomi: 2.7%; Munck 1975; Køie and Kreis 1977). In addition the oil composition is changed (Shewry et al. 1979). The increased oil content of these genotypes may partly be due to the bigger embryo (Nelson 1967, 1969b; Tallberg 1977). The *hl* gene of *sorghum* also increases the oil content of the seeds (Singh and Axtell 1973).

Mutation programs were started in *Arachis hypogaea* (Patil 1971), *Brassica juncea* (Norain and Prakash 1969), *Foeniculum vulgare* (Dnyansagar and Jahagirda 1979), *Linum usitatissimum* (Rath and Scharf 1968; Seetharam 1971; Srinivasachar and Malik 1971) and *Ricinus communis* (Chandola et al. 1971). Some genotypes with increased oil content were selected. In *Brassica napus,* mutants with seeds differing in the composition of the fatty acids with regard to the proportion of linolenic and linoleic acid were obtained (Rakow 1973; Röbbelen and Nitsch 1975).

21 Other Plant Substances

Most of the research work on the genetic regulation of amounts an composition of plant substances is done on seed proteins and carbohydrates. Only little information is available on genes influencing other plant substances of our crops. Tables 43 and 44

Table 43. Genetically conditioned quantitative and qualitative differences in plant substances others than seed proteins and seed carbohydrates

Species	Literature	Remarks
a) Non seed carbohydrates		
Beta vulgaris	Melzer and Sackewitz 1979	Mutation program to increase sugar content in turnips; mutagens: EMS, NMH
Ipomoea batatas	Kukimura and Takemata 1975	Greater variability in sugar content; mutagens: gamma rays, EI
Prunus spec.	Zwintzscher 1977	Increased sugar content in fruits of "Schattenmorelle"; mutagen: X rays
Saccharum spec.	Jagathesan 1976	Mutants with increased sugar content
Solanum tuberosum	Kolesnikova and Maksimova 1977	Dry matter and starch content improved by about 3% in the first vegetation period after treatment; different mutagens
	Khvostova 1967	Potato mutants with increased starch content
b) Phenols		
Gossypium spec.	Gutierrez et al. 1972	Spontaneous mutant with almost no gossypol
	King and Leffler 1979	Mutants without gossypol and pigment
c) Alkaloids		
Capsicum annuum	Bansal 1969	A mutant *(514)* with increased capsacin content and reduced pigmentation; different mutagens
Lupinus angustifolius	Zachow 1967	Mutant with reduced alkaloid content and normal fertility; mutagen: X rays
Lupinus digitatus	Byth 1975	Mutant with low alkaloid content; mutagen: EMS
Lupinus luteus	Jashowsky and Golovchenko 1967	Radio mutant with reduced alkaloid content (0,01%)
Lupinus mutabilis	Pakendorf 1974	Sweet mutant with low alkaloid content: mutagen: gamma rays
Melilotus albus	Scheibe and Micke 1963–1966	16 mutants with reduced coumarin content, reduction from 5% to 0.01%; mutagen: X rays
Papaver somniferum	Ivanova 1972	Mutants with high morphin content; mutagens: EI, NEU
Solanum viarum	Dnyansagar and Pingle 1977	A mutant with a higher amount of solasodine; mutagen: EMS. Solasodine is supposed to be directly related to the quantity of photosynthetic tissue

summarize some of the findings referring not only to seed substances, but also to components of other plant organs. In principle, they can be divided into two classes as follows:

- Substances, the increase of which is desired. This holds true for the sugar content in *sugar cane* (Jagathesan 1976) or the content of ascorbic acid in fruits of *black currants* (Gröber 1967).
- Substances, which should be removed. An impressive example is a toxic substance in *Lathyrus sativus* (BOAA; Nerkar 1971, 1972) or bad tasting alkaloids in *Lupinus*. As far as information is available, research work in this field seems to be worthwhile. Several mutants with reduced alcaloid content in the genus *Lupinus* are known (Zachow 1967; Pakendorf 1974; Byth 1975).

Table 44. Genetically conditioned quantitative and qualitative differences in further plant substances

Substance	Species	Literature	Remarks
Anthrachinon	*Rubia tinctoria*	Kondratenko 1975	Mutants with relatively high content of anthrachinon derivates; mutagen: NEU
Arginine	*Camellia sinensis*	Nakayama 1973	Bud mutants; one mutant with increased arginine content in leaves (arginine influences taste positively); mutagen: X rays
Ascorbic acid	*Ribes nigrum*	Gröber 1967	Increased ascorbic acid content in the fruits; mutagens: X rays, EMS
Diastase	*Hordeum vulgare*	Hayter and Allison 1975	Mutant with increased diastatic power in seeds
Mentha oil	*Mentha arvensis*	Kak and Kaul 1980	Dormant suckers irradiated: stable mutants with changed composition of oils
Oestrogenic isoflavons	*Trifolium subterraneum*	Francis and Millington 1965	Oestrogenic isoflavons in clover reduce lambing; it was supposed that this substance can be reduced by mutation
β-N-oxalyl-α,β-diamino-propionic acid (BOAA)	*Lathyrus sativus*	Nerkar 1971, 1972; Swaminathan 1969b	BOAA is a neurotoxin in seeds of *Lathyrus sativus*, cultivated because of high seed protein content and drought resistance. After application of mutagens greater variation in BOAA content (0.5%–2.5%); mutagens: gamma rays, EMS
Phytic acid	*Phaseolus* ssp.	Lolas and Markakis 1975	Phytic acid is able to chelate minerals, thus reducing the availability of them in animal nutrition; 50 varieties with different content of phytic acid

22 The Nutritional Value of Mutants

The aim of mutation breeding as far as seed storage substances are concerned consists in developing varieties of good nutritional quality. For energy supply in the body, carbohydrates, fats, and proteins are exchangeable. The special importance of proteins, however, lies in supplying amino acids; thus, they cannot be replaced by other substances. As proteins of various sources are differently composed, their nutritional value varies. Animal proteins are rich in essential and sulfur-containing amino acids, the amount of which is considerably smaller in seed proteins.

The improvement of seed proteins is possible by increasing the total protein content of the seed meal or by improving protein quality. For nutritional purposes, the increase of the total protein content is only relevant if the quality is not altered detrimentally. As already mentioned, late nitrogen fertilization, for example in normal *barley,* increases total seed protein content by increasing especially the prolamin fraction which is low in lysine. Under these conditions, the prolamin content per seed meal and the lysine content per protein are negatively correlated (Paulis et al. 1975a, *maize;* Eggum 1970, *barley*). Because of this, the increased protein content can be connected with a reduction of the nutritional value of the respective seeds. Such negative correlations are known for *maize* (Michael 1963), *barley* (Doll 1977), *wheat* (Nehring 1963) and *sorghum* (Waggle et al. 1966). On the other hand, the digestibility of the respective seed protein is increased because prolamins are better digestible than other seed proteins (Eggum 1970).

The relations between the biochemical conditions in the seeds and the nutritional characteristics are manifold. A judgement of the practical use of a genotype thus requires extensive knowledge of the seed properties or — if possible — feeding trials. Some experiences with such trials in connection with prospective mutants are already available; some details will be given in the following chapters.

22.1 Maize Mutants

Meals being composed of *opaque 2* and *floury 2* maize were fed to different animals. In both cases, maize genotypes were chosen exhibiting an increased total protein content due to an appropriate genetic background for both genes. From experiments with pigs it turned out that the average growth gain per day is more than doubled under the influence of *o 2* maize (daily gain with normal maize: 0.15 kg, with *o 2* maize: 0.36 kg). By using *fl 2* maize, there is an increase by a factor of 1.8 (Klein et al. 1971; Nelson and Mertz 1973). Thus, a smaller quantity of food is necessary for reaching the same degree of growth. The addition of free amino acids to normal maize diet in comparable amounts as *o 2* or *fl 2* maize would provide, has a similar but somewhat reduced positive effect in the case of the *o 2* amino acids, but no effect in the case of *fl 2* ones. The increased nutritional effectiveness of the *o 2* genotype is due to the higher amount of lysine and tryptophane provided. This was proved by adding these two amino acids to normal maize diet. The findings with *fl 2* indicate that — in contrast to *o 2* — the increased content of lysine and tryptophane is not the only reason for the improved nutritional value of this mutant; certain interactions between some other components seem to play a role in this case. Similar results have been obtained with rats (Nelson 1969a; Eggum 1973; Fig. 42) and chicken (Jarquin et al. 1970).

Feeding trials with rats using the recombinant *su 1/o 2* revealed a higher growth rate than with *o 2* maize alone. This seems to be due to the higher proportion of lysine as compared to the *o 2* genotype. N-retention and protein digestibility are similar to *o 2* (Sgarbieri et al. 1977).

After the first promising experiments with animals had been carried out, meals containing *o 2* maize were given to children. The nutritional value of *o 2* proteins turned out to be equal to that of milk proteins giving the same gain of body weight. If normal maize is used for the meals, other conditions were found. A negative protein balance was measured, indicating that normal maize proteins are not an adequate source for the growth of children (Nelson 1967, 1969a). According to calculations given by Bressoni and Elias (1979), the amount of an ideal protein should be 0.55 g/kg/d (FAO/WHO recommendations). This quantity provides all essential amino acids in required composition. If normal maize is used comparatively as protein source, 0.59 g/kg/d are needed. If *o 2* maize is used, only 0.44 g/kg/d are necessary, this is 25% less. Even if we consider a reduced yield of 10%–15%, *o 2* maize proves to be of advantage in nutrition.

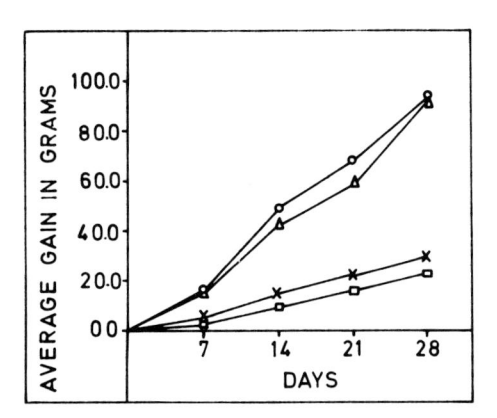

Fig. 42. Average weekly weight gain in weanling rats, fat rations composed of 94% whole grain, 2% vitamins and 4% minerals. (After Singh and Axtell 1973). ×———× normal maize; △———△ opaque 2 maize; ◻———◻ normal sorghum; ○———○ hl sorghum

22.2 Barley Mutants

From feeding trials with rats, pigs, and poultry, high lysine genotypes were found to have in general a higher nutritional quality than low lysine ones. Especially hiproly barley and the 1508 mutant of Bomi barley were tested in this respect (Eggum 1973, 1978; Doll et al. 1974). The various seed proteins of barley are characterized by different amino acid patterns with different amounts of lysine. Furthermore, their digestibility is not equal. While prolamine is highly digestible, albumins and globulins — the main sources for lysine — are not. Their low digestibility is conditioned by the fact that they are mainly deposited in the aleurone layer. The cells of this layer are thick-walled and cellulosic. In addition, some of the proteins are bound to the cell walls and thus are less available. As barley — in contrast to wheat for example — has a multiple aleurone layer, a considerable portion of the proteins is not available. Because of these conditions, the situation in seeds of the high lysine lines is somewhat contradictory: The biological value of the proteins is increased because of their high lysine content, the digestibility, however, is reduced as the lysine rich proteins are less digestible (Eggum 1977a, b, 1978).

22.3 Sorghum Genotypes

The digestibility of sorghum protein is low, obviously due to the tannin content of the seeds. These substances tend to bind to proteins making them less available for nutrition. Differences in tannin content are known for different varieties of the crop (Axtell et al. 1973, 1981; Axtell and Lafayette 1976; Chibber et al. 1978). Thus, improvement of the nutritional efficiency of sorghum proteins under the influence of mutant genes might be possible. The tannin content seems to be simply inherited, being controlled by one or two genes. The protein-rich P 721 mutant has not yet been investigated in this respect.

High lysine sorghum was tested in rat-feeding trials. The weight gain of the animals is considerably increased if high lysine seed is the basis of protein source (Fig. 42). This nutritional improvement, similar as in maize and barley, is due to the increased lysine content (Singh and Axtell 1973; Axtell et al. 1974). The protein quality of *hl* sorghum and of mutant P 721 proved to be equal. Moreover, the high lysine gene has an effect on the flavor of the sorghum seeds. Although high lysine lines from Ethiopia are low in yield, they are cultivated and consumed together with normal sorghum because of their good taste.

22.4 Pea Mutants

In legumes, only very little information in connection with feeding trials is available. Some few results exist in *Pisum*. Those mutants of our collection, which have an increased protein content, do not show any deviations in their amino acid composition, indicating that no alterations of the protein quality exist. Feeding trials with pea mutants have not yet been carried out to a broader extent. The high-yielding mutant 489C and its mother variety were comparably fed to rats. The results gave no differences in the nutritional value of the two genotypes (Eggum, pers. commun.). This result is in accordance with the unchanged amino acid composition of this mutant as compared to its initial line.

23 General Aspects of Mutation Breeding with Regard to the Improvement of Seed Storage Substances

A great number of mutants of different crops has been analyzed during the past 10 to 15 years with regard to any improvement of agronomically important substances, but the practical results are relatively poor. This holds particularly true with regard to seed protein improvement by means of mutant genes. A tremendous amount of work has been done in this field in the frame of national and international research programs resulting in the selection of a very low number of protein mutants. As a consequence of the disproportion between expenditure and results, some international programs have been cancelled. Our knowledge on the genetical and biochemical basic processes, on the other hand, has been widened considerably. Therefore, the work done in this field should not be judged too pessimistically. It is certain now that the amount of nutritionally valuable seed substances as well as their composition can be positively altered under the influence of mutant genes (Table 29). This is a promising result, which will be the starting point for further research. Some negative interactions between desired alteration of seed storage substances and distinct yielding criteria have been found, but also ways to overcome them are proposed. High protein content of seeds, combined with strongly reduced seed size, is undesirable and cannot be utilized agronomically. It is still too early to pass final judgement on the practical value of protein mutants and related genotypes, because yielding analyses over a longer period are not yet available to a great extent. The greatest part of the material shows that the increase in one group of substances is mostly reached at the expense of the amount of other substances. Increase of protein, free sugars, or changes in the proportion of amylose and amylopectin, mostly result in reduction of the starch content, being the reason for unsatisfactory seed size. Exceptions are reported for almost all the crops studied. But the number of practically used mutants of this category is very small (Table 45). Thus, those genotypes which show biochemically promising alterations of their seed substances are obviously not competitive with the varieties or strains already existing.

In some cases, the reduction in yield is accepted in combination with changed amounts of storage substances, because other criteria are of interest. In case of free sugars, for example, the reduction of seed yield is tolerated because the respective *maize* kernels have a favorable sweet taste. They are not consumed for energy supply but as an additional food of pleasant taste. Similarly, the low yielding *hl sorghum* is cultivated in Ethiopia because of the pleasant taste of its grains. Also in connection with antinutritional substances, the mutants can be judged with regard to taste components. The reduction of bitter-tasting *lupine* alkaloids, for example, makes the respective genotypes useful as forage.

Genotypes with the *o 2* gene of the *maize* and the 1508 gene of the *barley* genome, showing qualitative improvements of their seed proteins, are not yet used for other than scientific purposes. Their yielding potentialities are inferior to those of the varieties already existing. Numerous attempts were made to find modifier genes and/or appropriate genetic backgrounds for these genes in order to overcome negative correlations. Experiences with *o 2* inbred lines have shown that crops with increased and improved protein content but reduced seed yield are not accepted by the farmers. Thus, the nutritional improvement, actually existing in these genotypes, is not yet utilized practically.

Of special importance are the fundamental findings obtained, elucidating the pathways of seed protein and seed carbohydrate metabolism:

— Results are available on interrelations between nitrogen metabolism and seed protein content. Although most of these findings have been obtained in nonmutated material, it is probable that specific mutant genes become effective on the basis of enzymatic regulation.
— The findings reveal that seed proteins are composed of a considerable number of fractions and subfractions. According to the generally accepted concept of protein biosynthesis, specific genes are responsible for the synthesis of each of these polypeptides. Thus, seed protein production is controlled by a considerable number of genes.
— The investigations on seed carbohydrates have shown that specific enzymes, responsible for distinct steps of starch synthesis, are controlled by special genes.
— Quite a number of genes, involved in the control of these processes, have already been localized and their function as structural or regulatory genes is discussed.
— Recombinants, homozygous for several mutant genes, are a valuable material for studying certain interactions between the genes involved.
— An important factor, influencing amount and composition of storage substances, is the environment. Numerous investigations, using genotypes of the world collections and of assortments of mutants of different crops, reveal that environmental factors impede the selection of genotypes with improved characters, especially if the biochemical differences are small. So far, only some few single components of the complex character "environment" have been investigated in detail. The effect of nitrogen fertilization on seed proteins is relatively well understood, whereas the specific effectiveness of other factors is less known. The cultivation of appropriate genotypes under the controlled conditions of phytotrons will certainly contribute to obtaining further information.

The main advantage of the research work done in the whole field is the considerable gain in basic knowledge, whereas the gain with regard to the practical utilization of this knowledge is small. But many experiences are necessary — positive and negative ones — before a completely new field of research can be converted into practical application. Many of the mutants analyzed with regard to their storage substances have been selected because of distinct morphologically discernible characters or of favorable yielding properties. Any genetically conditioned alterations of their storage substances have been found later on not to be the reason for their selection. It can be

Table 45. List of officially released varieties with improved seed storage substances (Sigurbjörnsson and Micke 1974; Mutation Breeding Newsletter 1–19, 1972–1982)

Developed variety	Country, year of release	Mutagen	Characteristics
1. Cereals			
Hordeum vulgare			
Sharbati Sonora	India 1967	Gamma rays	Higher protein and lysine content, short straw
Blazer	USA 1974	Thermal neurons	Malting barley with increased α-amylase
Fakel	USSR 1974	EI	Higher protein content, short straw
Oryza sativa			
Iratom 38	East Pakistan 1970	Gamma rays	Higher protein content, earliness
Hybrid mutant 95	India 1973	Gamma rays	High protein content, high yielding potentials
Yuan-feng-tsao	P.R. China 1973	Gamma rays	High lysine content, high yield
2. Legumes			
Glycine max			
Vikram T 61	India 1973	X rays	High oil content, larger kernels
Colorado Irradiado	Argentina	X rays	High oil content, high yielding
Phaseolus vulgaris			
Alfa	CSSR 1972	EMS 1966	Improved protein and seed yield
Giza 80	Egypt 1980	Gamma rays	High protein content, high yielding, big seeds
Pisum sativum			
Moskovsky	USSR 1974	DES 1967	High protein content, larger seeds
Lupinus spec.			
Kievsky	USSR 1969	–	High protein and lysine content, alkaloidless, high grain and forage yield
3. Other crops			
Brassica spec.			
Elite F	Sweden 1952	X rays	Increased oil content, increased yield
Regina vårrupe	Sweden 1973	X rays	Increased oil content, increased yield
Elite A			
Capsicum annuum			
MDV 1	India 1976	Gamma rays	Higher capsicine content, good yield
Helianthus annuus			
Pervenets	USSR 1977	DMS 1965	Higher oil content, altered composition of fatty acids
Linum usitatissimum			
Redwood 65	Canada 1965	X rays	Higher oil content, good yield
Mentha arvensis			
Rose mint	Japan 1977	Gamma rays	Improved quantity of oil (fragrance similar to rose oil), good yield
Saccharum officinalis			
Nami	Japan 1981	Gamma rays	Higher sugar content

Table 45 (continued)

Developed variety	Country, year of release	Mutagen	Characteristics
Sinapis spec.			
Svalöf's Primex	Sweden 1950	X rays	Increased oil content, increased yield
RLM 198	India 1975	Gamma rays	Increased oil content, increased yield
RLM 514	India 1980	Gamma rays	Increased oil content, less erucic acid, high yield
Solanum khasianum			
RRL-20-2	India 1975	Gamma rays	Higher solasodine content

assumed, that the prospects of this branch of mutation breeding are considerably improved when better biochemical methods for analyzing great numbers of M_2 and M_3 plants are available.

References

IAEA means "International Atomic Energy Agency"

Aastveit K (1966) Use of induced barley mutants in a cross-breeding programme. Mutations in Plant Breeding, pp 7–14, IAEA, Vienna

Abdalla MMF, Hussein HAS (1977) Effects of single and combined treatments of gamma rays and EMS on M_2-quantitative variation in *Vicia faba* L. Z Pflanzenzücht 78:57–64

Abdalla MMF, Hussein HAS, Ibrahim AF, Hindi L, Sharaan AN (1980) Analysis of radiation-induced erectoid barley mutants and associated yield characters. Egypt J Genet Cytol 9:167–192

Abdel-Hafez AAGI, Röbbelen G (1979) Differences in partial resistance of barley to powdery mildew (*Erysiphe graminis* DC. f. sp. *hordei* Marchal) after chemomutagenesis. I. Screening of mutants under field conditions. Z Pflanzenzücht 83:321–339

Abdel-Hafez AAGI, Röbbelen G (1981) Differences in partial resistance of barley to powdery mildew (*Erysiphe graminis* DC. f. sp. *hordei* Marchal) after chemomutagenesis. III. Agronomic performance of the mutants. Z Pflanzenzücht 86:99–109

Abdel-Hak TM, Kamel AH (1977) Mutation breeding for disease resistance in wheat and field beans. Induced Mutations against Plant Diseases, pp 305–314, IAEA, Vienna

Abi-Antoun M (1974) An evaluation of disease-resistant mutants. Induced Mutations for Disease Resistance in Crop Plants, pp 147–148, IAEA, Vienna

Abo-Hegazi AM (1979) High-protein lines in field beans *Vicia faba* from a breeding programme using γ-rays. Seed Protein Improvement in Cereals and Grain Legumes II, pp 33–36, IAEA, Vienna

Abo-Hegazi AM (1980) Seed protein and other characters in M_4 generation of chickpea. Indian J Genet Plant Breed 40:122–126

Abo-Hegazi AM, Shoeb ZE, Salama FM, Hakam M (1973) Breeding for improved protein in pulses using radiation-induced mutations. Nuclear Techniques for Seed Protein Improvement, pp 265–268, IAEA, Vienna

Abraham V, Desai BM (1976) Radiation induced mutants in tuberose. Indian J Genet Plant Breed 36:328–331

Adam G (1975) Mutations affecting protein characters in wheat. Breeding for Seed Protein Improvement Using Nuclear Techniques, pp 35–37, IAEA, Vienna

Afsar Awan M, Konzak CF, Rutger JN, Nilan RA (1980) Mutagenic effects of sodium azide in rice. Crop Sci 20:663–668

Ahmad S, Godward MBE (1981) Comparison of a radioresistant with a radiosensitive cultivar of *Cicer arietinum* L. – II. Differences in the number of chromosome aberrations at the same dose. Environ Exp Bot 21:143–151

Ahokas H (1975) Male sterile mutants of barley. I. Inaperturate pollen of the msg6cf mutant. Ann Bot Fenn 12:17–21

Ahokas H (1976) Genetic and morphologic characteristics of the gigantic mutant of oats, *Avena sativa* L. J Sci Agric Soc Finl 48:90–105

Ahokas H (1977) Increase in protein content by partial fertility. Barley Genet Newslett 7:6–8

Åkerberg E (1966) Tenerife – a place for research on plant ecology. Acta Univ Lund Sect II, 33:1–16

Al Didi MA (1965) Development of new Egyptian cotton strains by seed irradiation. Use Induced Mutations Plant Breed, Radiat Bot Suppl 5:579–583

Alexander DE, Lambert RJ, Dudley JW (1969) Breeding problems and potentials of modified protein maize. New Approaches to Breeding for Improved Plant Protein, pp 55–65, IAEA, Vienna

References

Alexander LJ, Oakes GL, Jaberg CA (1971) The production of two needed mutations in tomato by irradiation. J Hered 62:311–315

Ali MAM, Okiror SO, Rasmusson DC (1978) Performance of semidwarf barley. Crop Sci 18:418–422

Allen SE, Terman GL (1978) Yield and protein content in rice as affected by rate, source, method and time of applied N. Agron J 70:238–242

Al-Rubeai MAF, Godward MBE (1981) Genetic control of radiosensitivity in *Phaseolus vulgaris* L. Environ Exp Bot 21:211–216

Andersen AJ (1977) Grain yield and protein content of high-lysine and normal barley in relation to rate and time of nitrogen application. In: Miflin BJ, Zoschke M (eds) Carbohydrate and Protein Synthesis. Publ Comm Eur EUR 6043:107–119

Andersen AJ, Køie B (1975) N fertilization and yield response of high lysine and normal barley. Agron J 67:695–698

Andrew RH, Brink RA, Neal NP (1944) Some effects of the *waxy* and *sugary* genes on endosperm in maize. J Agr Res 69:355–371

Andrews DJ, Webster OJ (1971) A new factor for genetic male sterility in *Sorghum bicolor* (L.) Moench. Crop Sci 11:308–309

Ankineedu G, Sharma KD, Kulkarni LG (1968) Effects of fast neutrons and gamma rays on castor. Indian J Genet Plant Breed 28:31–39

Annapurna S, Reddy GM (1971) Modified *opaque* maize for possible applied use. Curr Sci 40:581–582

Appa Rao S, Rao SP, Jana MK (1975) Induction of nondormant mutants in black gram. J Hered 66:388–389

Arias J, Frey KJ (1973) Grain yield mutations induced by ethyl methanesulfonate treatment of oat seeds. Radiat Bot 13:73–85

Arnholdt B, Schuster W (1981) Durch Umwelt und Genotyp bedingte Variabilität des Rohprotein- und Rohfettgehaltes in Rapssamen. Fette Seifen Anstrichm 83:49–54

Arnold JM, Baumann LF, Makonnen D (1977) Physical and chemical kernel characteristics of normal and *opaque 2* endosperm maize hybrids. Crop Sci 17:362–368

Athwal DS, Bhalla SK, Sandhu SS, Brar HS (1970) A fertile dwarf and three other mutants in *Cicer*. Indian J Genet Plant Breed 30:261–266

Auerbach C, Kilbey BJ (1971) Mutation in Eukaryotes. Annu Rev Genet 5:163–218

Autran JC, Branlard GP (1979) Characterization of some wheat genotypic modifications of gliadin electrophoresis. Seed Protein Improvement in Cereals and Grain Legumes II:428, IAEA, Vienna

Axtell JD, Lafayette W (1976) Naturally occurring and induced genotypes of high lysine sorghum. Evaluation of Seed Protein Alterations by Mutation Breeding, pp 45–53, IAEA, Vienna

Axtell JD, Oswalt DL, Mertz ET, Pickett RC, Jamhunathan R, Srinivasan G (1973) Components of nutritional quality in grain sorghum. High-Quality Protein Maize. Dowden, Hutchinson and Ross, Stroudsburg, pp 374–386

Axtell JD, Mohan DP, Cummings DP (1974) Genetic improvement of biological efficiency and protein quality in sorghum. Proc 29th Ann Corn Sorghum Res Conf, pp 29–39, Paper 5826, Purdue Univ Agric Exp Stn Dep Agron

Axtell JD, Scoyoc SW van, Christensen PJ, Ejeta G (1979) Current status of protein quality improvement in grain sorghum. Seed Protein Improvement in Cereals and Grain Legumes II, pp 357–365, IAEA, Vienna

Axtell JD, Kirleis AW, Hassen MN, D'Croz Mason N, Mertz ET, Munck L (1981) Digestibility of sorghum proteins. Proc Natl Acad Sci USA 78:1333–1335

Badr M, Etman M (1977) Gamma-radiation induced effects on the X_1-generation in carnation (*Dianthus caryophyllus*, L.). Egypt J Genet Cytol 6:32–43

Baenziger PS, Glover DV (1977) Protein body size and distribution and protein matrix morphology in various endosperm mutants of *Zea mays* L. Crop Sci 17:415–420

Baenziger PS, Glover DV (1979) Dry matter accumulation in maize hybrids near isogenic for endosperm mutants conditioning protein quality. Crop Sci 19:345–349

Bagnara D (1971) Mutagenesis and genetic improvement in wheat. Genet Agraria 25:243–378

Bagnara D, Porreca G (1977) Exploitation of induced mutations in *durum* wheat: The breeding of the cultivar "Augusto". Mutat Breed Newslett 10:2–4

Bagnara D, Rossi L, Porreca G (1971) Use of radiation induced mutants in hybridization programs: methods and results. Genet Agraria 25:177–230

Bahl PN, Singh SP, Ram H, Ragu DB, Jain HK (1979) Breeding for improved plant architecture and high protein yields. Seed Protein Improvement in Cereals and Grain Legumes I, pp 297–306, IAEA, Vienna

Balaravi SP, Bansal HC, Eggum BO, Bhasharan S (1976) Characterisation of induced high protein and high lysine mutants in barley. J Sci Food Agric 27:545–552

Bálint A, Menyhert Z, Sutka J, Kovacs M, Kurnik E (1970) Increasing the protein content of maize by means of induced mutants. Improving Plant Protein by Nuclear Techniques, pp 77–84, IAEA, Vienna

Bálint A, Bedö Z, Kiss E (1977) Examination of *WF9* mutant sublines for lodging and *Fusarium* resistance. Induced Mutations Against Plant Diseases, pp 489–494, IAEA, Vienna

Bandel G, Gottschalk W (1978) Recombinants from crosses between fasciated and non-fasciated pea mutants. II. Late flowering recombinants. Z Pflanzenzücht 81:60–76

Banerjee SK, Swaminathan MS (1966) X-ray induced variability for protein content in bread wheat. Indian J Genet Plant Breed 26:203–209

Bansal HC (1969) Induced chilli mutant containing high capsaicine and low carotenoids in fruit. Radiations and Radiomimetic Substances in Mutation Breeding, Bombay, pp 285–292

Bansal HC (1970) A new mutant induced in barley. Curr Sci 39:494

Bansal HC (1971) Induced polygenic variability and genetic advance for maturity in barley. Int Symp Use Isotopes Radiat Agric Anim Husbandry Res, New Delhi, pp 146–153

Bansal HC (1972) Induction of early dwarf mutants in barley. Indian J Genet Plant Breed 32:203–206

Bansal HC, Sharma RP, Narula PN (1972) Induced mutations for amber grain color in two varieties of wheat. Wheat Inform Serv 33, 34:13

Barabás Z (1965) Induced quantitative somatic mutants in *Sorghum*. Use Induced Mutations Plant Breed. Radiat Bot Suppl 5:515–520

Barabás Z (1977) Prevention of gene erosion of old wheat varieties by backcrossing and X-ray irradiation. Induced Mutations against Plant Diseases, pp 29–32, IAEA, Vienna

Baradjanegara AA (1980) Utilization of fast neutrons and gamma rays for soybean improvement. Induced Mutations for Improvement of Grain Legume Production, pp 41–43, IAEA, Vienna

Barbosa HM (1971) Genes and gene combinations associated with protein, lysine and carbohydrate content in the endosperm of maize *(Zea mays)*. Ph D Thesis, Purdue Univ

Basu AK, Basu RK (1970) The induction of grain size and colour mutations in rice (*Oryza sativa* L.) by radioisotopes. Theor Appl Genet 40:232–236

Basu RK (1966) The induction of early flowering mutants in *Corchorus olitorius* L. Radiat Bot 6:39–47

Bauer R (1974) Westra, an X-ray-induced erect-growing black-currant variety, and its use in breeding. Polyploidy and Induced Mutations in Plant Breeding, pp 13–20, IAEA, Vienna

Bazavluk IM, Enken VB (1976) Minor mutations as a method of increasing the content and improving the quality of protein in soya beans. Genetika USSR 12:46–54

Beauchamp EG, Kannenberg LW, Hunter RB (1976) Nitrogen accumulation and translocation in corn genotypes following silking. Agron J 68:418–422

Behera NC, Patnaik SN (1979) Viable mutations in *Amaranthus*. Indian J Genet Plant Breed 39:163–170

Bender MA (1970) Neutron-induced genetic effects: A review. Radiat Bot 10:225–247

Berg T, Frogner S, Åastveit K (1976) Recombination of induced mutant alleles for grain yield in barley. Barley Genet 3:203–214

Berhe T (1973) Prospects for improving *Eragrostis tef* by mutation breeding. Nuclear Techniques for Seed Protein Improvement, pp 297–303, IAEA, Vienna

Bernaux P, Marie R (1977) Mutants induits chez le riz (*Oryza sativa* L.) pour la reponse a *Sclerotium oryzae* Catt et *Sclerotium hydrophilum* Sacc. Induced Mutations against Plant Diseases, pp 157–170, IAEA, Vienna

Bewley JD, Black M (1979) Physiology and Biochemistry of Seeds. Springer, Berlin Heidelberg New York, 306 pp

Bhagwat SG, Bhatia CR, Gopalakrishna T, Joshua DC, Mitra RK, Narahari P, Pawar SE, Thakare RG (1979) Increasing protein production in cereals and grain legumes. Seed Protein Improvement in Cereals and Grain Legumes II, pp 225–236, IAEA, Vienna

Bhan AK, Kaul MLH (1974) Gamma-ray induced "long grain" mutant in rice variety Jhona 349. Seed Res 2:56–58

Bhaskara Rao EVV, Reddi VR (1975) A radiation-induced highly productive mutant in *Sorghum*. Radiat Bot 15:29–32

Bhatia CR, Rabson R (1976) Bioenergetic considerations in cereal breeding for protein improvement. Science 194:1418

Bhatia CR, Desai RM, Suseelan KN (1978) Attempts to combine high lysine and increased grain protein in wheat. Seed Protein Improvement by Nucelar Techniques, pp 51–57, IAEA, Vienna

Bhatia CR, Jagannath DR, Gopal-Ayengar AR (1970) Induced micromutations for major protein fractions in wheat. Improving Plant Protein by Nuclear Techniques, pp 99–105, IAEA, Vienna

Bilquez AF, Magne C, Martin JP (1965) Bilan de six années de recherches sur l'emploi des rayonnements ionisants pour l'amélioration des plantes au Sénégal. Use Induced Mutations Plant Breed, Radiat Bot Suppl 5:585–601

Bishr MA, Abdel-Bary AA, Zaitoon MA (1973) Effects of gamma rays on some characters in Egyptian cotton. Egypt J Genet Cytol 2:365–369

Bjarnason M, Pollmer WG (1972) The maize germ: Its role as a contributing factor to protein quantity and quality. Z Pflanzenzücht 68:83–89

Bjarnason M, Pollmer WG, Klein D (1976) Inheritance of modified endosperm structure and lysine content in *opaque-2* maize. I. Modified endosperm structure. Cereal Res Commun 4:401–410

Bjarnason M, Pollmer WG, Klein D (1977) Inheritance of modified structure and lysine content in *opaque-2* maize. II. Lysine content. Cereal Res Commun 5:49–58

Black RC, Loersch JD, McArdle FJ, Creech RG (1966) Genetic interactions affecting maize phytoglycogen and the phytoglycogen forming branching enzyme. Genetics 53:661–668

Blixt S (1972) Mutation genetics in *Pisum*. Agri Hort Genet 30:1–293

Blixt S (1976a) Mutation induction in *Phleum*. Agri Hort Genet 34:59–82

Blixt S (1976b) A crossing programme with mutants in peas. Induced Mutations in Cross-Breeding, pp 21–36, IAEA, Vienna

Blixt S (1979) Natural and induced variability for seed protein in temperate legumes. Seed Protein Improvement in Cereals and Grain Legumes II, pp 3–21, IAEA, Vienna

Blixt S, Gottschalk W (1975) Mutation in the *Leguminosae*. Agri Hort Genet 23:33–85

Bogyo TP, Scarascia-Mugnozza GT, Sigurbjörnsson B, Bagnara D (1969) Adaptation studies with radiation-induced *durum* wheat mutants. Induced Mutations in Plants, pp 699–717, IAEA, Vienna

Bollich CN, Webb BD (1968) The inheritance of protein content in a rice cross. Proc 12th Rice Tech Work Group, USDA, Washington, DC, 30 pp

Borojević K (1965) The effect of irradiation and selection after irradiation on the number of kernels in wheat. Use Induced Mutations Plant Breed. Radiat Bot Suppl 5:505–513

Borojević K (1966) Studies on radiation-induced mutations in quantitative characters of wheat *(Triticum vulgare)*. Mutations in Plant Breeding, pp 15–38, IAEA, Vienna

Borojević K (1967) Study of quantitative characters of bearded mutation in wheat induced by irradiation. Erwin Baur Ged Vorl 4:199–204

Borojević K (1975) Evaluating resistance to *Puccinia recondita tritici* in mutant lines selected in wheat after mutagenic treatments. Radiat Bot 15:367–374

Borojević K (1976) Type of infection, severity and tolerance to *Puccinia recondita tritici* in old mutant lines of wheat. Induced Mutations for Disease Resistance in Crop Plants (1975), pp 41–61, IAEA, Vienna

Borojević K (1977) Studies on resistance to *Puccinia recondita tritici* in wheat population after mutagenic treatments. Induced Mutations against Plant Diseases, pp 393–401, IAEA, Vienna

Borojević K (1978) Induced variability for leaf rust resistance in *Triticum aestivum*. Proc 5th Int Wheat Genet Symp, New Delhi, pp 559–564

Borojević K (1979) Modification of resistance to *Puccinia recondita tritici* in wheat population by mutagenic treatments. Inst Rad Breed Ohmiya, Japan. Gamma Field Symp 18:25–32

Borojević K, Borojević S (1969) Stabilization of induced genetic variability in irradiated populations of *vulgare* wheat. Induced Mutations in Plants, pp 399–432, IAEA, Vienna

Borojević K, Borojević S (1972) Mutation breeding in wheat. Induced Mutations and Plant Improvement, pp 237–251, IAEA, Vienna

Bose S (1968) Short-straw mutants in Aman paddy. Sci Cult 34:32–35

Boulter D (1979) Structure and biosynthesis of legume storage proteins. Seed Protein Improvement in Cereals and Grain Legumes I, pp 125–134, IAEA, Vienna

Bouma J (1966) Praktické využití mutačního šlechtění u jarního ječmene – nová odruda diamant. Vydalo Sdruženi šlechtitelských a semenařských podniku v Praze, Praha, pp 133–139

Bouma J (1967) New variety of spring barley "Diamant" in Czechoslovakia. Erwin Baur Ged Vorl 4:177–182

Bouma J (1977) Highly efficient spring barley varieties originating from the mutant cultivar "Diamant". Mutat Breed Newslett 10:6

Bovkis EN (1978) Some problems of using irradiated pollen in genetics and selection of winter soft wheat *(Triticum aestivum)*. Genetika USSR 14:1237–1426

Bowen HJM (1965) Mutations in horticultural chrysanthemums. Use Induced Mutations Plant Breed. Radiat Bot Suppl 5:695–700

Boyer CD, Preiss J (1978) Multiple forms of starch branching enzyme of maize. Evidence for independent genetic control. Biochem Biophys Res Commun 80:169–175

Boyer CD, Shannon JC, Garwood DL, Creech RG (1976) Changes in starch granule size and amylose percentage during kernel development in several *Zea mays* L. genotypes. Cereal Chem 53:327–337

Bozzini A (1965) Sphaerococcoid, a radiation-induced mutation in *Triticum durum* DESF. Use Induced Mutations Plant Breed. Radiat Bot Suppl 5:375–383

Bozzini A (1971) First results of bunt resistance analysis in mutants of *durum* wheat. Mutation Breeding for Disease Resistance, pp 131–138, IAEA, Vienna

Bozzini A (1974a) Breeding possibilities offered by induced mutations in *durum* wheat. Theor Appl Genet 44:304–310

Bozzini A (1974b) Radiation-induced male sterility in *durum* wheat. Polyploidy and Induced Mutations in Plant Breeding, pp 23–25, IAEA, Vienna

Bozzini A, Fossati A, Scarascia Mugnozza GT (1967) Recurrent X-irradiation in *durum* wheat: Induction of variability in some morphological and physiological characters. Genet Agrar 21:353–362

Brandt AB (1975) In vivo incorporation of ^{14}C lys into the endosperm proteins of wild type and high-lysine barley. FEBS Lett 52:288–291

Brandt AB (1976) Endosperm protein formation during kernel development of wild type and a high-lysine barley mutant. Cereal Chem 53:890–901

Brandt A, Ingversen J (1976) In vitro synthesis of barley endosperm proteins on wild type and mutant templates. Carlsberg Res Commun 41:311–320

Bressoni R (1972) Legumes in human diets and how they might be improved. Nutritional Improvement of Food Legumes by Breeding, pp 15–42, UN

Bressoni R, Elias LG (1979) The world protein and nutritional situation. Seed Protein Improvement in Cereals and Grain Legumes I, pp 3–22, IAEA, Vienna

Brim CA, Young MF (1971) Inheritance of a male-sterile character in soybeans. Crop Sci 11:564–566

Brock RD (1965) Induced mutations affecting quantitative characters. Use Induced Mutations Plant Breed. Radiat Bot Suppl 5:251–264

Brock RD (1970) Mutations in quantitatively inherited traits induced by neutron irradiation. Radiat Bot 10:209–223

Brock RD (1971) The role of induced mutations in plant improvement. Radiat Bot 11:181–186

Brock RD, Andrew WD, Kirchner R, Crawford EJ (1971) Early-flowering mutants of *Medicago polymorpha* var. *polymorpha*. Austr J Agric Res 22:215–222

Broertjes C (1966) Mutation breeding of chrysanthemums. Euphytica 15:156–162
Broertjes C (1969a) Mutation breeding of *Streptocarpus*. Euphytica 18:333–339
Broertjes C (1969b) Induced mutations and breeding methods in vegetatively propagated species. Induced Mutations in Plants, pp 325–329, IAEA, Vienna
Broertjes C (1969c) Mutation breeding of vegetatively propagated crops. Genet Agraria 23:139–165
Broertjes C (1972a) Mutation breeding of *Achimenes*. Euphytica 21:48–62
Broertjes C (1972b) Use in plant breeding of acute, chronic or fractionated doses of X-rays or fast neutrons as illustrated with leaves of *Saintpaulia*. Cent Agric Publ Doc, Wageningen, 74 pp
Broertjes C (1972c) Improvement of vegetatively propagated plants by ionizing radiation. Induced Mutations and Plant Improvement, pp 293–299, IAEA, Vienna
Broertjes C (1976a) Mutation breeding of autotetraploid *Achimenes* cultivars. Euphytica 25: 297–304
Broertjes C (1976b) Mutation breeding in autotetraploid *Achimenes* cultivars. Improvement of Vegetatively Propagated Plants and Tree Crops through Induced Mutations, pp 1–12, IAEA, Vienna
Broertjes C, Ballego JM (1967) Mutation breeding of *Dahlia variabilis*. Euphytica 16:171–176
Broertjes C, Harten AM van (1978) Application of Mutation Breeding Methods in the Improvement of Vegetatively Propagated Crops. Elsevier, Amsterdam, 316 pp
Broertjes C, Leffring L (1972) Mutation breeding of *Kalanchoë*. Euphytica 21:415–423
Broertjes C, Verboom H (1974) Mutation breeding of *Alstroemeria*. Euphytica 23:39–44
Broertjes C, Haccius B, Weidlich S (1968) Adventitious bud formation on isolated leaves and its significance for mutation breeding. Euphytica 17:321–344
Broertjes C, Roest S, Bokelmann GS (1976) Mutation breeding of *Chrysanthemum morifolium* Ram. using in vivo and in vitro adventitious bud techniques. Euphytica 25:11–19
Broertjes C, Koene P, Veen JWH van (1980) A mutant of a mutant of a mutant of a ...: Irradiation of progressive radiation-induced mutants in a mutation-breeding programme with *Chrysanthemum morifolium* Ram. Euphytica 29:525–530
Brönnimann A, Fossati A (1974) Tolerance à *Septoria nodorum* BERK. chez le blé: Methodes d'infection et selection par mutagenese. Induced Mutations for Disease Resistance in Crop Plants, pp 117–123, IAEA, Vienna
Brown RP, Creech RG, Johnson IJ (1971) Genetic control of starch granule morphology and physical structure in developing maize endosperms. Crop Sci 11:297–301
Brunori A, Mannino P, Ancora G, Bozzini A (1977) Protein accumulation, RNA and soluble amino nitrogen content in developing endosperm of two varieties of *Triticum aestivum* with low and high protein seed. Theor Appl Genet 50:73–77
Bugge G (1970) Versuche zur züchterischen Nutzbarmachung einer strahleninduzierten kurzröhrigen Mutante von *Trifolium pratense* (L.). Z Pflanzenzücht 63:196–208
Buiatti M, Ragazzini R, D'Amato F (1965) Somatic mutations in the carnation induced by gamma radiation. Use Induced Mutations Plant Breed. Radiat Bot Suppl 5:719–723
Buiatti M, Tesi R (1969) *Gladiolus* improvement through radiation induced somatic mutation. Genet Agrar 23:180–185
Bulgarian Academy of Sciences (1978) Experimental Mutagenesis in Plants. Sofia, 466 pp
Burr B, Nelson OE (1973) The phosphorylase of developing maize seeds. Ann NY Acad Sci 210: 129–138
Burr FA (1979) Zein synthesis and processing on zein protein body membranes. Seed Protein Improvement in Cereals and Grain Legumes I, pp 159–162, IAEA, Vienna
Burton GW (1974) Radiation breeding of warm season forage and turf grasses. Polyploidy and Induced Mutations in Plant Breeding, pp 35–39, IAEA, Vienna
Burton GW (1976) Using gamma irradiation to improve sterile turf and forage bermuda grasses. Improvement of Vegetatively Propagated Plants and Tree Crops through Induced Mutations, pp 25–32, IAEA, Vienna
Burton GW (1981) Tifway-2 bermudagrass. Mut Breed Newslett 18:8–10
Burton GW, Constantin MJ, Dobson JW, Hanna WW, Powell JB (1980) An induced mutant of Coastcross 1 Bermudagrass with improved winterhardiness. Environ Exp Bot 20:115–117
Butler L (1977) Viability estimates for sixty tomato mutants. Can J Genet Cytol 19:31–38

Byth DE (1975) Grain legumes in Australia. Induced Mutations for the Improvement of Grain Legumes in South East Asia, pp 111–122, IAEA, Vienna

Cameron JW (1947) A study of the genetic control of carbohydrates in maize endosperm. Genetics 32:80–81

Cameron JW, Teas HJ (1954) Carbohydrate relationships in developing and mature endosperms of *brittle* and related maize genotypes. Am J Bot 41:50–55

Campbell AI, Wilson D (1977) Prospects for the development of disease-resistant temperate fruit plants by mutation induction. Induced Mutations against Plant Diseases, pp 215–226, IAEA, Vienna

Casey R (1979) Genetic variability in the structure of the α-subunits of legumin from *Pisum*. – A twodimensional gel electrophoresis study. Heredity 43:265–272

Černý J (1970) Contribution to the problems of mutation breeding. Genet Slechtení 6:181–191

Chakrabarti SN, Sen S (1975) Improvement in rice for nitrogen response through induction of mutation. Indian J Genet Plant Breed 35:454–458

Chaleff RS (1980) Further characterization of picrolam-tolerant mutants of *Nicotiana tabacum*. Theor Appl Genet 58:91–95

Chandola RP, Bhathagar CP, Sudha Sah (1971) Nature of variability induced by radiations. Int Symp Use Isotopes Radiat Agric Anim Husbandry Res, New Delhi, pp 46–50

Chekalin NM (1977) Types of induced macromutations in *Lathyrus sativus* L. Genetika USSR 13:2116–2122

Chen R (1970) Elektrophorese der Samenproteine von neutroneninduzierten *Pisum*-Mutanten in Polyacrylamid-Gel. Z Naturforsch 25:1461–1464

Chen R, Gottschalk W (1970) Neutroneninduzierte Mutanten von *Pisum sativum* und ihr Vergleich mit röntgeninduzierten Genotypen. Angew Bot 44:325–342

Chery J (1979) Influence de la fertilization azotée tardive sur le rendement et la qualité du grain de differentes varietés d'orge. Seed Protein Improvement in Cereals and Grain Legumes I, pp 283–295, IAEA, Vienna

Chevalier P, Schrader LE (1977) Genotypic differences in nitrate absorption and partitioning of N among plant parts in maize. Crop Sci 17:897–901

Chibber BAK, Mertz ET, Axtell JD (1978) Effects of dehulling on tannin content, protein distribution and quality of high and low tannin sorghum. J Agric Food Chem 26:679–683

Chibber BAK, Voicu E, Mertz ET (1977) Studies on corn proteins. XI. Distribution of lysine during germination of normal and *opaque-2* maize. Cereal Chem 54:558–564

Chourey PS (1981) Genetic control of sucrose synthetase in maize endosperm. Mol Gen Genet 184:372–376

Chourey PS, Nelson OE (1979) Interallelic complementation at the *sh* locus in maize at the enzyme level. Genetics 91:317–325

Chu YE, Wu SC, Tsai KS, Li YS (1973) Mutation breeding for high-protein rice varieties. Nuclear Techniques for Seed Protein Improvement, pp 145–147, IAEA, Vienna

Chutima K (1979) Results of multi-location tests over several years for yield and seed protein content of indigenous thai rice varieties. Seed Protein Improvement in Cereals and Grain Legumes II, pp 279–290, IAEA, Vienna

Coffman WR, Juliano BO (1979) Seed protein improvement in rice. Seed Protein Improvement in Cereals and Grain Legumes II, pp 261–276, IAEA, Vienna

Conger BV, Carabia JV (1977) Mutagenic effectiveness and efficiency of sodium azide versus ethyl methanesulfonate in maize. Mutat Res 46:285–296

Cornu A, Cassini R, Berville A, Vuillaume E (1977) Recherche par mutagenese d'une resistance a *Helminthosporium maydis*, race T, chez les mais a cytoplasme male-sterile Texas. Induced Mutations against Plant Diseases, pp 479–488, IAEA, Vienna

Coutinho MP (1977) Utilisation des rayonnements pour l'amélioration de la vigne au point de vue de la resistance au mildiou. Induced Mutations against Plant Diseases, pp 233–240, IAEA, Vienna

Creech R (1965) Genetic control of carbohydrate synthesis in maize endosperm. Genetics 52:1175–1186

Creech R (1968) Carbohydrate synthesis in maize. Adv Agron 20:275–322

Creech R, McArdle F (1966) Gene interaction for quantitative changes in carbohydrates in maize kernels. Crop Sci 6:192–194

Crocomo OJ, Neto AT, Ando A, Blixt S, Boulter D (1978) Breeding for improved protein content and quality in the bean *(Phaseolus vulgaris)*. II. Seed Protein Improvement by Nuclear Techniques, pp 207–222, IAEA, Vienna

Crocomo OJ, Lee TSG, Derbyshire E, Boulter D (1979) Biochemical investigations on the seed proteins of a Brazilian variety and mutant of *Phaseolus vulgaris*. Seed Protein Improvement in Cereals and Grain Legumes I, pp 217–227, IAEA, Vienna

Culpepper CW, Magoon CA (1924) Studies upon the relative merits of sweet corn varieties for canning purposes and the relation of maturity of corn to the quality of the canned product. J Agric Res 28:403–443

Cummings DP, Stuthman DD, Green CE (1978) Morphological mutations induced with ethyl methanesulfonate in oats. J Hered 69:3–7

Dahiya BS (1973) Improvement of mung bean through induced mutations. Indian J Genet Plant Breed 33:460–468

Dalby A, Davies IAI (1967) Ribonuclease activity in the developing seeds of normal and *opaque-2* maize. Science 155:1573–1575

Danielsson CE (1949a) Investigations of vicilin and legumin. Acta Chem Scand 3:41–49

Danielsson CE (1949b) Seed globulins of the *Gramineae* and *Leguminosae*. Biochem J 44:387–400

Danielsson CE (1952) Differences in the chemical composition of some pea proteins. Acta Chem Scand 6:139–148

Darbishire AD (1908) On the result of crossing round with wrinkled peas, with especial reference to their starch-grains. Proc Roy Soc London Ser B 80:122–135

Das PK (1969) Chronic gamma irradiation on some horticultural crops with special reference to the isolation of somatic mutants. Radiations and Radiomimetic Substances in Mutation Breeding, Bombay, pp 215–221

Dasananda S, Khambanonda P (1970) Induction of mutations in Thai rice varieties and subsequent selection and testing of beneficial mutant lines. Rice Breeding with Induced Mutations II, pp 105–110, IAEA, Vienna

Dasgupta P (1979) Induced mutation in rice. Genet Agrar 33:139–156

Daskaloff S (1968) A male sterile pepper (*C. annuum* L.) mutant. Theor Appl Genet 38:370–372

Daskaloff S (1973) Investigation of induced mutants in *Capsicum annuum* L. III. Genet Plant Breed 6:419–429

Daskaloff S (1976) Seed setting of male sterile mutants in connection with heterosis breeding in pepper *Capsicum annuum* L. Genet Agrar 30:407–417

Daskaloff S (1978) Results of experimental mutagenesis in pepper. Experimental Mutagenesis in Plants, Sofia, pp 383–390

Davis LW, Williams WP, Crook L (1970) Interrelationships of the protein and amino acid content of inbred lines of corn. J Agr Food Chem 18:357–360

Decau J, Pollacsek M (1970) Influence de la fumure azotée et de l'irrigation sur le rendement et la production de proteines de mais-grain, portant ou non le gene *opaque-2*. Improving Plant Protein by Nuclear Techniques, pp 357–366, IAEA, Vienna

Denić M (1970) Role of the microsomal fraction in the regulation of protein synthesis in maize endosperm. Improving Plant Protein by Nuclear Techniques, pp 381–388, IAEA, Vienna

Denić M (1978) Some characteristics of proteins in mutant lines of hexaploid wheat. Seed Protein Improvement by Nuclear Techniques, pp 365–381, IAEA, Vienna

Denić M, Konstantinov K, Dumanović I (1979) Molecular basis of gene action in storage protein synthesis. Seed Protein Improvement in Cereals and Grain Legumes II:426, IAEA, Vienna

Derbyshire E, Wright DJ, Boulter D (1976) Legumin and vicilin, storage proteins of legume seeds. Phytochemistry 15:3–24

Detering R (1978) Der Stärke- und Amylosegehalt von Mutanten und Rekombinanten der Species *Pisum sativum*. Diplomarbeit, Univ Bonn

Devreux M, Donini B, Scarascia Mugnozza GT (1972) Genetic effects of gametophyte irradiation in barley II. Frequency and types of mutations induced. Radiat Bot 12:87–98

Dhaliwal HS (1977) Genetic variability and improvement of seed proteins in wheat. Theor Appl Genet 51:71–79

Dhonukshe BL, Bhowal JG (1976) Radiation induced polygenic variability in dwarf wheats. Indian J Genet Plant Breed 36:1–5

Dierks-Ventling C (1982) Storage protein chracteristics of proline-requiring mutants of *Zea mays* (L.). Theor Appl Genet 61:145–149

DiFonzo N, Stanca AM (1977) EMS derived barley mutants with increased lysine content. Genet Agrar 31:401–409

DiFonzo N, Fornasori E, Salamini F, Reggiani R, Soave C (1980) Interaction of maize mutants *floury-2* and *opaque-7* with *opaque-2* in the synthesis of endosperm proteins. J Heredity 71:397–402

Djelepov K (1973) Mutation in the length of internodes of wheat. Wheat Inform Serv 36:3–5

Djelepov K (1978) Induced short-stemmed winter common wheat mutants. Experimental Mutagenesis in Plants, Sofia, pp 177–185

Dnyansagar VR, Jahagirda HA (1979) Induced mutants of *Foeniculum vulgare* in relation to seed and essential oil content. Rec Res Plant Sci, New Delhi, pp 286–291

Dnyansagar VR, Pingle AR (1977) Correlation between macromutants and solasodine content in *Solanum viarum*. Planta Medica 31:21–25

Dolgih ST (1969) Obtaining mutations of vegetables with male sterility. Genetika USSR 5(11):45–55

Doll H (1966) Yield and variability of chlorophyll-mutant heterozygotes in barley. Hereditas 56:255–276

Doll H (1972) Variation in protein quantity and quality induced in barley by EMS treatment. Induced Mutations and Plant Improvement, pp 331–341, IAEA, Vienna

Doll H (1973) Inheritance of the high-lysine character of a barley mutant. Hereditas 74:293–294

Doll H (1975) Genetic studies of high lysine barley mutants. Barley Genetics III. Proc 3rd Int Barley Genetics Symp, Garching, pp 542–546

Doll H (1977) Storage proteins in cereals. In: Muhammed A, Aksel R, Borstel RC (eds) Genetic Diversity in Plants. Plenum Press, New York London, pp 337–347

Doll H (1980) A nearly non-functional mutant allele of the storage protein locus *Hor2* in barley. Hereditas 93:217–222

Doll H, Brown AHD (1979) Hordein variation in wild *(Hordeum spontaneum)* and cultivated *(H. vulgare)* barley. Can J Genet Cytol 21:391–404

Doll H, Køie B (1975) Evaluation of high lysine barley mutants. Breeding for Seed Protein Improvement Using Nuclear Techniques, pp 55–59, IAEA, Vienna

Doll H, Køie B (1978) Influence of the high-lysine gene from barley mutant *1508* on grain, carbohydrate and protein yield. Seed Protein Improvement by Nuclear Technqiues, pp 107–113, IAEA, Vienna

Doll H, Køie B, Eggum BO (1974) Induced high lysine mutants in barley. Radiat Bot 14:73–80

Dommergues P, Heslot H, Gillot J, Martin C (1967) L'induction de mutations chez les rosiers. Erwin Baur Ged Vorl 4:319–348

Donini B (1976) The use of radiations to induce useful mutations in fruit trees. Improvement of Vegetatively Propagated Plants and Tree Crops through Induced Mutations, pp 55–67, IAEA, Vienna

Donini B, Devreux M (1970) Mutazioni indotte per irraggiamento di gametofiti in orzo. Genet Agrar 24:208–209

Donini B, Roselli G (1972) Mutations induced by irradiation of olive rooted cuttings. Genet Agrar 26:149–160

Donovan GR, Lee JW, Hill RD (1977) Compositional changes in the developing grain of high- and low-protein wheats. II. Starch and protein synthetic capacity. Cereal Chem 54:646–656

Doorenbos J, Karper JJ (1975) X-ray induced mutations in *Begonia x hiemalis*. Euphytica 24:13–19

Dormling I, Gustafsson Å (1969) Phytotron cultivation of early barley mutants. Theor Appl Genet 39:51–61

Dormling I, Gustafsson Å, Jung HR, Wettstein D v (1966) Phytotron cultivation of Svalöfs' Bonus barley and its mutant Svalöfs' Mari. Hereditas 56:221–237

Driscoll CJ, Barlow KK (1976) Male sterility in plants. Induction, isolation and utilization. Induced Mutations in Cross-Breeding, pp 123–131, IAEA, Vienna

Dümke B (1973) Untersuchungen über die Eiweißzusammensetzung einiger Mutanten und Rekombinanten von *Pisum sativum*. Ph D Thesis, Univ Bonn

Dumanović J, Denić M, Ehrenberg L, Bergstrand KG (1968) Radiation-induced heritable variation of quantitative characters in wheat. Hereditas 62:221–238

Dumanović J, Ehrenberg L, Denić M (1970) Induced variation of protein content and composition in hexaploid wheat. Improving Plant Protein by Nuclear Techniques, pp 107–119, IAEA, Vienna

Dumanović J, Denić M, Jovanović C, Ehrenberg L (1973) Variation in content and composition of protein in wheat induced by mutation. Nuclear Techniques for Seed Protein Improvement, pp 153–161, IAEA, Vienna

Duvick DN (1961) Protein granules of maize endosperm cells. Cereal Chem 38:374–384

Edwards LH, Williams ND, Gough FJ, Lebsock KL (1969) A chemically induced mutation for stem rust resistance in "Little Club" wheat. Crop Sci 9:838–839

Egamberdiev A, Payziev P (1977) Induction of mutations in wild cotton species. Genetika USSR 13:1736 1741

Eggum BO (1970) Über die Abhängigkeit der Proteinqualität vom Stickstoffgehalt der Gerste. Z Tierphysiol Tierernähr Futtermittelk 26:265

Eggum BO (1973) Biological availability of amino acid constituents in grain protein. Nuclear Techniques for Seed Protein Improvement, pp 391–408, IAEA, Vienna

Eggum BO (1977a) The nutritive quality of cereals. Cereals Res Commun 5:153–157

Eggum BO (1977b) Nutritional aspects of cereal proteins. Genetic Diversity in Plants, Plenum Press, New York London, pp 349–369

Eggum BO (1978) Protein quality of induced high lysine mutants in barley. Nutritional Improvement of Food and Feed Proteins. Plenum Press, New York London, pp 317–341

Einfeld E, Abdel-Hafez AG, Fuchs WH, Heitefuss R, Röbbelen G (1976) Investigations on resistance of barley against mildew *(Erysiphe graminis)*. Induced Mutations for Disease Resistance in Crop Plants, pp 81–90, IAEA, Vienna

Elmore CD, Spurgeon WI, Thom WO (1978) Nitrogen fertilization increases N and alters amino acid concentration of cotton seed. Agron J 71:713–716

El-Sayed SA (1977) Phytoalexin-generating capacity in relation to late blight resistance in certain tomato mutants induced by gamma irradiation of seeds. Induced Mutations against Plant Diseases, pp 265–274, IAEA, Vienna

Elshuni KA, Khvostova VV (1966) Mutations obtained in spring wheat lutescens 62 treated with fast neutrons and gamma-rays followed by partial removal of damaging irradiation effect. Genetika USSR 1966(6):37–46

Elwinger K (1978) Effect of nitrogen fertilization on the nutritional value of barley. Swed J Agric Res 8:107–112

Enchev Y (1976) Induced mutations in winter brewing barley and their use. Barley Genet 3:190–196

Enken VB (1967) Manifestation of Vavilov's law of homologous series in hereditary variability in experimental mutagenesis. Erwin Baur Ged Vorl 4:123–129

Eriksson G (1969) The waxy character. Hereditas 63:180–204

Erlander SR (1960) The production of amylose and amylopectin in corn endosperms and in potato tubers. Cereal Chem 37:81–93

Erwin Baur Gedächtnisvorlesungen IV (1966) Induzierte Mutationen und ihre Nutzung. Akademie Verlag, Berlin, 468 S

Ewertson G (1977) Protein content and grain quality relations in barley. Agri Hort Genet 25:1–104

Favret EA (1965) Induced mutations in breeding for disease resistance. Use Induced Mutations Plant Breed. Radiat Bot Suppl 5:521–536

Favret EA (1971) Different categories of mutations for disease reaction in the host organism. Mutation Breeding for Disease Resistance, pp 107–116, IAEA, Vienna

Favret EA (1972) El mejoramiento de las plantas por induccion de mutaciones en Latinoamerica. Induced Mutations and Plant Improvement, pp 49–60, IAEA, Vienna

Favret EA (1976) Breeding for disease resistance using induced mutations. Induced Mutations in Cross-Breeding, pp 95–111, IAEA, Vienna

Favret EA, Ryan GS (1966) New useful mutants in plant breeding. Mutations in Plant Breeding, pp 49–61, IAEA, Vienna

Favret EA, Solari R, Manghers L, Avila A (1969) Genetic control of the qualitative and quantitative production of endosperm proteins in wheat and barley. New Approaches to Breeding for Improved Plant Protein, pp 87–107, IAEA, Vienna

Favret EA, Manghers L, Solari R, Avila A, Monesiglio JC (1970) Gene control of protein production in cereal seeds. Improving Plant Protein by Nuclear Techniques, pp 87–96, IAEA, Vienna

Favret EA, Sarasola J, Solari RM (1974) Discrimination between specific and non-specific reactions in the host: pathogen relationship. Induced Mutations for Disease Resistance in Crop Plants, pp 23–34, IAEA, Vienna

Favret EA, Solari RM, Manghers LE (1979) Induced mutations for protein quantity and quality in wheat. Seed Protein Improvement in Cereals and Grain Legumes II, pp 211–222, IAEA, Vienna

Feenstra WJ, Jacobsen E (1980) Isolation of a nitrate reductase deficient mutant of *Pisum sativum* by means of selection for chlorate. Theor Appl Genet 58:39–42

Fester T, Søgård B (1969) The localization of *eceriferum* loci in barley. Hereditas 61:327–337

Fierlinger P, Vlk J (1966) On the possibilities of utilizing mutation processes in the selection of leguminous plants. Genetika Slechteni 2:87–92

Fossati A, Brönnimann A (1976) Tolerance of *Septoria nodorum* Berk. in wheat: Inheritance and potential in breeding. Induced Mutations for Disease Resistance in Crop Plants, pp 91–100, IAEA, Vienna

Foster CA (1976) Natural and induced mutations in hybrid barley breeding. Barley Genet 3:774–784

Francis CM, Millington AJ (1965) Isoflavone mutations in subterranean clover. Aust J Agric Res 16:565–573

Frey KJ (1965) Mutation Breeding for quantitative attributes. Use Induced Mutations Plant Breed. Radiat Bot Suppl 5:465–475

Frey KH (1973) Improvement of the quantity and quality of cereal grain protein. Alternate Sources of Protein for Animal Production. Natl Acad Sci Wash, DC, pp 9–41

Frey KJ (1977) Protein of oats. Z Pflanzenzücht 78:185–215

Frey KJ, Chandhanamutta P (1975) Spontaneous mutations as a source of variation in diploid, tetraploid and hexaploid oats (*Avena* ssp.). Egypt J Genet Cytol 4:238–249

Frey KJ, Browning JA, Simons MD (1976) Crown rust control on oats. Induced Mutations for Disease Resistance in Crop Plants, pp 101–111, IAEA, Vienna

Fuchs WH, Heitefuss R, Röbbelen G (1974) Investigations on the horizontal resistance of barley and wheat against mildew *(Erysiphe graminis)* and yellow rust *(Puccinia striiformis)*. Induced Mutations for Disease Resistance in Crop Plants, pp 79–84, IAEA, Vienna

Fugino Y (1978) Rice lipids. Cereal Chem 55:559–571

Fursov VN, Konoplia SP (1967) Induced mutations of the extreme type of branching in fine-fibre cotton. Genetika USSR 1967(5):162–166

Galal S, Ibrahim AF, Abdel-Hamid AM, Mahmoud IM (1975) Morphogenetical studies on the M_2- and M_3-populations of wheat (*Triticum aestivum* ssp. *vulgare* L.) after seed irradiation with gamma rays. Z Pflanzenzücht 74:189–198

Ganashan P (1970) Induced mutation studies with *Brachiaria brizantha* Stapf. and some *indica* rice varieties from Ceylon. Rice Breeding with Induced Mutations II, pp 7–12, IAEA, Vienna

Gardner CO (1969) Genetic variation in irradiated and control populations of corn after ten cycles of mass selection for high grain yield. Induced Mutations in Plants, pp 469–477, IAEA, Vienna

Gaul E (1981) Untersuchungen über Proteingehalt und -zusammensetzung in Samen und Hülsen verschiedener Reifestadien bei *Pisum*-Mutanten. Ph D Thesis, Univ Bonn

Gaul H (1958) Present aspects of induced mutations in plant breeding. Euphytica 7:275–289

Gaul H (1961) Use of induced mutants in seed-propagated species. Mutation and Plant Breeding NAS-NRC 891:206–251

Gaul H (1963) Mutationen in der Pflanzenzüchtung. Z Pflanzenzücht 50:194–307

Gaul H (1964) Mutations in plant breeding. Radiat Bot 4:155–232

Gaul H (1965a) Induced mutations in plant breeding. Genetics Today, Proc 11th Int Congr Genetics, The Hague, 3:689–709

Gaul H (1965b) The concept of macro- and micro-mutations and results on induced micro-mutations in barley. Use Induced Mutations Plant Breed. Radiat Bot Suppl 5:407–428

References

Gaul H (1965c) Use of mutations for plant breeding in Europe. Proc 6th Japan Conf Radioisotopes, Tokyo 1964, pp 843–860

Gaul H (1966) Züchterische Bedeutung von Kleinmutationen. I. Z Pflanzenzücht 55:1–20

Gaul H (1967) Studies on populations of micro-mutants in barley and wheat without and with selection. Erwin Baur Ged Vorl 4:269–281

Gaul H, Grunewaldt J (1971) Independent variation of culm length and spike-internode length of a barley *erectoides* mutant. Barley Genet 2:106–118

Gaul H, Hesemann CU (1966) Züchterische Bedeutung von Großmutationen. I. Z Pflanzenzücht 55:225–237

Gaul H, Lind V (1976) Variation of the pleiotropy effect in a changed genetic background demonstrated with barley mutants. Induced Mutations in Cross-breeding, pp 55–69, IAEA, Vienna

Gaul H, Mittelstenscheid L (1960) Hinweise zur Herstellung von Mutationen durch ionisierende Strahlen in der Pflanzenzüchtung. Z Pflanzenzücht 43:404–422

Gaul H, Ulonska E (1967) Züchterische Bedeutung von Kleinmutationen. II. Z Pflanzenzücht 58:341–368

Gaul H, Grunewaldt J, Hesemann CU (1968) Variation of character expression of barley mutants in a changed genetic background. Mutations in Plant Breeding II, pp 77–95, IAEA, Vienna

Gaul H, Ulonska E, Zum Winkel C, Braker G (1969) Micro-mutations influencing yield in barley – Studies over nine generations. Induced Mutations in Plants, pp 375–398, IAEA, Vienna

Gaul H, Grunewaldt J, Ulsonksa E (1971) Macro- and micro-mutations, their significance in breeding of autogamous cultivated plants. Int Symp Use Isotopes Radiat Agric Anim Husbandry Res, New Delhi, pp 137–145

Gaul H, Frimmel G, Gichner T, Ulonska E (1972) Efficiency of mutagenesis. Induced Mutations and Plant Improvement, pp 121–139, IAEA, Vienna

Gaul H, Ulsonska E, Lind V, Walther H (1973) Studies on selection for high protein and lysine content in barley mutants. Nuclear Techniques for Seed Protein Improvement, pp 209–215, IAEA, Vienna

Ghosh N, Sen S (1974) Effect of irradiation in jute. Indian J Genet Plant Breed 34A:931–936

Ghosh Hajra N, Halder S (1981) *Bu 79* – A promising small grain mutant in rice. Genet Agrar 35:327–338

Ghosh Hajra N, Mallick EH, Bairagi P (1982) Studies on protein characteristics in rice mutants. Theor Appl Genet 61:23–26

Gianazza E, Righetti PG, Pioli F, Galante E, Soave C (1976) Size and charge heterogeneity of zein in normal and *opaque 2* endosperms. Maydica 21:1–17

Gill KS, Nanda GS, Karam Chand (1974) Induced polygenic variability in M_3 and M_4 generations of barley cultivar C 164 for plant height, spike length and number of spikelets per spike. Genet Agrar 28:232–241

Glazacheva LI, Sidorova KK (1973) Type of inheritance of modified characters in some mutant pea forms. Genetika USSR 9(1):46–53

Goldenberg JB (1965) *"Afila"*, a new mutation in pea (*Pisum sativum* L.) Bol Genet 1:27–28

Golovchenko WI, Solonenko LP, Chernij IW, Khvostova VV, Trofimova OS (1970) Amino acid composition of protein in radiation-induced varieties of lupin for fodder and in economically valuable mutants of spring wheat. Improving Plant Protein by Nuclear Techniques, pp 149–162, IAEA, Vienna

Gopal-Ayengar AR, Rao NS, Joshua DC, Thakare RG (1971) Induction of short culm mutations in rice with gamma rays and fast neutrons. Int Symp Use Isotopes Radiat Agric Anim Husbandry Res, New Delhi, pp 80–87

Górny A (1978) Studies on genetic variation of the root system characters of mutants of the spring barley (*Hordeum vulgare* L.). Genet Polon 19:447–456

Gottschalk W (1960a) Über züchterisch verwendbare strahleninduzierte Mutanten von *Pisum sativum*. Züchter 30:33–42

Gottschalk W (1960b) Untersuchungen über die Befruchtungsverhältnisse von *Vicia faba* mit Hilfe einer früh erkennbaren Mutante. Züchter 30:22–27

Gottschalk W (1962) Untersuchungen über den Selektionswert strahleninduzierter Mutanten. Z Vererbungsl 93:188–202

Gottschalk W (1964) Die Wirkung mutierter Gene auf die Morphologie und Funktion pflanzlicher Organe. Bot Stud 14:359pp

Gottschalk W (1965a) Der Einfluß der Penetranzverhältnisse mutierter Gene auf die Leistungsfähigkeit von Positivmutanten. Publ Europ Atom Energy Commun EUR 2150.d:1–14

Gottschalk W (1965b) A chromosome region in *Pisum* with an exceptionally high susceptibility to X-rays. Use Induced Mutations Plant Breed. Radiat Bot Suppl 5:385–391

Gottschalk W (1966) The yield capacity of useful mutants. A critical review of a collection of mutant types of *Pisum*. Mutations in Plant Breeding, pp 85–101, IAEA, Vienna

Gottschalk W (1967) Neue Aspekte zum Problem der pleiotropen Genwirkung. Ber Dtsch Bot Ges 80:545–553

Gottschalk W (1968a) Simultaneous mutation of closely linked genes. A contribution to the interpretation of "pleiotropic" gene action. Mutations in Plant Breeding II, pp 97–109, IAEA, Vienna

Gottschalk W (1968b) Investigations on the genetic control of meiosis. Nucleus, Seminar Vol, pp 345–361

Gottschalk W (1970) The productivity of some mutants of the pea (*Pisum sativum* L.) and their hybrids. A contribution to the heterosis problem in self-fertilizing species. Euphytica 19:91–97

Gottschalk W (1971a) Die Bedeutung der Genmutationen für die Evolution der Pflanzen. Fischer, Stuttgart, 296 S

Gottschalk W (1971b) Problems and results in improvement of grain legumes through mutation breeding. Intern Symp Use Isotopes Radiat Agric Anim Husbandry Res, New Delhi, pp 116–136

Gottschalk W (1972a) Die Kombination mutierter Gene. Biol Zbl 91:91–109

Gottschalk W (1972b) Harmonische und disharmonische Genkombinationen in der Mutationszüchtung. Z Pflanzenzücht 67:221–232

Gottschalk W (1972c) Combination of mutated genes as an additional tool in plant breeding. Induced Mutations and Plant Improvement, pp 199–218, IAEA, Vienna

Gottschalk W (1973) The evolutionary qualification of some leaf mutants and recombinants of *Pisum*. Egypt J Genet Cytol 2:219–238

Gottschalk W (1975a) Mutation. Progr Bot 37:219–246

Gottschalk W (1975b) The influence of mutated genes on quantity and quality of seed proteins. Indian Agric 19:205–223

Gottschalk W (1976a) Adaptability of mutants to diverse natural environmental conditions. Induced Mutations in Cross-Breeding, pp 37–44, IAEA, Vienna

Gottschalk W (1976b) Pleiotropy and close linkage of mutated genes. New examples of mutations of closely linked genes. Induced Mutations in Cross-Breeding, pp 71–78, IAEA, Vienna

Gottschalk W (1976c) Genetically conditioned male sterility. Induced Mutations in Cross-Breeding, pp 133–140, IAEA, Vienna

Gottschalk W (1976d) Monogenic heterosis. Induced Mutations in Cross-Breeding, pp 189–197, IAEA, Vienna

Gottschalk W (1977a) Fasciated peas – Unusual mutants for breeding and research. J Nuclear Agric Biol 6:27–33

Gottschalk W (1977b) Mutation. Progr Bot 39:153–172

Gottschalk W (1978a) The performance of translocation-homozygous *Pisum* lines in comparison with translocation-heterozygous plants. Nucleus 21:29–34

Gottschalk W (1978b) Gene-ecological studies in *Pisum* mutants and recombinants. Genetika Beograd 10:43–61

Gottschalk W (1978c) The dependence of the penetrance of mutant genes on environment and genotypic background. Genetica 49:21–29

Gottschalk W (1978d) A pea mutant showing an increase in fertility in subsequent generations. Pisum Newslett 10:15

Gottschalk W (1978e) The breeding system of *Vicia faba*. Legume Res 1:69–76

Gottschalk W (1978f) Prospects and limits of mutation breeding. Indian Agric 22:65–91

Gottschalk W (1979a) The utilization of induced mutations in plant breeding. A review. Proc 1st Mediterr Conf Genet Cairo, pp 765–782

Gottschalk W (1979b) Differential behaviour of a mutant gene in *Pisum* recombinants. Genetika Beograd 11:15–28

Gottschalk W (1979c) The genetic and breeding behavior of fasciated peas. Egypt J Genet Cytol Suppl 8:75–87

Gottschalk W (1979d) Mutation: Higher plants. Progr Bot 41:185–197

Gottschalk W (1980a) Gene-ecology, a modern branch of mutation research. Medio Ambiente 4:103–114

Gottschalk W (1980b) Induced mutations in plant breeding. Genetika Beograd 12:233–262

Gottschalk W (1981a) Genetic constitution of seven fasciated pea mutants. A mutator gene in *Pisum*? Pulse Crops Newslett 1(1):54–55

Gottschalk W (1981b) The behaviour of fasciated pea mutants under different ecological conditions. Pulse Crops Newslett 1(2):30–31

Gottschalk W (1981c) Induced mutations in gene-ecological studies. Induced Mutations – a Tool in Plant Research, pp 411–436, IAEA, Vienna

Gottschalk W (1981d) Mutation: Higher Plants. Progr Bot 43:139–152

Gottschalk W (1981e) The behaviour of a micromutant in the phytotron. Pisum Newslett 13:15

Gottschalk W (1981f) Genetics of seed size of fasciated pea mutants. Pulse Crops Newslett 1(4):19–21

Gottschalk W (1981g) Investigations on the heat tolerance of *Pisum* mutants. Pulse Crops Newslett 1(3):25–26

Gottschalk W (1981h) The suppression of gene actions through environmental factors. Egypt J Genet Cytol 10:159–174

Gottschalk W (1982a) The flowering behaviour of *Pisum* genotypes under phytotron and field conditions. Biol Zbl 101:249–260

Gottschalk W (1982b) The behaviour of gene *efr* for earliness in new recombinants under short-day phytotron conditions. Pisum Newslett 14:15–16

Gottschalk W (1982c) The short-day reaction of a fasciated recombinant. Pisum Newslett 14:14

Gottschalk W (1982d) Gene interactions in mutation breeding. Proc Int Symp New Genet Approaches Crop Improvement, Karachi (in press)

Gottschalk W (1983) Seed protein production of Pisum mutants and recombinants. In: Gottschalk W, Müller HP (eds) Seed Proteins: Biochemistry, Genetics, Nutritive Value. Nijhoff, The Hague, pp 377–402

Gottschalk W, Abou-Salha A (1983) Penetrance and seed production in *Pisum*. Pulse Crops Newslett (in press)

Gottschalk W, Bandel G (1978) Recombinants from crosses between fasciated and non-fasciated pea mutants. I. Early flowering recombinants. Z Pflanzenzücht 80:117–128

Gottschalk W, Baquar SR (1972) Breakdown of meiosis in a mutant of *Pisum sativum*. Cytobiol 5:42–50

Gottschalk W, Chen R (1969) Die Penetranz mutierter Gene als begrenzender Faktor in der Mutationszüchtung. Z Pflanzenzücht 62:293–304

Gottschalk W, Hasenberg E (1977) Untersuchungen über den Abbau von Reservestoffen in den Kotyledonen von Erbsen-Mutanten. Angew Bot 51:265–276

Gottschalk W, Hussein HAS (1975) The productivity of fasciated pea recombinants and the interaction of the mutant genes involved. Z Pflanzenzücht 74:265–278

Gottschalk W, Hussein HAS (1976) The seed production of recombinants selected from crosses between fasciated and non-fasciated *Pisum* mutants. I. Recombinants from sister-mutants. Egypt J Genet Cytol 5:312–330

Gottschalk W, Imam MM (1972) The yielding capacity of mutants under different climatic conditions. Ghana J Sci 13:63–71

Gottschalk W, Kaul MLH (1973) Investigations on the co-operation of mutated genes. II. Ber Dtsch Bot Ges 86:513–524

Gottschalk W, Kaul MLH (1974) The genetic control of microsporogenesis in higher plants. Nucleus 17:133–166

Gottschalk W, Kaul MLH (1975) Gene-ecological investigations in *Pisum* mutants. I. The influence of climatic factors upon quantitative and qualitative characters. Z Pflanzenzücht 75:182–191

Gottschalk W, Kaul MLH (1980) Gene-ecological investigations in *Pisum* mutants. Part 2: Comparative performance in Germany and North India. Theor Appl Genet 56:71-79

Gottschalk W, Klein HD (1976) The influence of mutant genes on sporogenesis. A survey on the genetic control of meiosis in *Pisum sativum*. Theor Appl Genet 48:23-34

Gottschalk W, Kumar S (1972) The response of pea mutants to moderate and semi-tropical conditions. Z Pflanzenzücht 67:95-102

Gottschalk W, Milutinović V (1973a) Untersuchungen zur Heterosis bei Selbstbefruchtern. I. Genetika Beograd 5:59-72

Gottschalk W, Milutinović V (1973b) Untersuchungen zur Heterosis bei Selbstbefruchtern. II. Genetika Beograd 5:117-134

Gottschalk W, Müller HP (1970) Monogenic alteration of seed protein content and protein pattern in X-ray induced *Pisum* mutants. Improving Plant Protein by Nuclear Techniques, pp 201-212, IAEA, Vienna

Gottschalk W, Müller HP (1974) Quantitative and qualitative investigations on the seed proteins of mutants and recombinants of *Pisum sativum*. Theor Appl Genet 45:7-20

Gottschalk W, Müller HP (1979) The reaction of an early-flowering *Pisum* recombinant to environment and genotypic background. Seed Protein Improvement in Cereals and Grain Legumes I, pp 259-272, IAEA, Vienna

Gottschalk W, Müller HP (1983) Seed Proteins: Biochemistry, Genetics, Nutritive Value. Nijhoff, The Hague, 531 pp

Gottschalk W, Patil SH (1971) The reaction of *Pisum* mutants to different climatic conditions. Indian J Genet Plant Breed 31:403-406

Gottschalk W, Wolff G (1977) Problems of mutation breeding in *Pisum*. Legume Res 1:1-16

Gottschalk W, Wolff G (1983) The behaviour of a protein-rich *Pisum* mutant in crossing experiments. In: Gottschalk W, Müller HP (eds) Seed Proteins: Biochemistry, Genetics, Nutritive Value. Nijhoff, The Hague, pp 403-425

Gottschalk W, Müller HP, Wolff G (1975a) Relations between protein production, protein quality and environmental factors in *Pisum* mutants. Breeding for Seed Protein Improvement Using Nuclear Techniques, pp 105-123, IAEA, Vienna

Gottschalk W, Müller HP, Wolff G (1975b) The genetic control of seed protein production and composition. Egypt J Genet Cytol 4:453-468

Gottschalk W, Müller HP, Wolff G (1976) Further investigations on the genetic control of seed protein production in *Pisum* mutants. Evaluation of Seed Protein Alterations by Mutation Breeding, pp 157-177, IAEA, Vienna

Gottschalk W, Bordia PC, Kumar S (1978) Comparison of the performance of *Pisum* mutants and recombinants in Germany and Western India. Legume Res 2:19-28

Goud JV (1967) Induced mutations in bread wheat. Indian J Genet Plant Breed 27:40-55

Goud JV (1968) Selection experiments for quantitative characters in wheat after treatment with mutagens. Genet Agrar 22:119-135

Goud JV (1972) Mutation studies in *Sorghum*. Genet Polon 13(3):33-40

Goud JV, Nair KMD, Rao MG (1971) Induced polygenic mutations in ragi. Indian J Genet Plant Breed 31:202-208

Goud JV, Nayar KMD, Rao MG (1970) Mutagenesis in *Sorghum*. Indian J Genet Plant Breed 30:81-90

Gregory RP (1903) The seed characters of *Pisum sativum*. New Phytol II:226-228

Gregory WC (1968) A radiation breeding experiment with peanuts. Radiat Bot 8:81-147

Gridley HE, Evans AM (1979) Prospects for combining high yield with increased protein production in *Phaseolus vulgaris* L. Seed Protein Improvement in Cereals and Grain Legumes II, pp 47-58

Griffiths DW, Lawes DA (1978) Variation in the crude protein content of field beans (*Vicia faba* L.) in relation to the possible improvement of the protein content of the crop. Euphytica 27:487-495

Gröber K (1967) Some results of mutation experiments in apples and black currants. Erwin Baur Ged Vorl 4:377-382

References

Grunewaldt J (1974a) Untersuchungen an dem *erectoides*-Komplex von Gerstenmutanten. I. Z Pflanzenzücht 71:193–207

Grunewaldt J (1974b) Untersuchungen an dem *erectoides*-Komplex von Gerstenmutanten. II. Z Pflanzenzücht 71:330–340

Guiragossian V, Chibber BAK, van Scoyoc St, Jambunathan R, Mertz ET, Axtell JD (1978) Characteristics of proteins from normal, high lysine and high tannin sorghum. Agric Food Chem 26: 219–223

Gupta MN (1969) Use of gamma irradiation in the production of new varieties of perennial *Portulaca*. Radiations and Radiomimetic Substances in Mutation Breeding, Bombay, pp 206–214

Gupta MN, Shukla R (1971) Mutation breeding of *Chrysanthemum*. I. Int Symp Use Isotopes Radiat Agric Anim Husbandry Res, New Delhi, pp 164–174

Gustafsson Å (1942) Mutationsforschung und Züchtung. Züchter 14:57–64

Gustfasson Å (1947) Mutations in agricultural plants. Hereditas 33:1–100

Gustafsson Å (1951) Induction of changes in genes and chromosomes. II. Mutations, environment and evolution. Cold Spring Harb Symp Quant Biol 16:263–281

Gustafsson Å (1954) Mutations, viability, and population structure. Acta Agric Scand 4:601–632

Gustafsson Å (1963) Productive mutations induced in barley by ionizing radiations and chemical mutagens. Hereditas 50:211–262

Gustafsson Å (1965) Characteristics and rates of high productive mutants in diploid barley. Use Induced Mutations Plant Breed. Radiat Bot Suppl 5:323–337

Gustafsson Å (1969) A study of induced mutations in plants. Induced Mutations in Plants, pp 9–31, IAEA, Vienna

Gustafsson Å (1972) The genetic architecture of phenotype patterns in barley. Induced Mutations and Plant Improvement, pp 7–12, IAEA, Vienna

Gustafsson Å, Dormling I (1971) Phytotron analysis of dominance expression and overdominance in monohybrid barley. Int Symp Use Isotopes Radiat Agric Anim Husbandry Res, New Delhi, pp 3–12

Gustafsson Å, Dormling I (1972) Dominance and overdominance in phytotron analysis of monohybrid barley. Hereditas 70:185–216

Gustafsson Å, Ekman G (1967) Yield efficiency of the X-ray mutant Svalöf's "Pallas barley". Züchter 37:42–46

Gustafsson Å, Gadd I (1965a) Mutations and crop improvement. II. The genus *Lupinus (Leguminosae)*. Hereditas 53:15–39

Gustafsson Å, Gadd I (1965b) Mutations and crop improvement. III. *Ipomoea batatas* (L.) POIR. *(Convolvulaceae)*. Hereditas 53:77–89

Gustafsson Å, Gadd I (1965c) Mutations and crop improvement. IV. *Poa pratensis* L. *(Gramineae)*. Hereditas 53:90–102

Gustafsson Å, Gadd I (1965d) Mutations and crop improvement. V. *Arachis hypogaea* L. *(Leguminosae)*. Hereditas 53:143–164

Gustafsson Å, Gadd I (1965e) Mutations and crop improvement. VI. The genus *Avena* L. *(Gramineae)*. Hereditas 53:327–373

Gustafsson Å, Gadd I (1966) Mutations and crop improvement. VII. The genus *Oryza* L. *(Gramineae)*. Hereditas 55:273–357

Gustafsson Å, Lundqvist U (1976) Controlled environment and short-day tolerance in barley mutants. Induced Mutations in Cross-Breeding, pp 45–53, IAEA, Vienna

Gustafsson Å, Wettstein D v (1958) Mutationen und Mutationszüchtung. In: Kappert H, Rudorf W (eds) Handb Pflanzenzücht I. Parey, Berlin Hamburg, pp 612–699

Gustafsson Å, Lundqvist U, Ekberg I (1966) The viability reaction of gene mutations and chromosome translocations in comparison. Mutations in Plant Breeding, pp 103–107, IAEA-FAO, Vienna

Gustafsson Å, Lundqvist U, Ekman G (1968) Yield analysis after repeated mutagenic treatment and selection in barley. Mutations in Plant Breeding II, pp 113–128, IAEA, Vienna

Gustafsson Å, Hagberg A, Persson G, Wiklund K (1971) Induced mutations and barley improvement. Theor Appl Genet 41:239–248

Gustafsson Å, Lundqvist U, Kucera J, Ghatnekar J (1972) Mutagenesis of a fluctuating character: Grain dormancy in Kristina barley. Induced Mutations and Plant Improvement, pp 343–348, IAEA, Vienna

Gustafsson Å, Dormling I, Ekman G (1973a) Phytotron ecology of mutant genes. I. Heterosis in mutant crossings of barley. Hereditas 74:119–126

Gustafsson Å, Dormling I, Ekman G (1973b) Phytotron ecology of mutant genes. II. Dynamics of heterosis in an intralocus mutant hybrid of barley. Hereditas 74:247–258

Gustafsson Å, Dormling I, Ekman G (1973c) Phytotron ecology of mutant genes. III. Growth reactions of two quantitative traits in barley. Hereditas 75:75–82

Gustafsson Å, Dormling I, Ekman G (1974) Phytotron ecology of mutant genes. V. Intra- and interlocus overdominance involving early mutants of Bonus barley. Hereditas 77:237–254

Gustafsson Å, Dormling I, Ekman G (1975) Phytotron ecology of mutant genes. VI. Clima reactions of the *eceriferum* mutations $cer\text{-}i^{16}$ and $cer\text{-}c^{36}$. Hereditas 80:279–290

Gustafsson Å, Ekman G, Dormling I (1977) Effects of the Pallas gene in barley: phene analysis, overdominance, variability. Hereditas 86:251–266

Gutierrez M, Vrdoljak J, Ricciardi A (1972) Development of gossypol-glandless strains of cotton. Induced Mutations and Plant Improvement, pp 397–404, IAEA, Vienna

Haahr V, Wettstein D v (1976) Studies of an induced, high-yielding dwarf-mutant of spring barley. Barley Genet III:215–218

Hadjichristodoulou A, Della A (1978) Induced micromutations in barley for protein content and quality in low-rainfall areas. Seed Protein Improvement by Nuclear Techniques, pp 145–156, IAEA, Vienna

Hadley HH, Freeman JE, Javier EQ (1965) Effects of height mutations on grain yield in *Sorghum*. Crop Sci 5:11–14

Hagberg A (1967) The use of induced mutations in practical barley breeding at Svalöf. Erwin Baur Ged Vorl 4:147–154

Hagberg A (1978) The Svalöf cereal protein quality programme. Seed Protein Improvement by Nuclear Techniques, pp 91–105, IAEA, Vienna

Hagberg A, Hagberg G (1971) Chromosome aberrations and their utilization. Barley Genet 2:65–71

Hagberg A, Karlsson KE (1969) Breeding for high protein content and quality in barley. New Approaches to Breeding for Improved Plant Protein, pp 17–21, IAEA, Vienna

Hagberg A, Karlsson KE, Munck L (1970) Use of *hiproly* in barley breeding. Improved Plant Protein by Nucelar Techniques, pp 121–132, IAEA, Vienna

Hagberg A, Persson G, Hagberg G (1972) Utilization of induced chromosomal aberrations, translocations, duplications and trisomics in barley. Induced Mutations and Plant Improvement, pp 173–182, IAEA, Vienna

Hagberg A, Persson G, Ekman R, Karlsson KE, Tallberg AM, Stoy V, Bertholdsson NO, Mounla M, Johansson H (1979) The Svalöf protein quality breeding programme. Seed Protein Improvement in Cereals and Grain Legumes II, pp 303–313, IAEA, Vienna

Hagberg P, Lehmann L, Hagberg G, Karlsson BA (1976) Induction and search for mutants which can be used in the development of hybrid barley. Barley Genet 3:771–773

Hahne J (1981) Untersuchungen zur N-Aufnahme von Winterweizentypen unterschiedlicher Ertragsleistungen im Keimpflanzenstadium. 1. Mitt. Einfluß der Saatguteigenschaften auf die N-Aufnahme von Keimpflanzensprossen. Arch Züchtungsforsch 11:93–104

Halim AH, Wassom CE, Mitchell HL (1973) Trypsin inhibitor in corn (*Zea mays* L.) as influenced by genotype and moisture stress. Crop Sci 13:405–407

Haniš M (1973) Results in mutation breeding of wheat. Genet Šlechtení 9:253–260

Haniš M (1974) Induced mutations for disease resistance in wheat and barley. Induced Mutations for Disease Resistance in Crop Plants, pp 49–56, IAEA, Vienna

Haniš M, Knytl V, Jech Z (1969) Induction of mutants resistant to stripe rust yellow in the "Diana I" winter wheat variety. Genet Šlechtení 5:257–262

Haniš M, Hanišová A, Černý J, Sašek A (1976) Practical value of induced mutations for disease resistance in wheat, *Triticum aestivum*. Induced Mutations for Disease Resistance in Crop Plants pp 113–128, IAEA, Vienna

Haniš M, Hanišová A, Knytl V, Cerny J, Benc S (1977) Induced mutations for disease resistance in wheat and barley. Induced Mutations against Plant Diseases, pp 347–357, IAEA, Vienna

Hänsel H (1966) Induction of mutations in barley: Some practical and theoretical results. Mutations in Plant Breeding, pp 117–137, IAEA, Vienna

Hänsel H (1971) Experience with a mildew-resistant mutant (mut 3502) of "Vollkorn" barley induced in 1952. Mutation Breeding for Disease Resistance, pp 125–129, IAEA, Vienna

Hänsel H, Simon W, Ehrendorfer K (1972) Mutation breeding for yield and kernel performance in spring barley. Induced Mutations and Plant Improvement, pp 221–235, IAEA, Vienna

Haq MA, Shakoor A (1980) Use of induced mutations for improving resistance against *Ascochyta* blight in chickpea *(Cicer arietinum)* and yellow mosaic virus in mungbean *(Vigna radiata)*. Induced Mutations for Improvement of Grain Legume Production, pp 63–67, IAEA, Vienna

Haq MS, Rahman MM, Mansur A, Islam R (1971) Breeding for early high-yielding and disease-resistant rice varieties through induced mutations. Rice Breeding with Induced Mutations III, pp 35–46, IAEA, Vienna

Haq MS, Rahman MM, Chowdhury MH (1973) Studies of the quality of induced mutants of rice. Nuclear Techniques for Seed Protein Improvement, pp 139–144, IAEA, Vienna

Haq MS, Rahman MM, Mia MM, Ahmed HU (1974) Disease resistance of some mutants induced by gamma rays. Induced Mutations for Disease Resistance in Crop Plants, 150, IAEA, Vienna

Harder DE, McKenzie RIH, Martens JW, Brown PD (1977) Strategies for improving rust resistance in oats. Induced Mutations against Plant Diseases, pp 495–498, IAEA, Vienna

Harn C (1976) Studies on the high-protein mutants of rice. Evaluation of Seed Protein Alterations by Mutation Breeding, 143, IAEA, Vienna

Harn C, Won JL (1978) Inheritance of high protein, short culm and early maturity in a rice mutant. Seed Protein Improvement by Nuclear Techniques, 203, IAEA, Vienna

Harn C, Won JL, Park CK, Yoo JY (1973) Mutation breeding for improvement of rice protein. Nuclear Techniques for Seed Protein Improvement, pp 115–137, IAEA, Vienna

Harn C, Won JL, Choi KT (1975) The agronomic characters of a high protein rice mutant. Breeding for Seed Protein Improvement Using Nuclear Techniques, pp 17–22, IAEA, Vienna

Hartmann K (1981) Untersuchungen über die Interaktion mutierter Gene in frühblühenden und gegabelten Rekombinanten von *Pisum sativum* L. Ph D Thesis, Univ Bonn

Hartwig E (1979) Breeding productive soybeans with a higher percentage of protein. Seed Protein Improvement in Cereals and Grain Legumes II, pp 59–66, IAEA, Vienna

Häuser J, Fischbeck G (1976) Untersuchungen zur Lokalisierung einiger Mutationen von Gerste *(Hordeum sativum)*. Z Pflanzenzücht 77:269–280

Hayter AM, Allison MJ (1975) Breeding for high diastatic power. Barley Genet 3:612–619

Heinz DJ (1973) Sugar-cane improvement through induced mutations using vegetative propagules and cell culture techniques. Induced Mutations in Vegetatively Propagated Plants, pp 53–59, IAEA, Vienna

Hejgaard J, Boisen S (1980) High-lysine proteins in *Hiproly* barley breeding: Identification, nutritional significance and new screening methods. Hereditas 93:311–320

Hekstra G, Broertjes C (1968) Mutation breeding in bulbous *Iris*. Euphytica 17:345–351

Hentrich W (1977) Tests for the selection of mildew-resistant mutants in spring barley. Induced Mutations against Plant Diseases, pp 333–341, IAEA, Vienna

Hentrich W (1978) Multiple allelism in the *ml-o* locus of powdery mildew resistant barley mutants. Experimental Mutagenesis in Plants, pp 208–216, Sofia

Hentrich W (1979) Allelwirkung und Pleiotropie mehltauresistenter Mutanten des *ml-o* Locus der Gerste. Arch Züchtungsforsch 9:283–291

Hermelin T, Adam G (1978) Selection for increased variability of seed protein and lysine contents in wheat. Seed Protein Improvement by Nuclear Techniques, pp 293–300, IAEA, Vienna

Hesemann CU, Gaul H (1967) Züchterische Bedeutung von Großmutationen. II. Z Pflanzenzücht 58:1–14

Heslot H (1966) Induction de mutations par des agents mutagenes physiques et chimiques. Mutations in Plant Breeding, pp 139–149, IAEA, Vienna

Hess D (1968) Biochemische Genetik. Springer, Berlin Heidelberg New York, 353 S

Hiemke Ch (1973) Untersuchungen über den Stickstoffgehalt der Samen einiger strahleninduzierter Mutanten von *Pisum sativum* in Abhängigkeit von verschiedenen Faktoren. Ph D Thesis, Univ Bonn

Hillerislambers D, Rutger IN, Qualset CO, Wiser WI (1973) Genetic and environmental variation in protein content of rice (*Oryza sativa* L.). Euphytica 22:264–273

Hiraiwa S, Fujimaki H (1977) Male-sterile mutants of rice. Technical News 18, Inst Radiat Breed Ohmiya, Japan

Hiraiwa S, Tanaka S (1978) Effects of successive irradiation and mass screening for seed size, density and protein content of soybean. Seed Protein Improvement by Nuclear Techniques, pp 265–274, IAEA, Vienna

Hiraiwa S, Tanaka S (1979) Glabrous mutants of rice. Technical News 22, Inst Radiat Breed Ohmiya, Japan

Hiraiwa S, Tanaka S, Nakamura S (1976) Induction of mutants with higher protein content in soybean. Evaluation of Seed Protein Alterations by Mutation Breeding, pp 185–196, IAEA, Vienna

Hoffmann W (1959) Neuere Möglichkeiten der Mutationszüchtung. Z Pflanzenzücht 41:371–394

Hoffmann W, Nover I (1959) Ausgangsmaterial für die Züchtung mehltauresistenter Gersten. Z Pflanzenzücht 42:68–78

Horner CE, Melouk HA (1977) Screening, selection and evaluation of irradiation-induced mutants of spearmint for resistance to *Verticillium* wilt. Induced Mutations against Plant Diseases, pp 253–262, IAEA, Vienna

Hossain M, Sen S (1978a) Radiation induced high fibre yielding chlorophyll mutant in jute. Z Pflanzenzücht 81:77–79

Hossain M, Sen S (1978b) Early and photoinsensitive mutants in jute. Indian J Genet Plant Breed 38:179–181

Hsieh SC, Chang TM (1975) Radiation induced variations in photoperiod-sensitivity, thermosensitivity and the number of days to heading in rice. Euphytica 24:487–496

Hu CH (1973) Evaluation of breeding semidwarf rice by induced mutation and hybridization. Euphytica 22:562–574

Hu CH, Wu HP, Li HW (1970) Present status of rice breeding by induced mutations in Taiwan, Republic of China. Rice Breeding with Induced Mutations II, pp 13–19, IAEA, Vienna

Hussein HAS (1976) Seed protein traits of fasciated pea recombinants and the role of the mutant gene involved. Theor Appl Genet 47:231–242

Hussein HAS (1979) Changes in protein and amino acids under the influence of mutant gene combinations in *Pisum sativum*. Egypt J Genet Cytol 8:1–16

Hussein HAS, Abdalla MMF (1978) Protein and yield traits of field bean mutants induced with γ-rays, EMS and their combination. Seed Protein Improvement by Nuclear Techniques, pp 253–264, IAEA, Vienna

Hussein HAS, Abdalla MMF (1979) Gamma-ray and EMS-induced mutations in *Vicia faba* L. Seed Protein Improvement in Cereals and Grain Legumes II, pp 23–31, IAEA, Vienna

Hussein HAS, Disouki IAM (1979) Mutation breeding experiments in *Phaseolus vulgaris* (L.). II. Egypt J Genet Cytol 8:181–197

Hussein HAS, Gottschalk W (1976) The seed production of recombinants selected from crosses between fasciated and non-fasciated *Pisum* mutants. II. Recombinants from non-sister mutants. Egypt J Genet Cytol 5:387–399

Hussein HAS, Abdalla MMF, Sharaan AN (1979) Genetic analysis of radiation-induced early flowering mutants in barley. Egypt J Genet Cytol 8:233–241

Hussein HAS, Abdalla MMF, Hindi L, Ibrahim AF, Sharaan AN (1980) Biometrical analysis of radiation-induced early flowering barley mutants and associated characters. Egypt J Genet Cytol 9:145–166

Ibrahim AF, Sharaan AN (1974a) Variability of character expression in barley M_3- and M_4-bulk populations after seed irradiation with gamma rays. Z Pflanzenzücht 72:212–225

Ibrahim AF, Sharaan AN (1974b) Studies on certain early barley mutants in M_3- and M_4-generations after seed irradiation with gamma rays. Z Pflanzenzücht 73:44–57

Ibrahim AFM, Röbbelen G, Scheibe A, Zoschke M (1966) Über die genetische Differenzierung der Nachkommenschaft einer chemisch induzierten Gerstenmutante. Z Pflanzenzücht 56:251–284

References

Ichikawa S (1965) Radiation-induced mutants in a Japanese wheat variety, Shinchunaga. Wheat Inform Serv 19, 20:14

Ikeda F (1974) Radiation-induced fruit color mutation in the apple var. Fuji. Inst Rad Breed Ohmiya, Japan, Techn News 15

Indian Agricultural Research Institute (1973) Internal Symposium on the Use of Isotopes and Radiation in Agriculture and Animal Husbandry Research, New Delhi, 829 pp

Ingversen J (1975) Structure and composition of protein bodies from wild-type and high-lysine barley endosperm. Hereditas 81:69–76

Ingversen J, Andersen AJ, Doll H, Køie B (1973a) Selection and properties of high lysine barley. Nuclear Techniques for Seed Protein Improvement, pp 193–198, IAEA, Vienna

Ingversen J, Køie B, Doll H (1973b) Induced seed protein mutant in barley. Experientia 29: 1151–1152

Institute of Radiation Breeding, Ohmiya, Japan (1978) Mutation Breeding for Disease Resistance. Gamma Field Symp 17:91 pp

Institute of Radiation Breeding, Ohmiya, Japan (1979a) Crop Improvement by Induced Mutation. Gamma Field Symp 18:154 pp

Institute of Radiation Breeding, Ohmiya, Japan (1979b) Use of Dwarf Mutation. Gamma Field Symp 16:109 pp

International Atomic Energy Agency (1966) Mutations in Plant Breeding. Vienna, 271 pp

International Atomic Energy Agency (1968a) Rice Breeding with Induced Mutations. Vienna, 155 pp

International Atomic Energy Agency (1968b) Mutations in Plant Breeding II. Vienna, 315 pp

International Atomic Energy Agency (1969a) Induced Mutations in Plants. Vienna, 748 pp

International Atomic Energy Agency (1969b) New Approaches to Breeding for Improved Plant Protein. Vienna, 193 pp

International Atomic Energy Agency (1970a) Rice Breeding with Induced Mutations II. Vienna, 124 pp

International Atomic Energy Agency (1970b) Improving Plant Protein by Nuclear Techniques. Vienna, 458 pp

International Atomic Energy Agency (1971) Mutation Breeding for Disease Resistance. Vienna, 249 pp

International Atomic Energy Agency (1972) Induced Mutations and Plant Improvement. Vienna, 554 pp

International Atomic Energy Agency (1972–1982) List of mutant varieties. Mutat Breed Newslett 1–19

International Atomic Energy Agency (1973a) Induced Mutations in Vegetatively Propagated Plants. Vienna, 222 pp

International Atomic Energy Agency (1973b) Nuclear Techniques for Seed Protein Improvement. Vienna, 422 pp

International Atomic Energy Agency (1974a) Induced Mutations for Resistance in Crop Plants. Vienna, 193 pp

International Atomic Energy Agency (1974b) Polyploidy and Induced Mutations in Plant Breeding. Vienna, 413 pp

International Atomic Energy Agency (1975a) Improvement of Vegetatively Propagated Plants through Induced Mutations. Vienna, 139 pp

International Atomic Energy Agency (1975b) Breeding for Seed Protein Improvement Using Nuclear Techniques. Vienna, 225 pp

International Atomic Energy Agency (1976a) Improvement of Vegetatively Propagated Plants and Tree Crops through Induced Mutations. Vienna, 186 pp

International Atomic Energy Agency (1976b) Induced Mutations in Cross-Breeding. Vienna, 255 pp

International Atomic Energy Agency (1976c) Induced Mutations for Disease Resistance in Crop Plants. Vienna, 203 pp

International Atomic Energy Agency (1976d) Evaluation of Seed Protein Alterations by Mutation Breeding. Vienna, 216 pp

International Atomic Energy Agency (1977a) Induced Mutations against Plant Diseases. Vienna, 580 pp
International Atomic Energy Agency (1977b) Manual on Mutation Breeding, 2nd edn. Vienna, 288 pp
International Atomic Energy Agency (1977c) Induced Mutations for the Improvement of Grain Legumes in South East Asia (1975). Vienna, 187 pp
International Atomic Energy Agency (1978) Seed Protein Improvement by Nuclear Techniques. Vienna, 582 pp
International Atomic Energy Agency (1979) Seed Protein Improvement in Cereals and Grain Legumes I, II. Vienna, 421 pp and 472 pp
International Atomic Energy Agency (1980) Induced Mutations for Improvement of Grain Legume Production. Vienna, 129 pp
International Atomic Energy Agency (1981) Induced Mutations — a Tool in Plant Research. Vienna, 538 pp
International Atomic Energy Agency (1982) Induced Mutants for Cereal Grain Protein Improvement. Vienna, 216 pp
Ismachin M (1976) Rice seed protein improvement through mutation breeding techniques. Evaluation of Seed Protein Alterations by Mutation Breeding, pp 143–144, IAEA, Vienna
Ismachin M, Mikaelsen K (1976) Early maturing mutants for rice breeding and their use in cross-breeding programmes. Results of work in Indonesia. Induced Mutations in Cross-Breeding, pp 119–121, IAEA, Vienna
Ismachin Kartoprawiro M (1977) Selection for bacterial leaf-blight (*Xanthomonas oryzae*) and sheath-blight *(Rhizoctonia oryzae)* resistant mutants in a collection of early rice mutants. Induced Mutations against Plant Diseases, pp 199–211, IAEA, Vienna
Ismachin Kartoprawiro M, Mugiono M, Rijanti AM (1978) Induced mutations for high protein content in rice. Seed Protein Improvement by Nuclear Techniques, pp 157–165, IAEA, Vienna
Ismail MA, Ahmad SS (1979) The possibility of solving an important breeding problem related to growing rice in Egypt through induced mutagenesis. Proc 1st Mediterr Conf Genet, Cairo, pp 701–711
Ismail MA, Heakal MY, Fayed A (1976) Improvement of yield through induced mutagenesis in broad beans. Indian J Genet Plant Breed 36:347–350
Ivanova RM (1972) Use of experimental mutagenesis in breeding of *Papaver somniferum* L. Genetika USSR 8(1):30–37
Jaffé WG (1972) Factors affecting the nutritional value of beans. Nutritional Improvement of Food Legumes by Breeding, pp 43–48, IAEA, Vienna
Jagathesan D (1976) Induction and isolation of mutants in sugarcane. Improvement of Vegetatively Propagated Plants and Tree Crops through Induced Mutations, pp 69–82, IAEA, Vienna
Jagathesan D, Ratnam R (1978) A vigorous mutant sugarcane (*Saccharum* sp.) clone Co 527. Theor Appl Genet 51:311–313
Jagathesan D, Sreenivasan TV (1970) Induced mutations in sugar-cane. Indian J Agric Sci 40: 165–172
Jagathesan D, Balasundaram N, Alexander KC (1974) Induced mutations for disease resistance in sugarcance. Induced Mutations for Disease Resistance in Crop Plants, pp 151–153, IAEA, Vienna
Jain HK (1976) Induced mutations and improved plant types in pulses. Evaluation of Seed Protein Alterations by Mutation Breeding, 209, IAEA, Vienna
Jalani BS, Frey KJ (1979) Variation in groat-protein percentage of oats (*Avena sativa* L.) following selfing and outcrossing of M_1 plants. Egypt J Genet Cytol 8:57–70
Jaranowski JK (1976) Gamma-ray induced mutations in *Pisum arvense* (L.ss.). Genet Polon 17: 479–495
Jaranowski JK (1977) New genotypes of *Pisum* sp. derived from hybridization of mutants and cultivars. Genet Polon 18:337–355
Jaranowski JK, Broda Z (1978) Leaf mutants in diploid red clover (*Trifolium pratense* L.). Theor Appl. Genet 53:97–103
Jarquin R, Albertazzi C, Bressoni R (1970) Value of *opaque-2* protein for chicks. J Agric Food Chem 18:268–272

References

Jashowsky IV, Golovchenko VI (1967) The use of induced mutations in plant breeding at the Ukrainian Scientific Research Institute of Agriculture. Erwin Baur Ged Vorl 4:237–239

Jech Z (1966a) Vyvolávání mutací u ječmene a jejich prakticke použiti. Vydalo Sdruženi šlechtitelských a semenářských podniku v Praze (Praha), pp 177–184

Jech Z (1966b) Mutace u pšenice. Vydalo Sdruženi šlechtitelských a semenářských podniku v Praze (Praha), pp 185–192

Jensen J (1979a) Chromosomal location of one dominant and four recessive high-lysine genes in barley mutants. Seed Protein Improvement in Cereals and Grain Legumes I, pp 89–96, IAEA, Vienna

Jensen J (1979b) Location of a high-lysine gene and the DDT-resistance gene on barley chromosome 7. Euphytica 28:47–56

Jensen J (1981) Location of the high-lysine locus *lys4d* on barley chromosome 5. Barley Genet Newslett 11:45–47

Jermyn WA, McKenzie SL, Slinkard AE (1976) Variability and heritability of methionine content in peas (*Pisum sativum* L.). Crop Sci Dep Saskatchewan

Jimenez JR (1966) Protein fractionation studies of high lysine corn. Proc High Lysine Corn Conf, Purdue Univ, pp 74–79

Johnson VA, Lay ChL (1974) Genetic improvement of plant protein. J Agric Food Chem 22:558–566

Johnson VA, Mattern PJ, Whited DA, Schmidt JW (1969) Breeding for high protein content and quality in wheat. New Approaches to Breeding for Improved Plant Protein, pp 29–40. IAEA, Vienna

Johnson VA, Mattern PJ, Schmidt JW, Stroike JE (1973) Genetic advances in wheat protein quantity and composition. Proc 4th Int Wheat Genet Symp, Univ Miss Columbia, 547

Johnson VA, Mattern PJ, Kuhr SL (1979) Genetic improvement of wheat protein. Seed Protein Improvement in Cereals and Grain Legumes II, pp 165–179, IAEA, Vienna

Jolivet E, Nicol M, Baudet J, Mossé J (1970) Differences dans l'incorporation du $^{14}CO_2$ a la lumière chez de jeunes mais normal et *opaque-2*. Improving Plant Protein by Nuclear Techniques, pp 391–400, IAEA, Vienna

Jonassen J (1980a) Characteristics of *hiproly* barley I. Isolation and characterisation of two water-soluble high-lysine proteins. Carlsberg Res Commun 45:47–58

Jonassen J (1980b) Characteristics of *hiproly* barley II. Quantification of two proteins contributing to its high lysine content. Carlsberg Res Commun 45:59–68

Jones RA (1978) Effect of the *floury-2* locus on zein accumulation and RNA metabolism during maize endosperm development. Biochem Genet 16:27–38

Jones RA, Tsai CY (1977) Changes in lysine and tryptophane content during germination of normal and mutant maize seed. Cereal Chem 54:565–571

Jörgensen JH (1971a) Comparison of induced mutant genes with spontaneous genes in barley conditioning resistance to powdery mildew. Mutation Breeding for Disease Resistance, pp 117–124, IAEA, Vienna

Jörgensen JH (1971b) An allelic series of mutant genes for powdery-mildew resistance in barley. Barley Genet II:475–477

Jörgensen JH (1974) Induced mutations for powdery mildew resistance in barley. Induced Mutations for Disease Resistance in Crop Plants, 67, IAEA, Vienna

Jörgensen JH (1975) Identification of powdery mildew resistant barley mutants and their allelic relationship. Barley Genet III:446–455

Jörgensen JH (1976) Studies on recombination between alleles in the *ml-o* locus of barley and on pleiotropic effects of the alleles. Induced Mutations for Disease Resistance in Crop Plants pp 129–140, IAEA, Vienna

Jörgensen JH (1977) Spectrum of resistance conferred by *ml-o* powdery mildew resistance genes in barley. Euphytica 26:55–62

Josan N, Nicolae I (1974) Promising mutants in soya. Genetika USSR 10(6):43–50

Joshua DC, Rao NS, Bhatia CR, Mistry KB, Bhujbal BM (1969) Characteristics of dwarf mutants induced in rice. Radiations and Radiomimetic Substances in Mutation Breeding, Bombay, pp 185–196

Jung-Heiliger H, Horn W (1980) Variation nach mutagener Behandlung von Stecklingen und in vitro-Kulturen bei *Chrysanthemum*. Z Pflanzenzücht 85:185–199

Kadam SS, Singh J, Mehta SL (1973) Changes in isoenzymes in embryo and endosperm of normal and *opaque-2 Zea mays* during inbibition. Phytochem 12:1221–1225

Kaicker US, Swarup V (1972) Induced mutations in roses. Indian J Genet Plant Breed 32:257–265

Kaicker US, Swarup V, Singh H (1971) Some radiation induced mutants in *Tithonia*. Indian J Genet Plant Breed 31:218–222

Kajjari NB, Puttarudrappa A, Goud JV (1969) Prospects of mutation breeding in *Sorghum*. Radiations and Radiomimetic Substances in Mutation Breeding, Bombay, pp 365–370

Kak SN, Kaul BL (1980) Radiation induced useful mutants of japanese mint (*Mentha arvensis* L.). Z Pflanzenzücht 85:170–174

Kakade ML (1974) Biochemical basis for the differences in plant protein utilization. J Agr Food Chem 22:550–555

Kalashnik NA (1972) Genotypic variability of certain quantitative characters in mutant strains of spring wheat grown under different environmental conditions. Genetika USSR 8(9):163–165

Kalashnik NA, Khvostova VV, Cherny IV (1972) Genetic nature of spring wheat mutants. Genetika USSR 8(11):5–12

Kappert H (1915) Untersuchungen an Mark-, Kneifel- und Zuckererbsen und ihren Bastarden. Z Vererbungsl 13:1–57

Karlsson KE (1975) Linkage studies on the *lys*-gene in relation to some marker genes and translocations. Barley Genet III:536–541

Karlsson KE (1977) Linkage studies on a gene for high lysine content in Risø barley mutant *1508*. Barley Genet Newslett 7

Kashi A, Zoschke M (1968) Untersuchungen zur Samengewinnung der Futtererbsen vom *fasciata*-Typus in Rein- und Stützpflanzenkultur. Z Acker Pflanzenbau 128:273–308

Kataoka K (1974) Induction of high protein mutants in rice with ethyle imine treatment. Improvement of Plant Protein by Mutation. Gamma Field Symp 13:49–57

Kaul AK (1973) Mutation breeding and crop protein improvement. Nuclear Techniques for Seed Protein Improvement, pp 1–106, IAEA, Vienna

Kaul MLH (1977) Radiation-genetic studies in the garden pea. I. Two early flowering and ripening induced mutants. Curr Sci 46:198–200

Kaul MLH (1978) Mutation genetic studies in rice. I. Induced mutants of breeding value. Genetika Beograd 10:63–77

Kaul MLH (1980a) Seed protein variability in rice. Z Pflanzenzücht 84:302–312

Kaul MLH (1980b) Radiation genetic studies in garden pea. 9. Non-allelism of early flowering mutants and heterosis. Z Pflanzenzücht 84:192–200

Kaul AK (1981) Improvement of rice through induced mutations. "A fine grain recombinant of Ihona – 349". Curr Sci 50:770–771

Kaul MLH, Matta NK (1976) Radiation-genetic studies in the garden pea. III. Morphological variability, intercorrelations and genetic parameters. Genetika Beograd 8:37–47

Kaur S, Padmanabhan SY, Kaur P (1977) Induction of ressistance to blast disease *(Pyricularia oryzae)* in the high yielding variety, Ratna (IR 8 x TKM6). Induced Mutations against Plant Diseases, pp 147–156, IAEA, Vienna

Kaur S, Padmanabhan SY, Rao M (1976) Induction of resistance to blast disease (*Pyricularia oryzae* Cav.) in the high yielding variety, Ratna (IR 8 x TKM 6). Induced Mutations for Disease Resistance in Crop Plants (1975), pp 141–147, IAEA, Vienna

Kawai T (1967) New crop varieties by mutation breeding. Jpn Agric Res Quart 2:8–12

Kawai T (1968) Genetic studies on short-grain mutants in rice. Mutations in Plant Breeding II, pp 161–182, IAEA, Vienna

Kawai T (1969) Relative effectiveness of physical and chemical mutagens. Induced Mutations in Plants, pp 137–152, IAEA, Vienna

Kawai T (1974) Induction of blast-resistant mutations in rice. Induced Mutations for Disease Resistance in Crop Plants, 153, IAEA, Vienna

Kawai T, Narahari P (1971) Pattern of reduction of internode lengths and changes of some other characters in short-culm mutants in rice. Indian J Genet Plant Breed 31:421–441

Kellenbarger S, Silveira V, McCready RM, Owens HS, Chapman JL (1951) Inheritance of starch content in peas. Agron J 43:337–339

Kesavan V, Khan TN (1977) Induced mutations in winged bean. Seed Protein Improvement by Nuclear Techniques, 276, IAEA, Vienna

Khalatkar AS, Kashikar SG (1980) Sodiumazide mutagenicity in *Petunia hybrida*. Mutat Res 79: 81–85

Khamankar YG, Singhal NC, Jain HK (1978) Induced mutations and components of yield in bread wheat. Experimental Mutagenesis in Plants, pp 166–176, Sofia

Khan AH (1973) Improvement of quality and yield of wheat by mutation breeding. Wheat Inform Serv 36:7–8

Khan MRI, Gatehouse JA, Boulter D (1980) The seed proteins of cowpea *(Vigna unguiculata L. Walp.)*. J Exp Bot 31:1599–1611

Khangildin WV (1966) A new gene *leaf* inducing absence of leaflets in peas. Interaction between genes *Leaf* and Tl^w. Genetika USSR 1966(6):88–96

Kharkwal MC (1981) Induced micromutations in chickpea. Pulse Crops Newslett 1:17–18

Khvostova VV (1967) Mutagenic treatment of agricultural plants. Erwin Baur Ged Vorl 4: 21–27

Khvostova VV (1978) Cytogenetical analysis of mutants and its importance for breeding. Experimental Mutagenesis in Plants, pp 92–99, Sofia

Khvostova VV, Mozhaeva VS, Aigaes NS, Valeva SA (1965) Mutants induced by ionizing radiations and ethyleneimine in winter wheat. Mutat Res 2:339–344

Khvostova VV, Mozhaeva VS, Cherny IV (1969) Experimental mutagenesis in wheat. Genetika USSR 5(11):178–189

Kim SI, Saur L, Mossé J (1979) Some features of the inheritance of avenins, the alcohol soluble proteins of oat. Theor Appl Genet 54:49–54

King EE, Leffler HR (1979) Cottonseed protein: Its status and potential for improvement. Seed Protein Improvement in Cereals and Grain Legumes II, pp 385–391, IAEA, Vienna

Kiraly Z, Barabás Z (1974) Report on mutation breeding for mildew and rust resistance of wheat in Hungary. Induced Mutations for Disease Resistance in Crop Plants, pp 85–88, IAEA, Vienna

Kishore H, Das B, Subramanyam KN, Chandra R, Upadhya MD (1975) Use of induced mutations for potato improvement. Improvement of Vegetatively Propagated Plants through Induced Mutations, pp 77–82, IAEA, Vienna

Kivi EI (1967) The transgressive effect of the early-a^8 locus of "Mari" barley in a cross with six-rowed "Otra" barley. Erwin Baur Ged Vorl 4:155–159

Kivi EI, Ramm-Schmidt C (1969) Selection for resistance to sprouting in ^{60}Co-irradiated wheat. Induced Mutations in Plants, pp 535–540, IAEA, Vienna

Kivi EI, Rekunen M, Varis E (1974) Use of induced mutations in solving problems of northern crop production. Polyploidy and Induced Mutations in Plant Breeding, pp 187–194, IAEA, Vienna

Kleijer G, Fossati A (1976) Chromosomal location of a gene for male sterility in wheat *(Triticum aestivum)*. Wheat Inform Serv 41–42:12–13

Klein D, Frölich WG, Pollmer WG, Eberhard D (1980) Performance of high-lysine maize hybrids under western European climatic conditions. Improvement of Quality Traits of Maize for Grain and Silage Use. Nijhoff, The Hague, pp 75–90

Klein HD (1969) Male sterility in *Pisum*. Nucleus 12:167–172

Klein RG, Beeson WM, Cline TR, Mertz ET (1971) *Opaque-2* and *floury-2* corn studies with growing swine. J Anim Sci 32:256–261

Kleinhofs A, Kuo T (1980) Characterization of nitrate reductase deficient barley mutants. Mol Gen Genet 177:421–425

Kleinhofs A, Warner RL, Muehlbauer FJ, Nilan RA (1978) Induction and selection of specific gene mutations in *Hordeum* and *Pisum*. Mutat Res 51:29–35

Knapp E (1967) Ein neuer, gärtnerisch interessanter *Antirrhinum*-Typ mit offener Blüte. Züchter 37:140–142

Knott DR (1977) Studies on general resistance to stem rust in wheat. Induced Mutations against Plant Diseases, pp 81–88, IAEA, Vienna

Kobayashi T (1965) Radiation-induced beneficial mutants of sesame cultivated in Japan. Use Induced Mutations Plant Breed. Radiat Bot Suppl 5:399–403

Køie B, Doll H (1979) Protein and carbohydrate components in the Risø high-lysine barley mutant. Seed Protein Improvement in Cereals and Grain Legumes I, pp 208–214, IAEA, Vienna

Køie B, Kreis M (1977) Hordein and starch synthesis in developing high-lysine and normal barley seeds at different N-fertilizer levels. In: Miflin BJ, Zoschke M (eds) Carbohydrate and Protein Synthesis. Publ Comm Eur EUR 6043:137–150

Køie B, Doll H, Kreis M (1976a) Evaluation of a high-lysine barley gene using chromosome-doubled monoploids. Genetika Beograd 8:177–182

Køie B, Ingversen J, Andersen AJ (1976b) Composition and nutritional quality of barley protein. Evaluation of Seed Protein Alteration by Mutation Breeding, pp 55–61, IAEA, Vienna

Kolesnikova LG, Maksimova AD (1977) Effect of chemical mutagens on the content of dry matter and starch in potato tubers. Genetika USSR 13:1742–1755

Kondratenko LM (1975) The use of experimental mutagenesis in breeding of *Rubia tinctoria*. Genetika USSR 11(7):34–39

Konishi T (1976) The nature and characteristics of EMS-induced dwarf mutants in barley. Barley Genet III:181–189

Konzak CF (1956) Induction of mutations for disease resistance in cereals. Brookhaven Symp Biol 9:157–176

Konzak CF (1957) Genetic effects of radiation on higher plants. Q Rev Biol 32:27–45

Konzak CF (1972) Induced mutation research in wheat. Induced Mutations and Plant Improvement, pp 323–329, IAEA, Vienna

Konzak CF (1976) A review of semidwarfing gene sources and a description of some new mutants useful for breeding short-stature wheats. Induced Mutations in Cross-Breeding, pp 79–93, IAEA, Vienna

Konzak CF, Ramirez AI, Woo SC (1968) Development of genetic methods for wheat improvement. Mutations in Plant Breeding II, pp 183–192, IAEA, Vienna

Konzak CF, Nilan RF, Wagner J, Foster RJ (1973) Using mutagens and mutations in wheat breeding and genetic research. Proc 4th Int Wheat Genet Symp, Columbia, pp 275–281

Konzak CF, Line RF, Allan RE, Schafer JF (1977) Guidelines for the production, evaluation, and use of induced resistance to stripe rust in wheat. Induced Mutations against Plant Diseases, pp 437–459, IAEA, Vienna

Konzak CF, Mung NV, Warner RL, Rubenthaler GL, Finney PL (1978) Advances in technology and in genetics information for breeding improvements in wheat protein potentials. Seed Protein Improvement by Nuclear Techniques, pp 519–531, IAEA, Vienna

Koo FKS (1972) Mutation breeding in soybeans. Induced Mutations and Plant Improvement, pp 285–292, IAEA, Vienna

Kooistra E (1962) On the differences between smooth and three types of wrinkled peas. Euphytica 11:357–373

Kotvics G (1976) Mutation breeding in soybean. Genetika Beograd 8:129–135

Kramer HH, Whistler RL (1949) Quantitative effects of certain genes on the amylose content of corn endosperm starch. Agron J 41:409–411

Krausse GW, Goetze D, Sternkopf G (1974) Mutationsversuche bei Gerste. VI. Zur Verbesserung von Kornertrag, Standfestigkeit, Rohproteingehalt und Aminosäuren-Komposition mit Hilfe chemischer Mutagene. Arch Züchtungsforsch 4:97–116

Krausse GW, Melzer R, Schönleiter R (1974) Mutationsversuche bei Gerste. V. Arch Züchtungsforsch 4:65–74

Kreft I, Javornik B, Milkovic N (1975) Studies on the gene control of electrophoretic protein patterns in barley. Barley Genet 3:30–35

Kreis M (1978) Starch and free sugars during kernel development of Bomi barley and its high-lysine mutant *1508*. Seed Protein Improvement by Nuclear Techniques, pp 115–120, IAEA, Vienna

Kreis M, Doll H (1980) Starch and prolamin level in single and double high-lysine barley mutants. Physiol Plant 48:139–143

Kuckuck H, Peters R (1979) Induzierte Mutationen in selbstfertilen und inzuchtfesten Kulturformen aus der Kreuzung *Secale cereale* L x *Secale vavilovii* Grossh. und ihr Verhalten in Kreuzungen mit panmiktischen Sorten und selbststerilen Spontanmutationen. Z Pflanzenzücht 82:97–115

Kujala V (1953) Felderbse, bei welcher die ganze Blattspreite in Ranken umgewandelt ist. Arch Soc Zool Bot Fenn Vanamo 8:44–45

Kukimura H, Takemata T (1975) Induced quantitative variation by γ-rays and ethylene-imine in tuber bearing plants. Gamma Field Symp 14:25–38

Kukimura H, Ikeda F, Fujita H, Maeta T, Nakajima K, Katagiri K, Nakahira K, Somegou M (1975) Genetical, cytological and physiological studies on the induced mutants with special regard to effective methods for obtaining useful mutants in perennial woody plants. Improvement of Vegetatively Propagated Plants through Induced Mutations, pp 83–104, IAEA, Vienna

Kulkarni LG (1969) Induction of useful mutations in castor. Radiations and Radiomimetic Substances in Mutation Breeding, Bombay, pp 293–299

Kulshrestha VP, Mathur VS (1978) Study on induced mutants resembling commercial varieties in bread wheat. Theor Appl Genet 53:125–128

Kumar D (1976) Yielding ability of the mutants induced from wheat var. K68 against released varieties at different levels of nitrogen. Wheat Inform Serv 43:13–15

Kumar D (1977a) Studies on Co^{60} gamma-ray induced variability in common wheat cultivar K68. Egypt J Genet Cytol 6:229–243

Kumar D (1977b) A promising semi-dwarf mutant in wheat variety K68. Environ Exp Bot 17:79–85

Kumar D (1978) Soft and hard grain mutants induced in common wheat variety K68. Wheat Inform Serv 45/46:5–7

Kumar H, Agrawal RK (1981) Combining ability analysis of certain productive traits in pea (*Pisum sativum* L.). Z Pflanzenzücht 86:110–116

Kumar PR (1972) Radiation induced variability in improvement of brown sarson. Radiat Bot 12:309–313

Kumar PR, Das K (1977) Induced quantitative variation in self-compatible and self-incompatible forms in *Brassica*. Indian J Genet Plant Breed 37:5–11

Kumar S, Bansal HC, Singh D, Swaminathan MS (1967) Pathways of height reduction in induced dwarf mutations in barley. Z Pflanzenzücht 57:317–324

Künzel G, Scholz F (1972) Über Nutzungsmöglichkeiten von induzierten Chromosomenmutationen bei Kulturpflanzen. Kulturpfl 20:225–262

Künzel G, Scholz F (1973) Mögliche Systeme zur Nutzung von *Ms/ms*-Genen für die Erzeugung von Hybridsaatgut bei Gerste. Tag Ber Akad Landwirtsch Wiss DDR 122:281–287

Kushihuchi K, Okada M, Watanabe Sh (1974) Breeding for high protein rice by the use of artificially induced mutants. Gamma Field Symp 13:37–45

Kwon SH (1974) Selection for blast-resistant mutants in irradiated rice populations. Induced Mutations for Disease Resistance in Crop Plants, pp 103–115, IAEA, Vienna

Kwon SH, Oh JH (1977) Selection for blast-resistant mutants in irradiated rice populations. Induced Mutations against Plant Diseases, pp 131–145, IAEA, Vienna

Lacey CND, Campbell AI (1979) The characteristics and stability of Cox's Orange Pippin apple mutants showing different growth habits. Euphytica 28:119–126

Lafiandra D, Dalignano GB, Colaprico G (1979) Protein content and amino acid composition of seed in varieties of *Vicia faba*. Z Pflanzenzücht 83:308–314

Lamprecht H (1952) Polymere Gene und Chromosomenstruktur bei *Pisum*. Agri Hort Genet 10:158–168

Lapins KO (1973) Induced mutations in fruit trees. Induced Mutations in Vegetatively Propagated Plants, pp 1–19, IAEA, Vienna

Lapins KO (1975) Polyploidy and mutations induced in apricot by colchicine treatment. Can J Genet Cytol 17:591–599

Larik AS (1975) Induced mutations in quantitative characters of *Triticum aestivum* L. Genet Agrar 29:241–250

Lata P (1980) Effect of ionizing radiation on roses: Induction of somatic mutations. Eviron Exp Bot 20:325–333

Laughnan JR (1953) The effect of the *sh2* factor on carbohydrate reserves in the mature endosperm of maize. Genetics 38:485–499

Lawrence GJ, Shepherd KW (1981a) Chromosomal location of genes controlling seed proteins in species related to wheat. Theor Appl Genet 59:25–31

Lawrence GJ, Shepherd KW (1981b) Inheritance of glutenin protein subunits of wheat. Theor Appl Genet 60:333–337

Lawrence JM (1976) Environmental influences on wheat lysine content. J Agric Food Chem 24: 356–358

Lee Choo Kiang, Halloran GM (1977) Production and characterisation of an early dwarf mutant in soybean [*Glycine max* (L.) Merrill]. Mutat Res 43:223–230

Lehmann ChrO, Nover I, Scholz F (1975) The Gatersleben barley collection and its evaluation. Barley Genet 3:64–69

Li HW, Hu CH, Wu HP (1968) Induced mutation breeding of rice in the Republic of China. Rice Breeding with Induced Mutations, pp 17–24, IAEA, Vienna

Li HW, Hu CH, Woo SC (1971) Development of rice mutation breeding in Taiwan. Rice Breeding with Induced Mutations III, pp 69–76, IAEA, Vienna

Lind V, Gaul H (1976) Studies of pleiotropic genes and their character complexes in *erectoides* mutants. Barley Genet 3:171–180

Line RF, Konzak CF, Allan RE (1974) Evaluating resistance to *Puccinia striiformis* in wheat. Induced Mutations for Disease Resistance in Crop Plants, pp 125–132, IAEA, Vienna

Little R (1971) An attempt to induce resistance to *Septoria nodorum* and *Puccinia graminis* in wheat using gamma rays, neutrons and EMS as mutagenic agents. Mutation Breeding for Disease Resistance, pp 139–149, IAEA, Vienna

Lodha ML, Mali PC, Agarwal AK, Metha SL (1974) Changes in soluble protein and isoenzymes in normal and *opaque-2 Zea mays* endosperm during grain development. Phytochemistry 13: 539–542

Loesch PJ, Wiser WJ, Booth GD (1978) Emergence comparisons between *opaque* and normal segregates in two maize synthetics. Crop Sci 18:802–805

Lolas GM, Markakis P (1975) Phytic acid and other phosphorus compounds of beans (*Phaseolus vulgaris* L.). J Agric Food Chem 23:13–15

Lönnig WE (1982) Dominance, overdominance and epistasis in *Pisum sativum* L. Theor Appl Genet 63:255–264

Loose R de (1979) Radiation induced chimeric rearrangement in flower structure of *Rhododendron simsii* Planch. (*Azalea indica* L.). Use of recurrent irradiation. Euphytica 28:105–113

Lowry KL, Caton IE, Foad DE (1974) Electrophoretic methods for detecting differences in seed proteins of soybean varieties and induced mutants. J Agric Food Chem 22:1043–1045

Lu YC (1970) Mutation breeding for rust resistance in soybeans. Improving Plant Protein by Nuclear Techniques, pp 185–187, IAEA, Vienna

Lundqvist U (1976) Locus distribution of induced *eceriferum* mutants in barley. Barley Genet 3:162–163

Lundqvist U (1978) Locus distribution of induced *eceriferum* mutants in barley. Experimental Mutagenesis in Plants, pp 263–265, Sofia

Lundqvist U, Wettstein-Knowles P v, Wettstein D v (1968) Induction of *eceriferum* mutants in barley by ionizing radiations and chemical mutagens. II. Hereditas 59:473–504

Luse RA, Rachie KO (1979) Seed protein improvement in tropical food legumes. Seed Protein Improvement in Cereals and Grain Legumes II, pp 87–102, IAEA, Vienna

Ma Y, Nelson OE (1975) Amino acid composition and storage proteins in two new high lysine mutants in maize. Cereal Chem 52:412–419

MacKey J (1956) Mutation breeding in Europe. Brookhaven Symp Biol 9:141–152

MacKey J (1961) Methods of utilizing induced mutation in crop improvement. Mutation and Plant Breeding NAS-NRC 891:336–364

Mackill DJ, Rutger JN (1979) The inheritance of induced-mutant semidwarfing genes in rice. J Hered 70:335–341

Madhusudana RG, Siddiq EA (1976) Studies on induced variability for amylose content with reference to yield components and protein characteristics in rice. Environ Exp Bot 16:177–188

Mahna S, Singh D (1975) Induced floral mutation in *Physalis ixocarpa* Brot. Curr Sci 44:21–22

Majumder B (1965) Induction of an early mutant in *Oryza sativa* Linn. by radioactive phosphorus. Trans Bose Res Inst Calcutta 28(2):57–64

Majumder B (1966) Cytogenetical studies on some X-rays and ^{32}P induced mutants of rice. J Indian Bot Soc 45:257–265

References

Majumder B (1969) Analysis of the rice mutants induced by X-rays and radioactive phosphorus. Genet Agrar 23:231–234

Malepszy S, Grunewaldt J, Muluszynski M (1977) Über die Selektion von Mutanten in Zellkulturen aus haploider *Nicotiana sylvestris* Spegazz et Comes. Z Pflanzenzücht 79:160–166

Malik CP, Mary TN (1971) Mutation breeding in some rye grass species. Intern Symp Use Isotopes Radiat Agric Animal Husbandry Res New Delhi, pp 39–45

Malik CP, Mary TN (1974) Induced mutations in *Lolium* species. Genet Polon 15:63–83

Mallick EH (1978) Induction of mutations in regard to some morphological and grain characteristics in photosensitive and photoinsensitive rice varieties. Genet Polon 19:39–47

Marie R (1967) Experience in the induction of rice mutants in France. Erwin Baur Ged Vorl 4:227–230

Marie R (1970) Rice breeding with induced mutations in France. Rice Breeding with Induced Mutations II, pp 21–24, IAEA, Vienna

Marie R, Tinarelli A (1974) Rice mutants and resistance to blast disease. Induced Mutations for Disease Resistance in Crop Plants, 154, IAEA, Vienna

Marx GA, Hagedorn DJ (1962) Fasciation in *Pisum*. J Hered 53:31–43

Mar'yushkin VF, Shkvarnikov PK, Sichkar VI (1977) Induced variation of winter wheat yield components. Genetika USSR 13:1701–1709

Matsuo T, Yamaguchi H (1967) Studies on the mutation breeding in Japan. Erwin Baur Ged Vorl 4:211–225

McKenzie RIH, Martens JW (1974) Breeding for stem rust resistance in oats. Induced Mutations for Disease Resistance in Crop Plants, pp 45–48, IAEA, Vienna

McKenzie RIH, Martens JW, Harder DE, Brown PD (1976) Screening oat populations for rust resistant mutants. Induced Mutations for Disease Resistance in Crop Plants (1975), pp 159–163, IAEA, Vienna

Mehta NP (1972) Use of induced mutations in breeding better branched wheats. Indian J Genet Plant Breed 32:441–444

Mehta SL, Dongre HB, Johari RP, Lodha ML, Naik MS (1979) Biochemical constraints that determine protein quality and grain yield in cereals. Seed Protein Improvement in Cereals and Grain Legumes I, pp 241–256, IAEA, Vienna

Melzer R, Sackewitz H (1979) Probleme der Mutationsauslösung und Nutzung von Mutationen bei der Zuckerrübe (*Beta vulgaris* L.) Arch Züchtungsforsch 9:65–72

Mertz ET, Bates LS, Nelson OE (1964) Mutant gene that changes protein composition and increases the lysine content of maize endosperm. Science 145:279–286

Meyer VG (1969) Some effects of genes, cytoplasm, and environment on male sterility in cotton *(Gossypium)*. Crop Sci 9:237–242

Miah AJ, Awan MA (1971) Induced mutations in rice. Rice Breeding with Induced Mutations III, pp 77–89, IAEA, Vienna

Miah AJ, Bhatti IM (1968) Evolution of new rice varieties by induced mutations to increase yield and resistance to diseases and to improve seed quality. Rice Breeding with Induced Mutations, pp 75–96, IAEA, Vienna

Miah AJ, Bhatti IM, Awan A, Bari G (1970) Improvement of rice varieties by induced mutations to increase yield per acre and resistance to diseases and to improve seed quality. Rice Breeding with Induced Mutations II, pp 69–76, IAEA, Vienna

Mian HR, Kuspira J, Walker GWR, Muntjewerff N (1974) Histological and cytochemical studies of five genetic male-sterile lines of barley *(Hordeum vulgare)*. Can J Genet Cytol 16:355–379

Michael G (1963) Einfluß der Düngung auf Eiweißqualität und Eiweißfraktion der Nahrungspflanzen. Qual Plant Mater Veg 10:248–264

Micke A (1969a) Improvement of low yielding sweet clover mutants by heterosis breeding. Induced Mutations in Plants, pp 541–550, IAEA, Vienna

Micke A (1969b) Utilization of induced mutants as hybrids. Genet Agrar 23:262–268

Micke A (1974) Heterosis bei Kreuzung von Mutanten derselben Ausgangsform. Ber Arb Tag Arb Gemeinsch Saatzuchtleiter Gumpenstein, pp 314–328

Micke A (1976) Hybrid vigour in mutant crosses. Prospects and problems of exploitation studied with mutants of sweet clover. Induced Mutations in Cross-Breeding, pp 199–218, IAEA, Vienna

Micke A (1978) Änderung von Vegetationsrhythmus und Reifezeit von Kulturpflanzen durch induzierte Mutationen. Ber Arb Tag Arb Gemeinsch Saatzuchtleiter Gumpenstein, pp 239–265

Micke A (1979) Use of mutation induction to alter the ontogenetic pattern of crop plants. Inst Rad Breed, Ohmiya, Japan. Gamma Field Symp 18:1–23

Mickhailov L, Popova D (1977) Effect of polygenic mutating on the degree of heterotic effect in hybrid progeny. Genetika USSR 13:217–221

Miedema P (1973) The use of adventitious buds to prevent chimerism in mutation breeding of potato. Euphytica 22:209–218

Miflin BJ, Shewry PR (1979) The biology and biochemistry of cereal seed prolamins. Seed Protein Improvement in Cereals and Grain Legumes I, pp 137–157, IAEA, Vienna

Mikaelsen K (1973) Studies on the inheritance of the high seed protein content of an *erectoides* mutant *(H-14)* in barley. Nuclear Techniques for Seed Protein Improvement, 217, IAEA, Vienna

Mikaelsen K, Kartoprawiro MI (1973) Studies on inter- and intraplant variation in seed protein content of rice. Nuclear Techniques for Seed Protein Improvement, pp 199–202, IAEA, Vienna

Mikaelsen K, Saja Z, Simon J (1971) An early maturing mutant. Its value in breeding for disease resistance in rice. Rice Breeding with Induced Mutations III, pp 97–101, IAEA, Vienna

Milenkov M (1978) Some results from gamma-irradiation of the cherry (*P. avium* L.). Experimental Mutagenesis in Plants, pp 408–416, Sofia

Milner M (1975) Need for improved plant proteins in world nutrition. J Agric Food Chem 22:548–549

Misra PS (1978) Corn endosperm mutants: Amino acid composition and protein distribution. Indian J Genet Plant Breed 38:186–192

Misra PS, Jambunathan R, Mertz ET, Glover DV, Barbosa HM, McWhirter KS (1972) Endosperm protein synthesis in maize mutants with increased lysine content. Science 176:1425–1426

Misra PS, Mertz ET, Glover DV (1975) Studies on corn proteins. VI. Endosperm protein changes in single and double endosperm mutants of maize. Cereal Chem 52:161

Misra PS, Mertz ET, Glover DV (1976) Studies on the corn proteins. X. Polypeptide molecular weight distributions in the Londry-Moureaux fractions of normal and mutant endosperm. Cereal Chem 53:705–711

Misra RN, Jachuck PJ, Sampath S (1971) Induced mutations for practical utility in rice. Intern Symp Use Isotopes Radiat Agric Anim Husbandry Res New Delhi, pp 230–236

Mitra GC (1977) Inheritance of a male sterile complex mutant in white jute (*Corchorus capsularis* L.). Genetica 47:71–72

Mitra RK, Bhatia CR (1979) Bioenergetic considerations in the improvement of oil content and quality in oil-seed crops. Theor Appl Genet 54:41–47

Mitra RK, Bhatia CR, Rabson R (1979) Bioenergetic cost of altering the amino acid composition of cereal grains. Cereal Chem 56:249–252

Moës A (1965) L'amélioration de la productivité de l'orge de printemps par hybridation et par mutation. Use Induced Mutations Plant Breed. Radiat Bot Suppl 5:559–571

Moës A (1969) Mutation breeding in tetraploid *Gladiolus* (*Gladiolus gandavensis* v.h.). Genet Agrar 23:176–179

Moës A (1977) Le mutant 5455, orge resistant a l'oidium. Induced Mutations against Plant Diseases, pp 343–346, IAEA, Vienna

Moh CC (1971) Mutation breeding in seed-coat color of beans (*Phaseolus vulgaris* L.). Euphytica 20:119–125

Moh CC (1972) Induced seed-coat colour mutations in beans and their significance for bean improvement. Induced Mutations and Plant Improvement, pp 67–72, IAEA, Vienna

Moh CC (1973) Induced mutations in coffee. Induced Mutations in Vegetatively Propagated Plants, pp 129–136, IAEA, Vienna

Mohamed HA, Omar AM, Barahamtoushy ME (1965) Effect of radioactive cobalt on characters of some wheat varieties. Wheat Inform Serv 19, 20:16–17

Monti LM (1972) Mutants of agronomic value induced in the tomato variety Moneymaker by gamete irradiation. Genet Agrar 26:111–118

Monti LM, Frusciante L (1978) Pea breeding by using genes drastically influencing the leaf morphology. Genet Agrar 32:365–373
Monti LM, Scarascia Mugnozza GT (1967) Mutazioni per precocita e ramosita indotte in pisello. Genet Agrar 21:301–312
Monti LM, Scarascia Mugnozza GT (1970) Impiego di una mutatione per precocita in un programma di incrocio per il miglioramento genetico del pisello da industria. Genet Agrar 24:195–206
Monti LM, Scarascia Mugnozza GT (1972) Ricerche per miglioramento genetico per precocita in pisello. Genet Agrar 26:119–127
Monyo JH, Sugiyama T (1976) Effect of chemical and physical mutagens on protein content and some agronomic characters in rice. Evaluation of Seed Protein Alterations by Mutation Breeding, 144, IAEA, Vienna
Monyo JH, Sugiyama T (1978) Improvement of seed protein in rice through mutation breeding. Seed Protein Improvement by Nuclear Techniques, pp 181–189, IAEA, Vienna
Monyo JH, Sugiyama T, Kihupi AN (1979) Potentially high-yielding and high protein rice in induced mutation breeding. Seed Protein Improvement in Cereals and Grain Legumes II, pp 293–301, IAEA, Vienna
Morrison WR (1978) Wheat lipid composition. Cereal Chem 55:548–558
Morsi LR, Abo-Elenein RA, Mahmoud IM (1977) Studies on the induction of new genetic variability for quantitative traits by gamma-rays and N-nitroso-N-methyl-urethane in barley. Egypt J Genet Cytol 6:244–258
Mortiniello P, Lorenzoni C, Stanca AM, Maggiore T, Gentinetta E, Salamini F (1978) Seed quality differences between normal, *floury-2* and *opaque-2* maize inbreds. Euphytica 27:411–417
Mosconi C (1967) Results of a three year test with mutants of *durum* wheat in Marche region. Genet Agrar 21:173–185
Mossé J (1966) Alcohol-soluble proteins of cereal grains. Fed Proc 25:1663–1669
Mossé J (1968) Progrès en chimie agricole et alimentaire. Hermann, Paris, pp 47–81
Muench SR, Lejeune AJ, Nilan RA, Kleinhofs A (1976) Evidence of two independent high lysine genes in barley. Crop Sci 16:283–285
Muhammed A, Shakoor A, Tahir Nadeem M, Ali A, Ifzal SM, Sadiq M (1976) Seed protein improvement in wheat by mutation breeding. Evaluation of Seed Protein Alterations by Mutation Breeding, pp 107–117, IAEA, Vienna
Müller HP (1976) Seed protein content of healthy and heavily virus-infected pea genotypes. Pisum Newslett 8:45
Müller HP (1978a) The seed proteins in gene-ecological investigations. Legume Res 2:29–40
Müller HP (1978b) Die genetische Steuerung der Zusammensetzung von Samenproteinen. Z Pflanzenkr Pflanzenschutz 85:210–217
Müller HP (1979) The genetic control of seed protein polymorphism in *Pisum*. Proc 1st Mediterr Conf Genet, Cairo, pp 747–764
Müller HP (1980) Biochemische Darstellung und gelelektrophoretische Charakterisierung der Samenproteine von Leguminosen. Gött Pflanzenzüchter Seminar 4:105–116
Müller HP, Gottschalk W (1973) Quantitative and qualitative situation of seed proteins in mutants and recombinants of *Pisum sativum*. Nuclear Techniques for Seed Protein Improvement, pp 235–253, IAEA, Vienna
Müller HP, Gottschalk W (1978) Gene-ecological investigations on the protein production of different *Pisum* genotypes. Seed Protein Improvement by Nuclear Techniques, pp 301–314, IAEA, Vienna
Müller HP, Werner S (1979) Seed protein characteristics of *Pisum* varieties, mutants and recombinants. Proc Symp Seed Proteins of Dicotyledonous Plants, Gatersleben 1977. Abh Akad Wiss DDR, pp 189–209
Munck L (1972) Improvement of nutritional value in cereals. Hereditas. 72:1–128
Munck L (1975) Developing the concept of quality in barley breeding programmes for malt and feed. Barley Genet 3:526–535
Munck L (1976) Aspects of selection design and use of high lysine cereals. Evaluation of Seed Protein Alterations by Mutation Breeding, pp 3–17, IAEA, Vienna

Munck L, Karlsson KE, Hagberg A, Eggum BO (1970) Gene for improved nutritional value in barley seed protein. Science 168:985–987

Munck L, Karlsson KE, Hagberg A (1979) Selection and characterization of a high-protein, high-lysine variety from the World Barley Collection. Barley Genet 2:544–558

Mutation Breeding Newsletter (1972–1982). FAO/IAEA, Issues 1–19

Müntz K (1975) Wege zur Steigerung des Proteinertrages von Kulturpflanzen. Arch Züchtungsforsch 5:253–273

Müntz K, Hammer K, Lehmann C, Meister A, Rudolph A, Scholz F (1979) Variability of protein and lysine content in barley and wheat specimens from the world collection of cultivated plants at Gatersleben. Seed Protein Improvement in Cereals and Grain Legumes II, pp 183–200, IAEA, Vienna

Müntzing A, Bose S (1969) Induced mutations in inbred lines of rye. I. EMS treatments. Hereditas 62:382–408

Murray MJ (1969) Successful use of irradiation breeding to obtain *Verticillium*-resistant strains of peppermint, *Mentha piperita* L. Induced Mutations in Plants, pp 345–371, IAEA, Vienna

Murray MJ (1971) Additional observations on mutation breeding to obtain *Verticillium*-resistant strains of peppermint. Mutation Breeding for Disease Resistance, pp 171–195, IAEA, Vienna

Murty BR (1974) Mutation breeding for resistance to downy mildew and ergot in *Pennisetum* and to *Ascochyta* in chickpea. Induced Mutations for Disease Resistance in Crop Plants (1975), pp 89–100, IAEA, Vienna

Murty BR (1976) Mutation breeding for resistance to downy mildew and ergot in *Pennisetum* and to *Ascochyta* in chickpea. Induced Mutations for Disease Resistance in Crop Plants, pp 165–181, IAEA, Vienna

Muszynski S, Darlewska M, Dabrowska A (1976) Mutation induction in rye (*Secale cereale* L.) with physical and chemical mutagens. Genetika Beograd 8:203–205

Nadeem MT, Shakoor A, Ali A, Sadiq M (1978) Seed protein improvement in wheat and pulses through induced mutation. Seed Protein Improvement by Nuclear Techniques, pp 59–67, IAEA, Vienna

Nagl K (1973) Mutation breeding for improved protein in *durum* wheat. Nuclear Techniques for Seed Protein Improvement, pp 181–192, IAEA, Vienna

Nagl K (1976) Studies on cross progenies (F_2) of two drastic morphological high-protein mutants in *durum* wheat. Evaluation of Seed Protein Alterations by Mutation Breeding, pp 137–138, IAEA, Vienna

Nagl K (1978a) An induced tetrapoloid *sphaerococcoid* semidwarf mutant as a useful cross-parent for stem height reduction and protein improvement in hexaploid wheat. Seed Protein Improvement by Nuclear Techniques, pp 69–78, IAEA, Vienna

Nagl K (1978b) Breeding value of radio-induced mutants of *Vicia faba* var. minor. Seed Portein Improvement by Nuclear Techniques, pp 243–252, IAEA, Vienna

Nakai H, Goto M (1977) Mutation breeding of rice for bacterial leaf-blight resistance. Induced Mutations against Plant Diseases, pp 171–186, IAEA, Vienna

Nakai H, Saito M (1979) Increasing mutagenic efficiency of gamma-radiation in rice by irradiation of seeds at extremely low temperature. Euphytica 28:697–704

Nakajima K (1970) Gamma ray induced sports from a rose variety (Peace). Inst Radiat Breed Techn News 4

Nakajima K (1973) Induction of useful mutations of mulberry and roses by gamma rays. Induced Mutations in Vegetatively Propagated Plants, pp 105–116, IAEA, Vienna

Nakajima K (1977) Studies on the effective methods for induction of mutations of vegetatively propagated plants by the use of the gamma field. Bull Inst Rad Breed 4:1–105, Ohmiya, Japan

Nakayama A (1973) Induction of the somatic mutations in tea plants by gamma irradiation. Gamma Field Symp 12:37–45

Nakornthap A (1965) Radiation-induced somatic mutations in the ornamental *Canna*. Use Induced Mutations Plant Breed. Radiat Bot Suppl 5:705–712

Nalampang A (1975) Grain legumes in Thailand. Induced Mutations for the Improvement of Grain Legumes in South East Asia, pp 77–83, IAEA, Vienna

References

Narahari P (1969) X-ray induced mutants in an *indica* rice variety. Radiations and Radiomimetic Substances in Mutation Breeding; Bombay, pp 172–183

Narahari P, Bhatia CR, Gopalakrishna T, Mitra RK (1976) Mutation induction of protein variability in wheat and rice. Evaluation of Seed Protein Alterations by Mutation Breeding, pp 119–127, IAEA, Vienna

Narain A, Prakash S (1969) High yield radiation induced dwarf mutant in mustard (*Brassica juncea* Coss.). Radiations and Radiomimetic Substances in Mutation Breeding, Bombay, pp 414–417

Nardi S, Brunori A (1980) Protein accumulation and total RNA content in the endosperm of developing grain of *Triticum durum* in response to nitrogen fertilization. Genet Agrar 34: 113–121

Nayak P, Padmanabhan SY (1974) Induced mutations for disease resistance in rice. Induced Mutations for Disease Resistance in Crop Plants, pp 154–155, IAEA, Vienna

Nayar GG, George KP (1969) X-ray induced early flowering, appressed pod mutant in *Brassica juncea* Coss. Radiations and Radiomimetic Substances in Mutation Breeding, Bombay, pp 409–413

Nehring K (1963) Wege zur Lösung des Eiweiß-Problems in der Landwirtschaft der Deutschen Demokratischen Republik. Dtsch Akad Landwirtschaftswiss, Berlin VI

Nelson OE (1967) A mutant gene affecting protein synthesis in the maize endosperm. Genet Agrar 21:209–230

Nelson OE (1969a) Genetic modifications of protein quality in plants. Adv Agron 21:171–194

Nelson OE (1969b) The modification by mutation of protein quality in maize. New Approaches to Breeding for Improved Plant Protein, pp 41:54, IAEA, Vienna

Nelson OE (1970) Improvement of plant protein quality. Improving Plant Protein by Nuclear Techniques, pp 43–49, IAEA, Vienna

Nelson OE (1979) Inheritance of amino acid content in cereals. Seed Protein Improvement in Cereals and Grain Legumes I, pp 79–87, IAEA, Vienna

Nelson OE (1980) Genetic control of polysaccharide and storage protein synthesis in the endosperms of barley, maize and sorghum. Adv Cereal Sci Technol 3:41–71

Nelson OE, Mertz ET (1973) Nutritive value of *floury-2* maize. Nuclear Techniques for Seed Protein Improvement, pp 321–328, IAEA, Vienna

Nelson OE, Rines HW (1962) The enzymatic deficiency in the *waxy* mutant of maize. Biochem Biophys Res Commun 9:297–300

Nelson OE, Tsai CY (1964) Glucose transfer from adenosine diphosphate-glucose to starch in preparations of waxy seeds. Science 145:1194–1195

Nelson OE, Mertz E, Bates L (1965) Second mutant gene affecting the amino acid pattern of maize endosperm proteins. Science 150:1469–1470

Nerkar YS (1971) Studies on induced variation, heritability and response to selection for low neurotoxin content in *Lathyrus sativus*, a protein rich pulse. Intern Symp Use Isotopes Radiat Agric Anim Husbandry Res New Delhi, pp 201–205

Nerkar YS (1972) Induced variation and response to selection for low neurotoxin content in *Lathyrus sativus*. Indian J Genet Plant Breed 32:175–180

Nerkar YS (1976) Mutation studies in *Lathyrus sativus*. Indian J Genet Plant Breed 36:223–229

Nettevich ED, Sergeev AV (1973) Induced stunted mutant of spring barley. Genetika USSR 9(8): 37–44

Nettevich ED, Tzukanov AF (1976) Genetic characteristics of the induced semidwarf mutant variety "Fakel" of spring barley. Genetika USSR 12(12):22–31

Nigmatullin FG, Muninshoeva Z (1977) Investigations of the genetic nature of mutants in barley. Genetika USSR 13:1149–1152

Nilan RA (1964) The cytology and genetics of barley, 1951–1962. Monograph Suppl No 3, Res Studies Washington State Univ 32(1):1–278

Nilan RA (1966) Barley cytogenetics and breeding. Mutations in Plant Breeding, pp 177–185, IAEA, Vienna

Nilan RA (1972) Induced mutations and winter barley improvement. Induced Mutations and Plant Improvement, pp 349–351, IAEA, Vienna

Nilan RA, Powell JB, Conger BV, Muir CE (1968) Induction and utilization of inversions and mutations in barley. Mutations in Plant Breeding II, pp 193–203, IAEA, Vienna

Nilan RA, Sideris EG, Kleinhofs A, Sander C, Konzak CF (1973) Azide – a potent mutagen. Mutation Res 17:142–144

Nishida T (1973) Induction of the somatic mutations in deciduous fruit trees by gamma irradiation. Gamma Field Symp 12:1–15, Ohmiya, Japan

Norain A, Prakash S (1969) High yielding radiation induced dwarf mutant in mustard *(Brassica juncea)*. Radiations and Radiomimetic Substances in Mutation Breeding, Bombay, pp 414–417

Nwasike ChC, Mertz ET, Pickett RC, Glover DV, Chibber BAK, van Scoyoc StW (1979) Lysine level in solvent fractions of pearl millet. J Agric Food Chem 27:1329–1331

Nybom N, Koch A (1965) Induced mutations and breeding methods in vegetatively propagated plants. Use Induced Mutations Plant Breed. Radiat Bot Suppl 5:661–678

Offutt MS, Riggs RD (1970) Radiation-induced resistance to root-knot nematodes in Korean lespedeza. Crop Sci 10:49–50

Ohba K (1971) Studies on the radiation breeding of forest trees. Bull Inst Rad Breed 2:1–102

Ojomo OA (1976) Breeding for high protein content and quality in cowpea by nuclear techniques. Evaluation of Seed Protein Alterations by Mutation Breeding, 210, IAEA, Vienna

Ojomo OA, Omueti O (1978) Induction of mutants with higher quantity and better quality protein in cowpea *(Vigna unguiculata)*. Seed Protein Improvement by Nuclear Techniques, 275, IAEA, Vienna

Okuno K, Glover DV (1980) Effect of *sugary-1, brittle-1* and double recessive mutant on carbohydrate in developing kernels of maize. Ann Rep Div Genet Nat Inst Agric Sci Japan, pp 34–35

Okuno K, Kawai T (1977) Induction of short-culm mutations and inheritance of induced short-culm mutants in rice. Gamma Field Symp 16:39–62, Ohmiya, Japan

Okuno K, Kawai T (1978a) Variations of internode length and other characters in induced long-culm mutants of rice. Jpn J Breed 28:243–250

Okuno K, Kawai T (1978b) Genetic analysis of induced long-culm mutants in rice. Jpn J Breed 28:336–342

Okuno K, Kawai T (1979) Effect of short-culm mutant genes in internode elongation in rice. Ann Rep Nat Inst Agric Sci Japan 1979:38–39

Oram RN, Doll H (1981) Yield improvement in high lysine barley. Austr J Agric Res (in press)

Orlyuk AP (1973) Variability and heritability of grain quality in mutants of winter wheat. Genetika USSR 9(10):7–15

Ottaviano E, Camussi A (1975) Selection of lines in relation to the effect of the *opaque-2* gene on endosperm structure in maize. Z Pflanzenzücht 74:119–129

Pacucci G, Scarascia Mugnozza GT (1974) Agronomic evaluation of the *durum* wheat mutant line Casteldelmonte. Polyploidy and Induced Mutations in Plant Breeding, pp 247–258, IAEA, Vienna

Pacucci G, Scarascia Mugnozza GT, Bozzini A, Cavazza L, Mosconi C (1966) Comparison between lines of *durum* wheat imported or newly formed, grown at Lazio and Puglia during 1965. Genet Agrar 20:85–97

Padma A, Reddy GM (1977) Genetic behavior of five induced dwarf mutants in an *indica* rice cultivar. Crop Sci 17:860–863

Padmanabhan SY, Kaur S, Rao M (1976) Induction of resistance to bacterial leaf blight *(Xanthomonas oryzae)* disease in the high yielding variety Vijaya (IR8 x T90). Induced Mutations for Disease Resistance in Crop Plants 1975, pp 183–194, IAEA, Vienna

Padmanabhan SY, Kaur S, Rao M (1977) Induction of resistance to bacterial leaf-blight *(Xanthomonas oryzae)* disease in the high yielding variety Vijaya (IR8 x T90). Induced Mutations against Plant Diseases, pp 187–198, IAEA, Vienna

Paez AV (1973) Protein quality and kernel properties of modified *opaque-2* endosperm corn involving a recessive allele at the *sugary-2* locus. Crop Sci 13:633–636

Pakendorf KW (1974) Studies on the use of mutagenic agents in *Lupinus*. II. Z Pflanzenzücht 72:152–159

Panda BS (1981) Success in mutation breeding of sesame. Mutation Breed Newslett 18:10–11

Panton CA (1975) Effect of γ-irradiation on the protein content of soya bean. Breeding for Seed Protein Improvement Using Nuclear Techniques, 176, IAEA, Vienna

Panton CA, Coke LB, Pierre RE (1973) Seed protein improvement in certain legumes through induced mutation. Nuclear Techniques for Seed Protein Improvement, pp 269–271, IAEA, Vienna

Pape G, I-Sun-Shen, Schön WJ (1969) Zur quantitativen Bestimmung der Reservekohlenhydrate in Pflanzenmaterial. Z Acker- Pflanzenbau 130:16–23

Parliman BJ, Stushnoff C (1979) Mutant induction through adventitious buds of *Kohleria*. Euphytica 28:521–530

Parodi PC (1975) Improvement of protein content in wheat through mutation breeding. Breeding for Seed Protein Improvement Using Nuclear Techniques, pp 31–34, IAEA, Vienna

Parodi PC, Nebreda M (1977) Reaccion a enfermedades de seis genotipos de trigo (*Triticum* ssp.) tratados con rayos gamma. Induced Mutations against Plant Diseases, pp 375–383, IAEA, Vienna

Parodi PC, Nebreda M (1978) Mutation breeding to increase protein content in wheat (*Triticum* ssp.). Seed Protein Improvement by Nuclear Techniques, pp 33–39, IAEA, Vienna

Parodi PC, Nebreda M (1979) Protein and yield response of six wheat (*Triticum* ssp.) genotypes to gamma radiation. Seed Protein Improvement in Cereals and Grain Legumes II, pp 201–209, IAEA, Vienna

Parodi PC, Diaz MS, Nebreda IM (1976) Improvement of seed protein content in wheat (*Triticum* ssp.) by mutation breeding. Evaluation of Seed Protein Alterations by Mutation Breeding, 137, IAEA, Vienna

Pásztor K (1978) Effect of mutagens on the variability of morphological characters in maize. Acta Agron Acad Sci Hung 27:481–488

Pásztor K (1979) Applicability of morphological maize mutants in plant breeding. Acta Agron Acad Sci Hung 28:452–458

Patil SH (1971) Use of induced mutations in breeding for quantitative characters in groundnut. Intern Symp Use Isotopes Radiat Agric Anim Husbandry Res New Delhi, pp 154–163

Patil SH (1972) Induced mutations for improving quantitative characters of groundnut. Indian J Genet Plant Breed 32:451–459

Patil SH (1975) Mutation breeding in improving groundnut cultivars. Induced Mutations for the Improvement of Grain Legumes in South East Asia, pp 145–147, IAEA, Vienna

Patil SH (1980) Mutation breeding of groundnut at Trombay. Induced Mutations for Improvement of Grain Legume Production, pp 109–110, IAEA, Vienna

Patil SH, Thakare RG (1969) Yield potential of X-ray induced Trombay groundnut mutants. Radiations and Radiomimetic Substances in Mutation Breeding, Bombay, pp 375–386

Paulis JW, Wall JS (1979) Distribution and electrophoretic properties of alcohol-soluble proteins in normal and high-lysine sorghum. Cereal Chem 56:20–23

Paulis JW, Wall JS, Kwolek WK, Donaldson GL (1975a) Selection for high-lysine corns with varied kernel characteristics and compositions by a rapid turbidimetric assay for zein. J Agric Food Chem 22:318–323

Paulis JW, Beitz JA, Wall JS (1975b) Corn protein subunits: Molecular weights determined by sodium dodecyl sulfate polyacrylamide gel electrophoresis. J Agric Food Chem 23:197–201

Paulis JW, Wall JS, Sanderson I (1978) Origin of high methionine content in *sugary-1* corn endosperm. Cereal Chem 55:705–712

Pavithran K, Mohandas C (1976) Genetics of male sterility in rice. J Hered 67:252

Pawar MS (1971) Present status of rice breeding by induced mutations in Guyana, S. America. Rice Breeding with Induced Mutations III, pp 117–129, IAEA, Vienna

Peixoto Gomes E (1972) Informaciones sobre la induccion de mutaciones en trigo mediante la irradiacion de semillas. Induced Mutations and Plant Improvement, pp 443–445, IAEA, Vienna

Perez CM, Cagampang GB, Esmama BV, Monserrate RK, Juliano BO (1973) Protein metabolism in leaves and developing grains of rices, differing in grain protein content. Plant Physiol 51:537–542

Pernollet JC (1978) Proteinbodies of seeds: ultrastructure, biochemistry, biosynthesis and degradation. Phytochemistry 17:1473–1480

Persson G, Hagberg A (1969) Induced variation in a quantitative character in barley. Morphology and cytogenetics of *erectoides* mutants. Hereditas 61:115–178

Pfeifer RP (1965) The use of an induced mutation to develop a winter barley variety. Use Induced Mutations Plant Breed. Radiat Bot Suppl 5:573–578
Pisum Newsletter (1969–1982) Vol 1–14, Geneva, USA
Plotnikov VA (1973) Mutagenic effect of dimethylsulphate in sunflower (inbred line). Genetika USSR 9(5):15–22
Pohlheim F (1972) Untersuchungen zur Sproßvariation der *Cupressaceae*. 4. Arch Züchtungsforsch 2:223–235
Poláček J (1967) A contribution to the mutation breeding of tomatoes. Genet Slechtení 3:1–4
Poll L (1974) The production and growth of young apple compact mutants induced by ionizing radiaiton. Euphytica 23:521–533
Pollacsek M (1970) Modification chez le mais de l'expression du gène *opaque-2* par un gène suppresseur dominant. Ann Amélior Plantes 20:337–343
Pollacsek M, Caenen M, Rousset M (1972) Mise en évidence d'un deuxième gène suppresseur du gène *opaque-2* chez le mais. Ann Génét 15:173–176
Pollhamer E (1966) Mutation experiments with summer barley. V. Breeding and production value of the mutants of the variety MFB 104. Növénytermelés 15:217–226
Pollhamer E (1967) Importance of induced mutants in the research and improving of barley at Martonvásár. Erwin Baur Ged Vorl 4:169–175
Pollmer WG, Klein D, Bjarnason M, Eberhard D (1974) Die Verbesserung der Proteinqualität bei Mais als Züchtungsaufgabe. Z Acker-Pflanzenbau 140:241–254
Pollmer WG, Klein D, Dhillon BS (1977) Evaluation of lysine and protein content of maize inbred lines. Cereal Res Commun 5:369–376
Pomeranz Y, Robbins GS, Gilbertson JT, Booth GD (1977) Effect of nitrogen fertilization on lysine, threonine and methionine of hulled and hull-less barley cultivars. Cereal Chem 54: 1034–1042
Popova D, Mikhailov L (1973) Monogenic heterosis in tomato (*Lycopersicon esculentum* Mill.). Genetika USSR 9(12):34–39
Popova J (1975) Possibilities of inducing economically useful mutations in maize by seed treatment with ethyl methanesulphonate and N-nitroso-N-methylurethan. Genet Plant Breed 8: 282–290
Popova Y, Popov A (1971) A gamma-induced maize mutant resistant to *Helminthosporium turcicum* pass. Genet Plant Breed 4:361–364
Popović AO, Zečević LM, Koburović ML (1969) Use of gamma rays to induce genetic variability for quantitative traits in some wheat varieties (*Triticum aest.* ssp. *vulgare*). Genet Agrar 23: 247–261
Porsche W (1967) Results of X-irradiation in breeding of *Lupinus albus*. Erwin Baur Ged Vorl 4:241–244
Pospíšil B (1974) The importance of chemomutagenes in the breeding of flax. Genet Slechtení 10:187–196
Powell JB (1976) Induced mutations in highly heterozygous vegetatively propagated grasses. Induced Mutations in Cross-Breeding, pp 219–224, IAEA, Vienna
Powell JB, Toler RW (1980) Induced mutants in "Floratam" St. Augustinegrass. Crop Sci 20: 644–646
Powell JB, Burton GW, Young JR (1974) Mutations induced in vegetatively propagated turf bermudagrasses by gamma radiation. Crop Sci 14:327–330
Prakken R (1959) Induced mutation. Euphytica 8:270–322
Prasad AB, Das AK (1980) Morphological variants in khesari. Indian J Genet Plant Breed 40: 172–175
Prasad MVR (1976) Induced mutants in green gram. Indian J Genet Plant Breed 36:218–222
Preeatchenkoo A, Greenvald K (1967) Valubale mutants in irradiated barley. Genetika USSR 1967 (3):49–54
Prešev N, Jelenić D, Šukalović V (1973) Some genetic and biochemical characteristics of a new source of maize *opaque-2* mutant. Theor Appl Genet 43:23–26
Privalov GF (1968) Investigations of experimental mutagenesis in arboreous plants. Genetika USSR 4(6):144–158

Przybylska J (1976) Problems involved in breeding high lysine barley considered in the light of genetic, biochemical and nutritional studies. Genet Polon 17:83–109

Quednau HD (1972) Der Einfluß mutierter Gene auf die frühesten Stadien der Ontogenese. Theor Appl Genet 42:357–362

Quednau HD, Wolff G (1978a) Biometric evaluation of seed protein production of *Pisum* mutants and recombinants. Seed Protein Improvement by Nuclear Techniques, pp 315–329, IAEA, Vienna

Quednau HD, Wolff G (1978b) Investigations of the seed protein content of several pea genotypes grown in two different years. Theor Appl Genet 53:181–190

Rabson R, Burton GW, Hanna WW, Axmann H, Cross B (1978) The development of procedures for the selection of genotypes for improved protein of pearl millet from mutagenized populations of inbreds. Seed Protein Improvement by Nuclear Techniques, pp 279–291, IAEA, Vienna

Rabson R, Hanna WW, Burton GW, Axmann H (1979) Potential for improving the protein content of pearl millet grain using induced mutations. Seed Protein Improvement in Cereals and Grain Legumes II, pp 367–375, IAEA, Vienna

Raghuvanshi SS, Pathak CS, Singh AK (1978) Effects of preirradiation colchicine treatment on mutation spectrum of *Phaseolus aureus*. Roxb. Cytologia 43:143–151

Rai M, Das K (1976) Potentiality and genetic variability in irradiated population of linseed. Indian J Genet Plant Breed 36:20–25

Raini R, Sharma GS, Singh RB (1978) Genetics of certain grain quality characters in normal and EMS-treated populations of bread wheat. Indian J Genet Plant Breed 38:202–206

Rakha A, Sahrigy MA, Salama FA (1974) Transfer of genes controlling higher percentage of protein and essential amino acids into the Egyptian local lines of maize. Egypt J Gen Cytol 3:155–171

Rakow G (1973) Selektion auf Linol- und Linolensäuregehalt in Rapssamen nach mutagener Behandlung. Z Pflanzenzücht 69:62–82

Ram M (1974) Useful induced mutations in rice. Induced Mutations for Disease Resistance in Crop Plants, pp 161–164, IAEA, Vienna

Ramirez IA, Allan RE, Konzak CF, Becker WA (1969) Combining ability of winter wheat. Induced Mutations in Plants, pp 445–455, IAEA, Vienna

Ramirez Araya I, Sanz de Cortazar C, Konzak CF (1972) Mutaciones inducidas y programa de mejoramiento del trigo en Chile. Induced Mutations and Plant Improvement, pp 425–433, IAEA, Vienna

Rana RS (1965) Radiation-induced variation in ray-floret characteristics of annual *Chrysanthemum*. Euphytica 14:296–300

Rao CH, Tickoo JL, Ram H, Jain KH (1975) Improvement of pulse crops through induced mutation. Reconstruction of plant type. Breeding for Seed Protein Improvement Using Nuclear Techniques, pp 125–131, IAEA, Vienna

Rao GM, Siddiq EA (1977) Induced variation for yield and its components in rice. Indian J Genet Plant Breed 37:12–21

Rao KP, Rains DW, Qualset CO, Huffaker RC (1977) Nitrogen nutrition and grain protein in two spring wheat genotypes differing in nitrate reductase activity. Crop Sci 17:283–286

Rao SC, Croy LI (1972) Protease and nitrate reductase seasonal patterns and their relation to grain protein production of "high" vs "low" protein wheat varieties. J Agric Food Chem 20:1138–1141

Rao TJ, Srinivasan KV, Alexander KC (1966) A red rot resistant mutant of sugar-cane induced by gamma irradiation. Proc Indian Acad Sci 64:224

Rath L, Scharf H (1968) Ölgehalt und Sättigungsgrad der Öle sowie korrelative Beziehungen zwischen einigen Merkmalen bei Leinmutanten. Theor Appl Genet 38:280–288

Raut RN, Jain HK, Panwar RS (1971) Induced mutants of economic value in cotton. Intern Symp Use Isotopes Radiat Agric Anim Husbandry Res New Delhi, pp 223–229

Raut RN, Sharma B, Pokhriyal SC, Singh MP, Jain HK (1974) Induced mutations for mildew resistance in Bajra *(Pennisetum typhoides)* and rust resistance in wheat *(Triticum aestivum)*. Induced Mutations for Disease Resistance in Crop Plants, 165, IAEA, Vienna

Reddy GM (1975) Induced grain type and dwarf mutants in rice breeding. J Cytol Genet Congr Suppl, pp 143–145

Reddy GM (1977) Ethyl methane sulphonate induced grain shape mutations in some local cultivars of rice. Riv Il Riso 26:187–192

Reddy GM, Padma A (1976) Some induced dwarfing genes non-allelic to Dee-geo-woo-gen gene in rice, variety Tellakattera. Theor Appl Genet 47:115–118

Reddy GM, Reddy TP (1971) Induced semidwarf Basumati rice mutant for commercial use. Intern Symp Use Isotopes Radiat Agric Anim Husbandry Res New Delhi, pp 237–241

Reddy GM, Reddy TP (1972) Induction of fine grain mutants in the rice variety IR-8. Sabrao Newslett 4:139–142

Reddy GM, Reddy TP (1973a) Induction of some grain shape and morphological mutations in rice, variety IR-8. Radiat Bot 13:181–184

Reddy GM, Reddy TP (1973b) Induced grain shape mutation in some varieties of rice. Indian J Genet Plant Breed 34A:321–330

Reddy GM, Sarala AK (1979) Studies on the amylose content and gelatinisation temperature in certain local cultivars and induced grain shape mutants in rice. Euphytica 28:665–674

Reddy TP, Padma A, Reddy GM (1975) Short-culm mutations induced in rice. Indian J Genet Plant Breed 35:31–37

Ree JH (1968) Rice breeding problems in Korea. Rice Breeding with Induced Mutations, pp 119–125, IAEA, Vienna

Ree JH (1971) Induced mutations for rice improvement in Korea. Rice Breeding with Induced Mutations III, pp 131–147, IAEA, Vienna

Ree JH, Chung HS, Chung BC (1974) Reaction of the rice mutant MR 515-17 to some races of *Pyricularia oryzae* in Korea. Induced Mutations for Disease Resistance in Crop Plants, pp 166–168, IAEA, Vienna

Reed C, Penner D (1978) Peptidases and trypsin inhibitor in the developing endosperm of *opaque-2* and normal corn. Agron J 70:337–340

Reinhold M (1980a) Genetische Charakterisierung von Mutanten mit abweichender Mehltauresistenz aus einer Gerstensorte mit mittlerer Resistenz. 1. Quantitative Erbanalyse im Freiland. Z Pflanzenzücht 84:63–77

Reinhold M (1980b) 2. Erbanalyse nach Infektion mit Pathotypen unter kontrollierten Bedingungen. Z Pflanzenzücht 84:89–99

Remussi C, Gutierrez HP (1965) Obtenócin de una linea precoz de girasol (*Helianthus annuus* L.) por tratamiento de "semillas" con rayons X. Use Induced Mutations Plant Breed. Radiat Bot Suppl 5:603–609

Rendig VV, Broadhent FE (1978) Proteins and amino acids in grain of maize grown with various levels of N. Agron J 71:509–512

Richards JE, Soper RJ (1978) Effect of N fertilizer on yield, protein content and symbiotic N fixation in Fababeans. Agron J 71:807–811

Richter E (1974) Knöllchenbildung, Ertrag und Eiweißgehalt der Samen bei *Pisum sativum*. Landw Forsch 27:330–342

Richter E (1976) Merkmalsdifferenzen zwischen Pal- und Markerbsen. I. Gehalt reifer Samen an primären Inhaltsstoffen. Gartenbauwissenschaft 42:72–78

Rick CM, Boynton JE (1967) A temperature-sensitive, male-sterile mutant of the tomato. Am J Bot. 54:601–611

Röbbelen G (1959) 15 Jahre Mutationsauslösung durch Chemikalien. Züchter 29:92–95

Röbbelen G (1979) The challenge of breeding for improved protein crops. Seed Protein Improvement in Cereals and Grain Legumes I, pp 27–41, IAEA, Vienna

Röbbelen G, Nitsch A (1975) Genetical and physiological investigations on mutants for polyenoic fatty acids in rapeseed, *Brassica napus*. Z Pflanzenzücht 75:93–105

Röbbelen GP, Abdel-Hafez AG, Reinhold M (1977) Use of mutants to study host/pathogen relations. Induced Mutations against Plant Diseases, pp 359–374, IAEA, Vienna

Robbins GS, Pomeranz Y, Briggle LW (1971) Amino acid composition of oat groats. J Agric Food Chem 19:536–539

Roby F (1972) Doce mutaciones en el peral Williams obtenidas por injertos de ramitas irradiadas. Rev Invest Agropec INTA Bs Aires, Ser 2,9:55–64

Rod J, Vagnerova V (1970) Beitrag zur Vererbung des *fasciata*-Typus bei Erbse. Acta Univ Agric Brne, Fac Agron, Rada A 18:9–15
Römer F-W (1973) Heterosis und Kombinationseignung in diallelen Kreuzungen von strahleninduzierten Steinklee-Mutanten. Z Pflanzenzücht 70:323–348
Römer F-W, Micke A (1974) Combining ability and heterosis of radiation-induced mutants of *Melilotus albus* DES. Polyploidy and Induced Mutations in Plant Breeding, pp 275–276, IAEA, Vienna
Rooney LW (1978) Sorghum and pearl millet lipids. Cereal Chem 55:584–590
Rossi L (1972) Mutagenesis in *durum* wheat: isolating and agronomic traits of early mutants in Capeiti variety. Genet Agrar 26:291–304
Rossi L, Bagnara D (1972) Mutagenesis in *durum* wheat: isolating and agronomic traits of short straw mutants in Capeiti variety. Genet Agrar 26:305–320
Roy K, Jana MK (1975) On the nature and utilization of mutation in rice. Indian Agric 19:267–271
Roy Davies D, Hedley CL (1975) The induction by mutation of all-year-round flowering in *Streptocarpus*. Euphytica 24:269–275
Rubaihayo PR (1975) The use of γ-ray induced mutations in *Phaseolus vulgaris* (L.). Z Pflanzenzücht 75:257–261
Rubaihayo PR (1976a) Interrelationship among some yield characters and the productivity of mutants of three grain legumes. Evaluation of Seed Protein Alterations by Mutation Breeding, pp 179–184, IAEA, Vienna
Rubaihayo PR (1976b) Utilization of γ-rays for soybean improvement. Egypt J Genet Cytol 5:136–140
Rubaihayo PR (1978) The performance of gamma-ray induced mutants of three pulse crops. Seed Protein Improvement by Nuclear Techniques, pp 235–241, IAEA, Vienna
Rubaihayo PR, Leakey CLA (1973) Protein improvement in beans and soybeans by mutation breeding. Nuclear Techniques for Seed Protein Improvement, pp 291–296, IAEA, Vienna
Ruddick DP, Marsters M (1977) Mutant barley varieties in the U.K. Mutat Breed Newslett 10:5
Ruebenbauer T, Kaleta S (1980) Inheritance of protein content in kernels of inbred lines of rye (*Secale cereale* L.). Genet Polon 21:1–15
Rutger JN, Peterson ML, Hu CH, Lehman WF (1976) Induction of useful short stature and early maturing mutants in two *japonica* rice cultivars. Crop Sci 16:631–635
Saini SS, Sharma D (1970) Radiation induced variation in rice improvement. Indian J Genet Plant Breed 30:569–578
Sajó Z (1978) High-protein rice selections following mutagenic treatment. Seed Protein Improvement by Nuclear Techniques, pp 202–203, IAEA, Vienna
Salamini F, Borghi B, Lorenzoni C (1970) The effect of the *opaque-2* gene on yield in maize. Euphytica 19:531–538
Salamini F, DiFonzo N, Gentinetta E, Soave C (1979) A dominant mutation interferring with protein accumulation in maize seeds. Seed Protein Improvement in Cereals and Grain Legumes I, pp 97–108, IAEA, Vienna
Samoto S (1975) Short stature mutants induced by gamma-ray irradiation to the Japanese rice variety Koshihikari. Gamma Field Symp 14:11–23. Inst Radiat Breed Ohmiya, Japan
Sampath S, Jachuck PJ (1969) The uses of wild rices in mutation breeding. Radiations and Radiomimetic Substances in Mutation Breeding, Bombay, pp 263–270
Sander C, Muehlbauer FJ (1977) Mutagenic effects of sodium azide and gamma irradiation in *Pisum*. Environ Exp Bot 17:43–47
Sandhu TS, Bhullor BS, Cheema HS, Gill AS (1979) Variability and inter-relationships among grain protein yield and yield components in mungbean. Indian J Genet Plant Breed 39:480–484
Santos IS (1969) Induction of mutations in mungbean (*Phaseolus aureus* Roxb.) and genetic studies of some of the mutants. Induced Mutations in Plants, pp 169–179, IAEA, Vienna
Santos IS, Fukusawa CA, Elec JV, De la Rosa AM (1970) Acclimatization and improvement of a Lincoln variety soybean through mutation breeding. Improving Plant Protein by Nuclear Techniques, pp 189–196, IAEA, Vienna

Sarala AK, Reddy GM (1979) Role of local germplasm and induced mutations in the improvement of the protein content in rice. Theor Appl Genet 54:75–79

Sasakuma T, Maan SS, Williams ND (1978) EMS-induced male-sterile mutants in euplasmic and alloplasmic common wheat. Crop Sci 18:850–853

Sašek A, Kubánek J, Haniš M, Černý J (1977) Application of electrophoretic analysis of the grain proteins for the identification of the wheat radiomutant *ST 7750*.Genet Šlechtení 13:161–168

Satory M (1975) Chrysanthemenzüchtung mit Hilfe künstlicher Mutationsauslösung. Gartenwelt 75:433–435

Sawhney RN, Sharma JR (1979) Induced variability for polygenic traits in wheat (*Triticum aestivum* L.). Z Pflanzenzücht 83:57–63

Sawhney RN, Jain HK, Joshi BC, Singh D (1971) Induction of mutations at loci determining qualitative and quantitative traits in wheat. Int Symp Use Isotopes Radiat Agric Anim Husbandry Res New Delhi, pp 214–222

Sawhney RN, Chopra VL, Mohindro HR, Kumar R (1982) Improved grain mutant in the wheat, variety Arjun (HD 2009). Wheat Inform Serv 54:32–34

Scarascia-Mugnozza GT (1965) Induced mutations in breeding for lodging resistance. Use Induced Mutations Plant Breed. Radiat Bot Suppl 5:537–558

Scarascia-Mugnozza GT (1966a) Research on mutation breeding in *durum* wheat. Mutations in Plant Breeding, pp 191–196, IAEA, Vienna

Scarascia-Mugnozza GT (1966b) Use of induced mutations for the genetic improvement of agricultural plants. Genet Agrar 20:140–178

Scarascia-Mugnozza GT (1967) Mutation breeding in *Triticum durum*. Erwin Baur Ged Vorl 4: 205–210

Scarascia-Mugnozza GT (1969a) Problems in using experimental mutagenesis for breeding purposes. Induced Mutations in Plants, pp 485–499, IAEA, Vienna

Scarascia-Mugnozza GT (1969b) Mutation breeding in sexually propagated plants. Genet Agrar 23:187–219

Scarascia-Mugnozza GT, Bagnara D, Bozzini A, Mosconi C (1965) Nuovi dati sulla risposta alla concimazione azotata di mutanti di frumento duro. Genet Agrar 19:195–198

Scarascia-Mugnozza GT, Pacucci G, Bozzini A, Cavazza L, Mosconi C (1966) Evolution of new lines of *durum* wheat obtained by mutation and by crosses. Tests performed in Lazio and Puglia in the period 1961–1964. Genet Agrar 20:66–84

Scarascia-Mugnozza GT, Bagnara D, Bozzini A (1972) Mutagenesis applied to *durum* wheat. Results and perspectives. Induced Mutations and Plant Improvement, pp 183–198, IAEA, Vienna

Scheibe A (1954) Der *fasciata*-Typus bei *Pisum*, seine pflanzenbauliche und züchterische Bedeutung. Z Pflanzenzücht 33:31–58

Scheibe A (1965) Die neue Mähdresch-Futtererbse 'Ornamenta'. Saatgutwirtschaft 17:116–117

Scheibe A (1968) Der *fasciata*-Erbsentypus im Rahmen der Saatenanerkennung. Saatgutwirtschaft 20:126–128

Scheibe A (1971) Futtererbsen mit neuen Eigenschaften. Flur und Furche (1971), pp 2–3

Scheibe A (1977) Zum Problem der Genotypenkonstanz nach Mutationszüchtung, dargestellt an Hand langjähriger Erfahrungen an *Pisum sativum*. Ber Arbeitstagung Saatzuchtleiter Gumpenstein, pp 87–94

Scheibe A, Bruns A (1953) Eine kurzröhrige weißblühende Mutante bei *Trifolium pratense* nach Röntgenbestrahlung. Angew Bot 27:70–74

Scheibe A, Micke A (1967) Experimentally induced mutations in leguminous forage plants and their agronomic value. Erwin Baur Ged Vorl 4:231–236

Schneider A (1951) Untersuchungen über die Eignung von Erbsensorten für Zwecke der Naßkonservierung. II. Qualitative Unterschiede von Schal- und Markerbsenstärke und ihre Einflüsse auf die Aufgußflüssigkeit der Naßkonserven. Züchter 21:275–281

Scholz F (1965) Experiments on the use of induced mutants to hybridization breeding in barley. Induction of Mutations and the Mutation Process, Praha, pp 73–79

Scholz F (1967) Utility of induced mutants of barley in hybridization. Erwin Baur Ged Vorl 4:161–168

Scholz F (1971) Utilization of induced mutations in barley. Barley Genet 2:94–105

Scholz F (1972) Induced high-protein mutants of barley-Problems in breeding for protein content. Proc Symp Breed Product Barley, Kroměříž, pp 255–265

Scholz F (1975) Problems of breeding for high protein yield in barley. Barley Genet 3:548–555

Scholz F (1976a) Zur Frage der Kombination von hohem Eiweißgehalt mit hohem Kornertrag bei Gerste. Tag Ber Akad Landwirtsch Wiss 143:173–189

Scholz F (1976b) Experience and opinions on using induced mutants in cross-breeding. Induced Mutations in Cross-Breeding, pp 5–19, IAEA, Vienna

Schuster W, Marquard R (1973) Über den Einfluß des Standortes und das Anbaujahr auf Protein- und Fettgehalt sowie das Fettsäuremuster bei unterschiedlichen Sojabohnensorten. Fette Seifen Anstrichm 75:289–298

Schwarzbach E (1967) Recessive total resistance of barley to mildew (*Erysiphe graminis* D.C.f.sp. *hordei* Marchal) as a mutation induced by ethylmethansulfonate. Genetika Šlechtení 3:159–162

Scossiroli RE (1965) Value of induced mutations for quantitative characters in plant breeding. Use Induced Mutations Plant Breed. Radiat Bot Suppl 5:443–450

Scossiroli RE, Sarti A (1967) Analysis of genetic variability in alfalfa after X-ray treatment of seeds. Genet Agrar 21:340–352

Seetharam A (1971) Changes in oil content and seed colour associated with a mutation for yellow seed coat colour in *Linum usitatissimum* L. Z Pflanzenzücht 66:331–334

Seetharaman R, Srivastava DP (1971) Inheritance of height and other characters in a rice mutant. Indian J Genet Plant Breed 31:237–242

Sethi GS (1974) Long-peduncled dwarf: A new mutant type induced in barley. Euphytica 23:237–239

Sethi GS (1975) Induced mutations of plant breeding significance in barley. Indian J Genet Plant Breed 35:109–114

Sethi GS, Bhateria SD (1977) Morphology and cytogenetics of some new macromutants in barley. Indian J Genet Plant Breed 37:73–79

Sethi GS, Gill KS, Kalia HR (1969) Comparative mutagenicity of P^{32} and S^{35} in barley, *Hordeum sativum* Jess. Radiations and Radiomimetic Substances in Mutation Breeding, Bombay, pp 135–145

Sgarbieri VC, da Silva WJ, Antunes PL, Amaya-F I (1977) Chemical composition and nutritional properties of a *sugary-1/opaque-2 (su1/o2)* variety of maize (*Zea mays* L.). J Agric Food Chem 25:1098–1101

Shaikh MAQ, Miah AJ, Mia MM, Rahman A, Hanif MA (1976) Evolution of rice varieties with improved quality through induced mutations. Evaluation of Seed Protein Alterations by Mutation Breeding, pp 138–139, IAEA, Vienna

Shaikh MAQ, Kaul AK, Mia MM, Choudhury MH, Bhuiya AD (1978) Screening for natural variants and induced mutants of some legumes for protein content and yielding potential. Seed Protein Improvement by Nuclear Techniques, pp 223–233, IAEA, Vienna

Shaikh MAQ, Ahmed ZU, Khan AI, Majid MA (1980) An anatomical screening approach to selection of high yielding mutants of jute (*Corchorus capsularis* L.). Environ Exp Bot 20:287–296

Shakoor A, Haq MA (1980) Improvement of plant architecture in chickpea and mungbean. Induced Mutations for Improvement of Grain Legume Production, pp 59–62, IAEA, Vienna

Shakoor A, Ahsanul-Haq M, Sadiq M, Sarwar G (1977) Induction of resistance to yellow mosaic virus in mungbean through induced mutations. Induced Mutations against Plant Diseases, pp 293–302, IAEA, Vienna

Sharma D, Lal GS, Tawar ML, Shrivastava MN (1974) EMS induced variation for heading date and the performance of early flowering mutants in rice. Indian J Genet Plant Breed 34:216–220

Sharma RK, Reinbergs E (1976) Male sterility genes in barley and their sensitivity to light and temperature intensity. Indian J Genet Plant Breed 36:59–63

Sharma SK, Sharma B (1978a) Induction of male sterility in lentil *(Lens culinaris medic.)*. Legume Res 2:45–48

Sharma SK, Sharma B (1978b) Induction of tendril mutations in lentil *(Lens culinaris medic.)*. Curr Sci 47:864–866

Sharma SK, Sharma B (1979a) Pattern of induced mutability in different genotypes of lentil (*Lens culinaris* Medik.). Z Pflanzenzücht 83:315–320

Sharma SK, Sharma B (1979b) Induced alterations in seed colour of lentil. Indian J Agric Sci 49: 174–176

Shevtzov VM (1969a) Induction of mutants in oats. Genetika USSR 5(2) 5–11

Shevtzov VM (1969b) The effect of mutagenes on the mutation frequency in barley. Genetika USSR 5(3):33–42

Shewry PR, Pratt HM, Leggatt MM, Miflin BJ (1979) Protein metabolism in developing endosperms of high-lysine and normal barley. Cereal Chem 56:110–117

Shewry PR, Faulks AJ, Pickering RA, Jones IT, Finch RA, Miflin BJ (1980) The genetic analysis of barley storage proteins. Heredity 44:383–389

Shifriss C (1973) Additional spontaneous male-sterile mutant in *Capsicum annuum* L. Euphytica 22:527–529

Shkvarnikov PK, Morgun VV (1974) Mutations in maize induced by chemical mutagens. Polyploidy and Induced Mutations in Plant Breeding, pp 295–302, IAEA, Vienna

Shumny VK (1978) Use of mutants for the production of heterosis. Analysis of cases of monohybrid heterosis in plants. Experimental Mutagenesis in Plants, Sofia, pp 376–382

Shumny VK, Sidorova KK, Belova LA (1970) Study of the heterozygous state in nine mutant genes in pea. Genetika USSR 6(8):12–19

Shumny VK, Belova LI, Sharova LA (1971) Cases of monohybrid heterosis in pea. Genetika USSR 7(9):36–41

Sichkar VI (1976) Improvement of cereals and legumes with respect to protein content and quality by means of experimental mutagenesis. Genetika USSR 12(2):145–153

Sichkar VI, Shkvarnikov PK, Mar'yushkin VF (1975) Variation of quantitative characters in winter wheat induced by chemical mutagens. Genetika USSR 11(2):5–13

Sichkar VI, Shkvarnikov PK, Mar'yushkin VF, Suvorinov AM, Podmogilniy VV (1980) Investigation of the productivity of winter wheat mutants. Genetika USSR 16:564–565

Siddiq EA, Swaminathan MS (1968) Induced mutations in relation to the breeding and phylogenetic differentiation of *Oryza sativa*. Rice Breeding with Induced Mutations, pp 25–51, IAEA, Vienna

Siddiq EA, Kaul AK, Puri RP, Singh VP, Swaminathan MS (1970) Mutagen-induced variability in protein characters in *Oryza sativa*. Mutat Res 10:81–84

Siddiqui JA (1971) Polygenic variation following irradiation in interspecific crosses of cotton. Indian J Genet Plant Breed 31:461–470

Siddiqui KA, Arain AG (1974) Performance and selection for yield of wheat mutants derived from different cultivars. Euphytica 23:585–590

Siddiqui KA, Doll H (1973) Screening for improved protein quality mutants in wheat. Z Pflanzenzücht 70:143–147

Siddiqui KA, Siddiqui HA (1974) Reactions of wheat mutants to *Puccinia graminis tritici* and *Puccinia recondita*. Induced Mutations for Disease Resistance in Crop Plants, pp 169–172, IAEA, Vienna

Siddiqui KA, Arain MA, Jafri KA (1981) Evaluation of a high yielding wheat variety through fast neutron treatment. Mutat Breed Newslett 18:6–7

Sidorova KK (1970) The study of allelism in phenotypically identical mutants of pea in connection with the law of homologous series in hereditary variability. Genetika USSR 6(11):23–35

Sidorova KK (1975) Influence of different types of mutations on development of the organism in the homo- and heterozygous states (in the example of mutants of *Pisum sativum* L.). Genetika USSR 11(1):35–46

Sidorvoa KK (1981a) Effect of different mutation types in the homo- and heterozygous state in *Pisum sativum* L. Pulse Crops Newslett 1:52–53

Sidorova KK (1981b) Influence of genotypic background on the expressivity of mutant genes of pea. Pulse Crops Newslett 1(3):23–24

Sidorova KK, Bobodzhanov VA (1977) Ecological study of allelic pea mutants. Genetika USSR 13(4):583–592

Sidorova KK, Khvostova VV (1972) Investigation of the ecology of the mutant gene. Induced Mutations and Plant Improvement, pp 277–284, IAEA, Vienna

Sidorova KK, Uzhintzeva LP (1969) Ecological studies on induced pea mutants. Genetika USSR 5(8):46–51
Sidorova KK, Khanghildin VV, Debely GA (1969) The experimental mutagenesis in pea breeding. Agrobiologia (USSR) 4(4):538–543
Sidorova KK, Kalinina NP, Bobodjanov VA (1972) Ecology of mutant gene in homo- and heterozygous conditions. Genetika USSR 8(1):23–29
Sigurbjörnsson B (1968) Induced mutations as a tool for improving world food sources and international cooperation in their use. Hereditas 59:375–395
Sigurbjörnsson B (1976) The improvement of barley through induced mutation. Barley Genet 3:84–95
Sigurbjörnsson B, Micke A (1969) Progress in mutation breeding. Induced Mutations in Plants, pp 673–698, IAEA, Vienna
Sigurbjörnsson B, Micke A (1973) List of varieties of vegetatively propagated plants developed by utilizing induced mutations. Induced Mutations in Vegetatively Propagated Plants, pp 195–202, IAEA, Vienna
Sigurbjörnsson B, Micke A (1974) Philosphy and accomplishments of mutation breeding. Polyploidy and Induced Mutations in Plant Breeding, pp 303–343, IAEA, Vienna
Simon J, Sajo Z (1976) A new radiomutant rice variety in Hungarian rice production. Genetika Beograd 8:223–226
Simon J, Sajo Z (1977) A new radiomutant rice variety in the Hungarian rice production. Improving Crop and Animal Productivity, pp 192–195, ISNA, New Delhi
Simons MD, Frey KJ (1977) Induced mutations for tolerance of oats to crown rust. Induced Mutations against Plant Diseases, pp 499–512, IAEA, Vienna
Simons MD, Youngs VL, Booth GD, Forsberg RA (1979) Effect of crown rust on protein and groat percentages of oat grain. Crop Sci 19:703–706
Singh BV, Singh J (1977) Development and evaluation in an *opaque-2* maize composite at three plant population densities. Crop Sci 17:515–516
Singh CB (1973) Variation in quantitative characters in the F_2 population derived from unirradiated and irradiated hybrids of rice. Indian J Genet Plant Breed 33:369–372
Singh CB, Rao YP (1971) Association between resistance to *Xanthomonas oryzae* and morphological and quality characters in induced mutants of *indica* and *japonica* varieties of rice. Indian J Genet Plant Breed 31:369–373
Singh DP (1971) Radiation induced polygenic mutations in jute. Intern Symp Use Isotopes Radiat Agric Animal Husbandry Res New Delhi, 58–64
Singh DP (1981) Fasciated mutant in greengram (*Vigna radiata* L. Wilczek). Mutat Breed Newslett 18:5
Singh DP, Sharma BK, Banerjee SC (1973) X-ray induced mutations in jute (*Corchorus capsularis* L. and *Corchorus olitorius* L.). Genet Agrar 27:115–147
Singh R, Axtell JD (1973) High lysine mutant gene *(hl)* that improves protein quality and biological value of grain sorghum. Crop Sci 13:535–539
Singh RJ, Ikehashi H (1981) Monogenic male-sterility in rice: Induction, identification and inheritance. Crop Sci 21:286–289
Singh SP, Drolsom PN (1974) Induced early-maturing mutation in *Sorghum.* Crop Sci 14:377–380
Singh SP, Drolsom PN (1977) Genetic analyses of four diethyl sulfate-induced culm height mutants of *Sorghum.* Crop Sci 17:617–621
Singh V, Sastry LVS (1977a) Studies on the proteins of the mutants of barley grain. I. Extraction and electrophoretic characterization. Cereal Chem 54:1–12
Singh V, Sastry LVS (1977b) Studies on the proteins of the mutants of barley grain. II. Fractionation and characterization of the alcohol-soluble proteins. J Agric Food Chem 25:912–917
Singh VP, Chaturvedi SN (1981a) Gamma-ray induced male sterile mutants in mung bean. Pulse Crops Newslett 1:41
Singh VP, Chaturvedi SN (1981b) The productivity of some mutants of the moong bean. II. Variation in size and number of pods. Genet Agrar 35:295–300
Singhal NL, Jain HK, Austin A (1978) Induced variability for protein content in bread wheat. Seed Protein Improvement by Nuclear Techniques, pp 41–50, IAEA, Vienna

Sinhamahapatra SP, Rakshit SC (1981) Effect of selection for plant height in X-ray treated population of jute *(Corchorus capsularis)*. Z Pflanzenzücht 86:329–335

Sjödin J, Martensson P, Magyarosi T (1981a) Selection for increased protein quantity in field bean (*Vicia faba* L.). Z Pflanzenzücht 86:210–220

Sjödin J, Martensson P, Magyarosi T (1981b) Selection for antinutritional substances in field bean (*Vicia faba* L.). Z Pflanzenzücht 86:231–247

Skirvin RM (1978) Natural and induced variation in tissue culture. Euphytica 27:241–266

Skorda EA (1977) Stem and stripe rust resistance in wheat induced by gamma rays and thermal neutrons. Induced Mutations against Plant Diseases, pp 385–392, IAEA, Vienna

Slinkard AE (1972) Field pea research. Crop Dev Centre, Univ Saskatchewan, Canada

Slinkard AE (1974) Production, utilization and marketing of field peas. Ann Rep 1 Crop Dev Centre, Univ Saskatchewan, Canada

Smith HH (1958) Radiation in the production of useful mutations. Bot Rev 24:1–24

Smith HH (1972) Comparative genetic effects of different physical mutagens in higher plants. Induced Mutations and Plant Improvement, pp 75–93, IAEA, Vienna

Smutkupt S, Gymantasire P (1975) Improvement of soya bean protein by mutation breeding. Breed Seed Protein Improvement Using Nuclear Techniques, 177, IAEA, Vienna

Snoad B (1974) A preliminary assessment of "leafless peas". Euphytica 23:257–265

Snoad B (1975) Alteration of plant architecture and the development of the leafless pea. Induced Mutations for the Improvement of Grain Legumes in South East Asia, pp 123–132, IAEA, Vienna

Snoad B (1981) Development and breeding of leafless dried peas at the John Innes Institute, U.K. Pulse Crops Newslett 1(1):58–59

Snoad B, Hedley CL (1981) Potential for redesigning the pea crop using spontaneous and induced mutations. Induced Mutations – a Tool in Plant Research, pp 111–126, IAEA, Vienna

Soave C, Viotti A, DiFonzo N, Salamini F (1979) Maize prolamin: Synthesis and genetic regulation. Seed Protein Improvement in Cereals and Grain Legumes I, pp 165–173, IAEA, Vienna

Soave C, Reggioni R, DiFonzo N, Salamini F (1981) Clustering of genes for 20 kd zein subunits in the short arm of maize chromosome 7. Genetics 97:363–377

Sodek L, Wilson CM (1970) Incorporation of leucine-^{14}C and lysine-^{14}C into protein in the developing endosperm of normal and *opaque-2* corn. Arch Biochem Biophys 140:29–38

Sodek L, Wilson CM (1971) Amino acid composition of proteins isolated from normal, *opaque-2* and *floury-2* corn endosperms by a modified Osborne procedure. J Agric Food Chem 19:1144–1150

Soendsen I, Martin B (1980) Characteristics of *hiproly* barley. III. Amino acid sequences of two lysine-rich proteins. Carlsberg Res Commun 45:79–85

Solovjeva VK (1958) New forms of a podded vegetable pea. Agrobiolog 1958(5):124–126 (Russian without English summary)

Sozinov AA, Netsvetaev VP, Poperelya FA, Navolotsky VD (1976) Recovery of Pr-α locus activity in barley (*Hordeum vulgare* L.) mutant hybridization and its dose effect. Genetika USSR 12(10):55–59

Sparrow AH, Konzak CF (1958) The use of ionizing radiation in plant breeding; accomplishments and prospects. Camellia Culture, pp 425–452

Spiegel-Roy P, Kochba J (1975) Production of solid mutants in *Citrus* utilizing new approaches and techniques. Improvement of Vegetatively Propagated Plants through Induced Mutations, pp 113–127, IAEA, Vienna

Sprague GF, Brimhall B, Kraut RM (1943) Some effects of the *waxy* gene in corn on properties of the endosperm starch. Agron J 35:817–822

Sraon HS, Reeves DL, Rumbaugh MO (1975) Quantitative gene action for protein content in oats. Crop Sci 15:668–670

Sree Ramulu K (1968) Induction of chlorophyll and viable mutations in *Sorghum*. Genet Agrar 22:320–334

Sree Ramulu K (1974a) Mutational response pattern in different genotypes of grain *Sorghum*. Z Pflanzenzücht 73:58–70

Sree Ramulu K (1974b) Induced polygenic variability in *Sorghum*. Genet Agrar 28:278–291

References

Sree Ramulu K (1975) Mutation breeding in *Sorghum*. Z Pflanzenzücht 74:1–17

Srinivasachar D, Malik RS (1971) Possibilities of improvement of linseed, *Linum usitatissimum* L., for seed yield, oil content and iodine value of oil by ionized radiations. Intern Symp Use Isotopes Radiat Agric Anim Husbandry Res, New Delhi, pp 242–249

Stefanov T, Gorastev H, Gramatikova M (1978) Use of short-stem mutants in winter barley breeding. Experimental Mutagenesis in Plants, pp 266–274, Sofia

Steinbrück G (1977) Der Einfluß mutierter Gene auf die Speicherung von Zuckern und Stärke und die Kohlenhydratentwicklung während der ersten Keimwochen bei *Pisum sativum*. Ph D Thesis, Univ Bonn

Stephanov T, Gorastev Ch (1976) Using of induced mutagenesis and intervarietal hybridization in winter two-rowed barley breeding. Barley Genet 3:197–202

Stoilov M, Daskaloff S (1976) Some results on the combined use of induced mutations and heterosis breeding. Induced Mutations in Cross-Breeding, pp 179–188, IAEA, Veinna

Stoilov M, Hristova J (1978) Possibilities for inducing maize mutations with valuable biological qualities. Experim Mutagenesis in Plants, pp 333–339, Sofia

Stoilov M, Popov A (1976) Inducing disease resistant maize mutations via ionizing radiation. Genet Plant Breed Sofia 9:416–419

Stoilov M, Popova J (1974/75) Radiosensitivity of maize and changes of its protein content induced by gamma-irradiation. Genet Selek 7–8:186

Streitberg H (1966a) Schaffung wirtschaftlich wertvoller Sproßvarianten bei Rosen durch Behandlung mit Röntgenstrahlen. Z Pflanzenzücht 55:165–182

Streitberg H (1966b) Schaffung von Sproßvarianten bei Azaleen durch Behandlung mit Röntgenstrahlen. Z Pflanzenzücht 56:70–87

Streitberg H (1967) Production of economically valuable variations in roses and azaleas by means of X-irradiation. Erwin Baur Ged Vorl 4:359–362

Stubbe H (1959) Some results and problems of theoretical and applied mutation research. Indian J Genet Plant Breed 19:13–29

Stubbe H (1967) On the relationships between the spontaneous and experimentally induced form diversity and on some experiments on the evolution of cultivated plants. Erwin Baur Ged Vorl 4:99–121

Subramanian D (1980) Effects of gamma-radiation in *Vigna*. Indian J Genet Plant Breed 40:187–194

Sutcliffe JF, Pate JS (1977) The Physiology of the Garden Pea. Academic Press, London New York San Francisco, 500 pp

Světlík V (1967) Mutation breeding in smooth-stalked meadow grass *(Poa pratensis)*. Erwin Baur Ged Vorl 4:255–262

Swaminathan MS (1963) Evaluation of the use of induced micro- and macro-mutations in breeding of polyploid crop plants. Proc Symp Energ Nucl Agric Rome, pp 243–277

Swaminathan MS (1965) A comparison of mutation induction in diploids and polyploids. Use Induced Mutations Plant Breed. Radiat Bot Suppl 5:619–641

Swaminathan MS (1969a) The role of mutation breeding in a changing agriculture. Radiations and Radiomimetic Substances in Mutation Breeding, Bombay, pp 427–441

Swaminathan MS (1969b) Role of mutation breeding in a changing agriculture. Induced Mutations in Plants, pp 719–736, IAEA, Vienna

Swaminathan MS (1971) Mutation breeding and agricultural progress. J Indian Bot Soc 50A: 416–429

Swaminathan MS, Austin A, Kaul AK, Naik MS (1969) Genetic and agronomic enrichment of the quantity and quality of proteins in cereals and pulses. New Approaches to Breeding for Improved Plant Protein, pp 71–86, IAEA, Vienna

Swaminathan MS, Siddiq EA, Singh CB, Pai RA (1970) Mutation breeding in rice in India. Rice Breeding with Induced Mutations II, pp 25–43, IAEA, Vienna

Swarup V, Gill HS (1968) X-ray induced mutations in French bean. Indian J Genet Plant Breed 28:44–58

Swarup V, Raghava SPS (1974) Induced mutation for resistance to leaf-curl virus and its inheritance in garden zinnia. Indian J Genet Plant Breed 34:17–21

Swarup V, Kaicker US, Gill HS (1971) Induced mutations in French bean and roses. Intern Symp Use Isotopes Radiat Agric Animal Husbandry Res, New Delhi, pp 13–25

Takagi Y (1970) Monogenic recessive male sterility in oil rape (*Brassia napus* L.) induced by gamma irradiation. Z Pflanzenzücht 64:242–247

Takenaka S (1969) Some useful mutations induced by gamma irradiation in rice. Induced Mutations in Plants, pp 517–527, IAEA, Vienna

Tallberg A (1977) The amino-acid composition in endosperm and embryo of a barley variety and its high lysine mutant. Hereditas 87:43–46

Tallberg A (1981a) Protein and lysine content in high-lysine double-recessives of barley. I. Combinations between mutant *1508* and a *Hiproly* back-cross. Hereditas 94:253–260

Tallberg A (1981b) Protein and lysine content in high-lysine double-recessives of barley. II. Combination between mutant *7* and a *Hiproly* back-cross. Hereditas 94:261–268

Tallberg A (1982) Characterization of high-lysine barley genotypes. Hereditas 96:229–245

Tamrazian EE (1968) Physiological and genetic effects of ethylene imine and ethyl-sulphate on *Rudbeckia*. Genetika USSR 4(4):31–37

Tanaka S (1968) Radiation-induced mutations in rice. Rice Breeding with Induced Mutations, pp 53–64, IAEA, Vienna

Tanaka S (1969) Some useful mutations induced by γ-irradiation in rice. Induced Mutations in Plants, pp 517–527, IAEA, Vienna

Tanaka S (1973) Varietal differences in protein content of rice. Nuclear Techniques for Seed Protein Improvement, pp 107–113, IAEA, Vienna

Tanaka S (1974) Induction of mutations in protein content of rice. Gamma Field Symp 13:61–68

Tanaka S (1975) Induction of mutations in protein content of rice. Breeding for Seed Protein Improvement Using Nuclear Techniques, pp 176–177, IAEA, Vienna

Tanaka S (1976) Induction of mutations in protein content of rice. Evaluation of Seed Protein Alterations by Mutation Breeding, pp 139–140, IAEA, Vienna

Tanaka S (1978) Differences in nitrogen absorption between a radiation-induced high-protein rice mutant and its original variety. Seed Protein Improvement by Nuclear Techniques, pp 199–201, IAEA, Vienna

Tanaka S, Hiraiwa S (1978) Induction of high-protein mutants in rice. Seed Protein Improvement by Nuclear Techniques, pp 191–198, IAEA, Vienna

Tanaka S, Takagi Y (1970) Protein content of rice mutants. Improving Plant Protein by Nuclear Techniques, pp 55–62, IAEA, Vienna

Tanaka S, Tamura S (1968) A short report on γ-ray induced rice mutants having high protein content. Jpn Agric Res Q 3(3):1–4

Tarasenko ND (1977a) Obtaining potato mutants of the variety Lorch resistant to *Synchytrium endobioticum*. Induced Mutations against Plant Diseases, pp 247–251, IAEA, Vienna

Tarasenko ND (1977b) Mutagenic efficiency of high-energy protons in potato. Z Pflanzenzücht 79:79–81

Tavčar A (1965) Gamma-ray irradiation of seeds of wheat, barley and inbreds of maize and the formation of some useful point mutations. Use Induced Mutations Plant Breed. Radiat Bot Suppl 5:159–174

Tessi J, Scarascia-Mugnozza GT, Sigurbjörnsson B, Bagnara D (1968) First-year results in the FAO-IAEA Near East uniform regional trials of radio-induced *durum* wheat mutants. Mutations in Plant Breeding II, pp 251–272, IAEA, Vienna

Thakare RG, Joshua DC, Rao NS (1973) Induced viable mutations in *Corchorus olitorius* L. Indian J Genet Plant Breed 33:204–228

Thierfeldt HJ (1972) Qualitative und quantitative Untersuchungen über den Einfluß mutierter Gene auf das Zuckerspektrum der Samen strahleninduzierter Mutanten und Rekombinanten von *Pisum sativum*. Ph D Thesis, Univ Bonn

Thomas E, King PJ, Potrykus I (1979) Improvement of crop plants via single cells in vitro – an assessment. Z Pflanzenzücht 82:1–30

Thomson JA, Doll H (1979) Genetics and evolution of seed storage proteins. Seed Protein Improvement in Cereals and Grain Legumes I, pp 109–123, IAEA, Vienna

References

Thomson JA, Millerd A, Schroeder HE (1979) Genotype-dependent patterns of accumulation on seed storage proteins in *Pisum*. Seed Protein Improvement in Cereals and Grain Legumes I, pp 231–240, IAEA, Vienna

Toft Viuf B (1969) Breeding of barley varieties with high protein content with respect to quality. New Approaches to Breeding for Improved Plant Protein, pp 23–28, IAEA, Vienna

Tong WF, Chu YE, Li HW (1970) Variation in protein and amino acid contents among genetic stocks of rice. Improving Plant Protein by Nuclear Techniques, pp 71–76, IAEA, Vienna

Trofimovskaya AJ, Zhukovsky PM (1967) EMS-induced mutagenesis in winter barley. Genetika USSR 1967 (4):13–28

Trujillo-Figueroa R (1968) Indirekte Frühselektion auf induzierte genetische Variabilität in Ertragsmerkmalen nach EMS-Behandlung von Weizen. Z Pflanzenzücht 60:327–348

Tsai CY, Glover DV (1974) Effect of the *brittle-1 sugary-1* double mutant combination on carbohydrate and postharvest quality of sweet corn. Crop Sci 14:808–810

Tsai CY, Nelson OE (1966) Starch-deficient maize mutant lacking adenosine diphosphate glucose pyrophosphorylase activity. Science 151:341–343

Tsai CY, Larkins BA, Glover DV (1978) Interaction of the *opaque-2* gene with starch-forming mutant genes on the synthesis of zein in maize endosperm. Biochem Genet 16:883–896

Tulmann Neto A, Menten JOM, Ando A, Costa AS, Alberini J (1980) Induced mutations for disease resistance in beans (*Phaseolus vulgaris* L.). Induced Mutations for Improvement of Grain Legume Production, pp 89–95, IAEA, Vienna

Uhlik J, Urban J (1976) Evaluation of the effect of thermal neutrons, fast neutrons, and ENU on the M_3 and M_4 generation of *Lens culinaris*. Genet Šlechteni 12:129–138

Ukai Y, Yamashita A (1979) Mutant of barley resistant to barley yellow mosaic virus (BYMV). Technical News 21, Inst Rad Breed Ohmiya, Japan

Ukai Y, Yamashita A (1980) Induced mutation for resistance to barley yellow mosaic virus. Jpn J Breed 30:125–130

Ullrich SE, Eslick RF (1978a) Lysine and protein characterization of spontaneous shrunken endosperm mutants of barley. Crop Sci 18:809–812

Ullrich SE, Eslick RF (1978b) Inheritance of the associated kernel characters, high lysine and shrunken endosperm of the barley mutant Bomi, Risø *1508*. Crop Sci 18:828–831

Ulonska E, Baumer M (1976) Studies on the effect of conditions during growth on the success of selection for protein quantity and quality in barley. Evaluation of Seed Protein Alterations by Mutation Breeding, pp 95–106, IAEA, Vienna

Ulonska E, Gaul H, Baumer M, Fritz A (1973) Breeding for protein quantity and quality in barley. Nuclear Techniques for Seed Protein Improvement, pp 219–234, IAEA, Vienna

Ulonska E, Gaul H, Baumer M (1975) Investigation of selection methods in mutation breeding of barley for protein quality and quantity. Breeding for Seed Protein Improvement Using Nuclear Techniques, pp 61–77, IAEA, Vienna

Upadhya MD, Purohit AN (1971) Mutation induction for the day neutral reaction in potato. Intern Symp Use Isotopes Radiat Agric Animal Husbandry Res, New Delhi, pp 51–57

Upadhya MD, Purohit AN (1973) Mutation induction and screening procedure for physiological efficiency in potato. Induced Mutations in Vegetatively Propagated Plants, pp 61–66, IAEA, Vienna

Upadhya MD, Dayal TR, Dev B, Chaudhri VP, Sharda RT, Chandra R (1974) Chemical mutagenesis for day-neutral mutations in potato. Polyploidy and Induced Mutations in Plant Breeding, pp 379–383, IAEA, Vienna

Upadhya MD, Chandra R, Abraham MJ (1976) Mutagenic treatments towards increasing the frequency of day-neutral mutations and standardisation of procedures for tissue culture in potato. Improvement of Vegetatively Propagated Plants and Tree Crops through Induced Mutation, pp 151–170, IAEA, Vienna

Uzhintseva LP, Sidorova KK (1979) Genetic nature of pea mutants characterized by early flowering. Genetika USSR 15:1076–1082

Vasal SK (1975) Use of genetic modifiers to obtain normal-type kernels with the *opaque-2* gene. High Protein Quality Maize. Dowden, Hutchinson and Ross, Stroudsbury, pp 197–201

Vasal SK, Villegas E, Bauer R (1979) Present status of breeding quality protein maize. Seed Protein Improvement in Cereals and Grain Legumes II, pp 127–150, IAEA, Vienna

Vassileva M (1976) Obtaining forms with useful traits and high protein content in *Pisum sativum* L. Genetika Beograd 8:143–153

Vassileva M (1978) Induced genetic variety in *P. sativum*. Experimental Mutagenesis in Plants (Sofia), pp 440–447

Vassileva M, Milanova GP (1977) New mutant forms in *Pisum sativum* L. C R Acad Bulg Sci 30(6): 929–932

Vasudevan KN, Singh D, Nerwal SK (1969) Effect of irradiation on some quantitative characters in barley. Indian J Genet Plant Breed 29:36–41

Velikovsky V (1978) Evaluation and use of mutant forms of oats. Genet Šlechteni 14:117–125

Vershinin AV, Sokolov VA, Shumny VK (1976) Physiological and biochemical aspects of monohybrid heterosis derived from pea chlorophyll mutant. I. Pigment composition and proteins of chloroplasts. Genetika USSR 12(2):52–58

Vershinin AV, Sokolov VA, Shumny VK (1979) Physiological and biochemical aspects of monohybrid heterosis derived from pea chlorophyll mutants. III. The growth analysis. Genetika USSR 15:2006–2012

Viado GB (1968) Studies on induction of mutations in rice. Their nature and use for rice improvement. Rice Breeding with Induced Mutations, pp 99–108, IAEA, Vienna

Vilawan S, Siddiq EA (1973) Study on mutational manipulation of protein characteristics in rice. Theor Appl Genet 43:276–280

Virupakshappa K, Mahishi DM, Shivashankar G (1980) Variation in cowpea following hybridization and mutagenesis. Indian J Genet Plant Breed 40:396–398

Visser T (1973) Methods and results of mutation breeding in deciduous fruits, with special reference to the induction of compact and fruit mutations in apple. Induced Mutations in Vegetatively Propagated Plants, pp 21–33, IAEA, Vienna

Visser T, Verhaegh JJ, de Vries DP (1971) Pre-selection of compact mutants induced by X-ray treatment in apple and pear. Euphytica 20:195–207

Votava V, Našinec J (1974) Induction and utilization of short-stalked mutations in oats. Genet Šlechteni 10:179–186

Waggle D, Porrish DB, Deyae CW (1966) Nutritive value of protein in high and low protein content sorghum grain as measured by rat performance and amino acid assays. J Nutr 88:370

Wagner M (1969) Beobachtungen an Winterweizenmutanten unter Einbeziehung von Qualitätsmerkmalen. Z Pflanzenzücht 61:288–296

Walther H (1975) Selection of improved protein and lysine mutants induced in spring barley by EMS and X-ray treatment. Barley Genetics 3:557–564

Walther H, Seibold KH (1978) Induced variation in protein mutants after multiple EMS and X-ray treatment. Seed Protein Improvement by Nuclear Techniques, pp 131–144, IAEA, Vienna

Walther H, Seibold KH (1979) Improved protein mutants selected from barley after multiple EMS and X-ray treatment. Seed Protein Improvement in Cereals and Grain Legumes II, pp 327–343, IAEA, Vienna

Walther H, Gaul H, Ulonska E, Seibold KH (1975) Variation and selection of protein and lysine mutants in spring barley. Breeding for Seed Protein Improvement Using Nuclear Techniques, pp 79–89, IAEA, Vienna

Weaver JB, Ashley T (1971) Analysis of a dominant gene for male-sterility in upland cotton, *Gossypium hirsutum* L. Crop Sci 11:596–598

Weber EJ (1978) Corn lipids. Cereal Chem 55:572–584

Weber E, Gottschalk W (1973) Die Beziehungen zwischen Zellgröße und Internodienlänge bei strahleninduzierten *Pisum*-Mutanten. Beitr Biol Pfl 49:101–126

Weber G, Lark KG (1979) An efficient plating system for rapid isolation of mutants from plant cell suspensions. Theor Appl Genet 55:81–86

Weiling F, Gottschalk W (1961) Die genetische Konstitution der X_1-Pflanzen nach Röntgenbestrahlung ruhender Samen. Biol Zbl 80:579–612

Wellensiek SJ (1965) The origin of early-flowering neutron-induced mutants in peas. Use Induced Mutations Plant Breed. Radiat Bot Suppl 5:393–397

Wellensiek SJ, van Brenk G, Buiskool R (1975) The restoration of fertility in sterile M_2 pea mutants. Mutat Res 27:327–330

Wenzel U (1981) Untersuchungen über die Samenproteine sowie deren Fraktionen bei verbänderten Genotypen von *Pisum sativum*. Ph D Thesis, Univ Bonn

Wettstein D v (1954) The pleiotropic effects of *erectoides* factors and their bearing on the property of straw-stiffness. Acta Agric Scand 4:491–506

Wheatherwax P (1922) A rare carbohydrate in maize. Genetics 7:568–572

Whistler RL, Johnson C (1964) Amylose and amylopectin. Methods in Carbohydrate Chemistry, vol 4. New York, London, pp 25–28

White OE (1948) Fasciation. Bot Rev 14:319–358

Wiberg A (1973) Mutants of barley with induced resistance to powdery mildew. Hereditas 75:83–100

Wiberg A (1977) Mutation work done at Svalöf, Sweden, for improving disease resistance in barley. Induced Mutations against Plant Diseases, pp 317–332, IAEA, Vienna

Wilson CM (1973) Plant nucleases. IV. Genetic control of ribonuclease activity in corn endosperm. Biochem Genet 9:53–62

Wilson CM, Shewry PR, Miflin BJ (1981) Maize endosperm proteins compared by sodium dodecyl sulfate gel electrophoresis and isoelectric focusing. Cereal Chem 58:275–281

Winkler U, Schön WJ (1980) Amino acid composition of the kernel proteins in barley resulting from nitrogen fertilization at different stages of development. Z Acker Pflanzenbau 149:506–512

Wojciechowska W (1973) Mutagenic effect of X-raying and of N-nitroso-N-methylurea (NMH) treatment on serradella (*Ornithopus sativus* Brot.). Genet Polon 14:269–294

Wolf MJ, Khoo V, Seckinger HL (1967) Subcellular structure of endosperm protein in high lysine and normal corn. Science 157:556–557

Wolff G (1975) Quantitative Untersuchungen über den Proteingehalt von Samen von *Pisum sativum*. Z Pflanzenzücht 75:43–54

Wolff G (1979) The alteration of seed proteins by induced and spontaneous mutation. Proc 1st Mediterr Conf Genet, Cairo, pp 783–798

Wolff G (1980a) Investigations on the relations within the family *Papilionaceae* on the basis of electrophoretic banding patterns. Theor Appl Genet 57:225–232

Wolff G (1980b) Electrophoretical investigations on seed albumins of *Pisum sativum*: Identification of several enzymes. Pisum Newslett 12:76

Woo SC, Konzak CF (1969) Genetic analysis of short-culm mutants induced by ethyl methane sulphonate in *Triticum aestivum* L. Induced Mutations in Plants, pp 551–555, IAEA, Vienna

Woo SC, Chi-Ming Ng (1974) Preliminary results of blast resistance test for mutant TP 309-EMS-1. Induced Mutations for Disease Resistance in Crop Plants, 173, IAEA, Vienna

Yamaguchi H, Tano S, Tatara A, Hirai S, Hasegawa K, Hiraki M (1974) Mutations induced in germinating barley seed by diethyl sulphate treatment at the interphase. Polyploidy and Induced Mutations in Plant Breeding, pp 393–399, IAEA, Vienna

Yamaguchi I, Yamashita A (1979) Resistant mutants of two-rowed barley to powdery mildew (*Erysiphe graminis* f.sp. *hordei*). Techn News 20, Inst Rad Breed, Ohmiya, Japan

Yamakawa K (1969) Induction of male-sterile tomato mutants by gamma irradiation. Techn News 3, Inst Rad Breed, Ohmiya, Japan

Yamakawa K (1970) Radiation-induced mutants of *Chrysanthemum* and their somatic chromosome number. Techn News 6, Inst Rad Breed, Ohmiya, Japan

Yamakawa K, Nagata N (1975) Three tomato lines obtained by the use of chronic gamma radiation with combined resistance to TMV and *Fusarium* race J-3. Techn News 16, Inst Rad Breed, Ohmiya, Japan

Yamasaki Y, Kawai T (1968) Artificial induction of blast-resistant mutations in rice. Rice Breeding with Induced Mutations, pp 65–73, IAEA, Vienna

Yamashita A, Ukai Y, Yamaguchi I (1972) Comparison of genetic effects of gamma-ray irradiation and treatments of chemical mutagens in a six-rowed barley. Gamma Field Symp 11:73–91

Yonezawa K (1975) Method and efficiency of mutation breeding for quantitative characters. Gamma Field Symp 14:39–58

Yonezawa K, Yamagata H (1977) On the optimum mutation rate and optimum dose for practical mutation breeding. Euphytica 26:413–426

Yordanov M, Zagorcheva L, Stoyanova Z (1977) Results from induced mutagenesis in the tomato line XXIV-a. Genet Plant Breed 10:13–23

Young OH, Warner JRL, Kleinhofs A (1980) Effect of nitrate reductase deficiency upon growth, yield, and protein in barley. Crop Sci 20:487–490

Youngs VL (1978) Oat lipids. Cereal Chem 55:591–597

Youngs VL, Forsberg RA (1979) Protein-oil relationships in oats. Crop Sci 19:798–802

Zacharias M (1967) The yields of early ripening soybean mutants in relation to the climate conditions. Erwin Baur Ged Vorl 4:245–249

Zachow F (1967) Ein neues Gen für Alkaloidarmut bei *Lupinus angustifolius*. Züchter 37:35–38

Zagaja SW, Przybyla A (1973) Gamma-ray mutants in apples. Induced Mutations in Vegetatively Propagated Plants, pp 35–40, IAEA, Vienna

Zagaja SW, Przybyla A (1976) Compact type mutants in apple and sour cherries. Improvement of Vegetatively Propagated Plants and Tree Crops through Induced Mutations, pp 171–184, IAEA, Vienna

Zagorcheva L, Yordanov M (1978) Results from induced mutagenesis in tomato. Experimental Mutagenesis in Plants, pp 359–369, Sofia

Zeller A (1957) Physiologie der Fettbildung und Fettspeicherung bei höheren Pflanzen. In: Ruhland W (Hrsg) Handb Pflanzenphysiol, Bd VII. Springer, Berlin Göttingen Heidelberg, 280 pp

Zima KI, Khajinov MI, Normov AA (1972) Changes in seed weight, protein and lysine content in *opaque-2* mutant of maize in backcross generations. Genetika USSR 8(7):5–8

Zink F (1979) Effect of late application of nitrogen on the grain protein of three high-yielding maize hybrids with different protein or lysine content. Seed Protein Improvement in Cereals and Grain Legumes I, pp 273–280, IAEA, Vienna

Zuber MS, Grogan CO, Deatherage WL, Hubbard JE, Schulze WE, MacMasters MM (1958) Breeding high-amylose corn. Agron J 50:9–12

Zwintzscher M (1967) On the variability of isolates of the sour cherry variety "Schattenmorelle" from a mutation trial. Erwin Baur Ged Vorl 4:363–367

Subject Index

F = Figure
T = Table

Achimenes 53, 85, 86, T11, T12, T18
adventitious bud technique 86
African violet 85, 86
albumins 131, 132, 135, 143, 153, 164, 165, F41, T26, T31, T37
aleurone layer 132, 149, 176, T29
alkaloids 173, 178, T43
Alstroemeria 67, 85, 86, T12
Amaranthus hypochondriacus 58
amylopectin 8, 165, 167, 168, 170, 178, T39
amylose 8, 165–170, 178, T29, T39, T41, T42
anthocyanidins 67
anthrachinon T44
Antirrhinum majus 68
apical dominance 52, T9
Arachis hypogaea 1, 9, 29, 33, 34, 82, 126, 170, 171, T3, T8
Ascochyta rabei T23
ascorbic acid 173, T44
autotetraploids 86
Avena sativa 8, 28, 31, 61, 118, 158, T2
  disease resistance 97, T22
  dwarfs 50, T7
  heterosis 90
  lodging resistance T7
  seed lipids 170
  seed proteins 131, 135, 158, T26–T28
avenin T26
Azalea T11, T12

barley: see *Hordeum vulgare*
Begonia 31, 71, 86, T11
Bermuda grasses 5, 11, 85–87, 102–104, T22
beta rays 10, T1
Beta vulgaris 12, T43
black currant 53, 173
BOAA 8, 173, T44
Botrytis fabae T23
Brachiaria brizantha 87
branching 53, 54
Brassica campestris 13, 54, 104, T3, T18
Brassica juncea 52, 54, 80, 104, 171, T9, T18, T45

Brassica napus 69, 171, T13
broad bean: see *Phaseolus vulgaris*

Camellia sinensis T44
Canna 71, T11, T12
capsicine T45
Capsicum annuum 68, 69, 90, T8, T43, T45
capsule borer 102
carbohydrates: see seed carbohydrates
carnation: see *Dianthus*
Cassava 85
castor: see *Ricinus communis*
cell cultures 11
Celosia argentea 58
Cercosporella herpotrichoides T21
cereals 1, 7, 23, 31, 43ff., 54, 142ff., F1, F35, T2, T4–T7, T13–T16, T19–T22;
  see also under the various crops
cherries 52, 70, 85, T43
chimerism of M_1 plants 14ff., 86, F2
chlorophyll mutants 10, 87
Chrysanthemum 10, 31, 53, 86, T11, T12
Cicer arietinum 4, 29, F1, T8, T23
  seed proteins 159
Citrus sinensis T18
Claviceps microcephala 97
cockscomb 58
Coffea arabica T9
colchicine 11
Compositae 66
Corchorus capsularis 53, 64, 75, 76, T3, T13, T18, T24
Corchorus olitorius 53, 64, 75, 76, T3, T9, T10, T18, T24
cotton: see *Gossypium*
coumarine 13, 88, F17, T43
crossbreeding 1, 26, 31, 32ff., 45, 46, 48, 50, 53, 81, 82, 91, 96, T4–t6, T14, T17, T23
Cryptomeria japonica T9
Cynodon 11, 85–87

Dahlia 31, 67, 85, T11, T12
Datura innoxia 11
deficiencies 3, 4

deficit of mutants 14, 16
Dianthus caryophyllus 53, 71, 86, T11, T12
Diaporthe nomurai T24
diastase T44
dichotomy: see stem bifurcation (*Pisum*)
diethyl sulfate 4, 10, T2, T4–T8, T14–T17, T45
dimethyl sulfate T13, T45
diplontic selection 14, 86
Drosophila 4
drought resistance 30, 103, T17
dwarfs 12, 43ff., 64, 102, T4–T9; see also individual cereals and legumes

earliness 12, 13, 29, 30, 54, 69, 75ff., 87, 101, 103, T9, T14–T18; see also individual crops
Eleusine coracana T2
Empoasca devastans 102
environment, influence on gene expression 6, 7, 62, 116ff.; see also gene-ecology and phytotron experiments
epistasis 35, 48, 57, 62, 63, 89, 90
Eragrostis tef 158
erectoides types 43ff., T4–T9; see also under *Hordeum, Oryza, Triticum*
Erysiphe graminis 91, 96, T19, T21
ethyleneimine 3, 10, 71, 83, 152, T2, T4–T6, T8, T10, T11, T13–T15, T17, T19–T21, T43, T45
ethyl methane sulfonate 3, 10, 12, 28, 30, 69, 81, 86, 91, 93, 97, 103, 129, 152, T2–T7, T10–T17, T19–T23, T43–T45

flavonols 67
flower color 65ff., 87, T12
flower shape 65, 87, T11
flower size T11
flowering behaviour 6, 75ff.; see also the individual crops
Foeniculum vulgare 171
fruit color 87
fruit shape 87
fruit trees 52, 85, 87, 91, T9
Fusarium T22, T23

gall fly 102
gall midge 102
gamma rays 3–5, 10–13, 27–29, 31, 46, 52, 61, 64, 71, 74, 76, 81, 83, 86, 87, 93, 94, 101, 103, 126, 128, 155, T1–T24, T43, T45
garlic 85
gene-ecology 37, 62, 82, 106, 107, 114, 116ff., 123ff., F19, F24–F27, T25; see also phytotron experiments

genotypic background 105, 179
germination rates following mutagenic treatments 16, F3
Gladiolus T12
gliadin T26
globulins 131, 132, 135, 143, 153, 164, 165, T26, T31, T37
glutelins 131, 143, 156, 157, T26, T29, T31, T37
Glycine max 8, 9, 11, 23, 29, 62, 83, 118, F1, T10, T13
  disease resistance T23
  earliness 81–83, T17
  erectoides types T8
  lodging resistance T8
  seed lipids 170, T45
  seed proteins 134, 136, 158, 159
  shattering resistance 104
Golden mosaic virus T23
Gossypium 54, 68, 71, 81, 83, F1, T3, T43
  seed proteins 135
Gossypium barbadense 29
Gossypium hirsutum 75, 83, 102, T13
gossypol T43
grain hardness 23
grapes 85

haploids as initial material 11, 87
haricot bean 23, 101, T23
heat susceptibility 103
heat tolerance 103
Helianthus annuus 13, 69, T13, T18, T45
Helminthosporium maydis T22
Helminthosporium oryzae T20
Helminthosporium sativum 94, T19
Helminthosporium turcicum T22
herbicide tolerance 102
heterosis 54, 62, 68, 69, 88ff., 101
Hibiscus T11, T12
hordein 153, T26, T35
Hordeum vulgare 1–8, 21, 26, 27, 30, 32, 34, 68, 69, F1, T2
  albumins 176, T37
  desynaptic mutants 150
  diastase content T44
  disease resistance 33, 91ff., T19
  dormancy 129
  dwarfs 27, 45, 74, T4
  earliness 33, 34, 75, 80–82, 90, 94, 119, 123, T4, T14
  eceriferum 5, 6, 123
  erectoides 5, 6, 32, 34, 44ff., 74, 105, 106, 116, 123, 155, T4
  free amino acids 153, T35

Subject Index

Hordeum vulgare
  free sugars 168, T40
  gene-ecology 119; see also phytotron experiments
  gigas 5
  globulins 176, T37
  glutelins T37
  heterosis 90
  hiproly 150ff., 154, 176, F35
  hordein 153, T35
  lodging resistance 27, 32, 33, 45, 46, 54, T4
  lysine content 150ff., 152, 176, F35, T35, T36, T45
  male sterility T13
  mildew resistance 33
  ml-o locus 91–93, T19
  multiple alleles 5, 92, 152
  mutant 1508: 150, 152ff., 168, 171, 179, T35–T37, T40
  notch 154, T35
  nutritional value 176
  phytotron experiments 83, 90, 123, 134, T14
  pleiotropism 105, 106
  prolamins 154, 155, 176, T37
  protein bodies 153
  protein mutants 150ff., F35, T35–T38, T45; see also seed proteins
  recombinants 40, 168
  Risø mutants 152ff., 168
  seed carbohydrates 168, T40
  seed proteins 131, 134–136, 142, 174, T26–T28, T45
  seed quality 128
  seed size 80, 93
  semidwarfs T4
  starch content 153, 168, T40
  tillering 45, 54
  translocations 8
  tryptophane content 152
  vine 5
  waxy endosperm 5
  winterhardiness 45, 103
hydroxylamine T2
hypostasis 35, 39, 57, 62, 63, 90, 121, 122

inbred lines 12
inflorescences 67, 68
initial cells 14ff., F2
Ipomoea batatas 9, 85, T43
Iresine herbestii 71
Iris 68, 85, 87, T11
irradiation, recurrent 10
isoflavons T44

jassids 102
Jute: see *Corchorus*

kafirin T26
Kalanchoë 53, 71, 86, T11, T12
Kohleria 52, T12

lateness 82, 87
Lathyrus sativus 8, 29, 173, T13, T17, T44
leaf color 71
leaf-curl virus T24
leaf mutants 71ff.
leaves, multifoliate 13
legumes 1, 7, 9, 51, 52, 100, 158ff., F36–F41, T3, T8, T9, T13, T17, T23; see also under the various crops
legumin 132, 165
Lens culinaris 4, 73, 128, 159, T3, T13
Lespedeza stipulacea T23
Lilium T12
Linum usitatissimum 64, 126, 171, T3, T45
lodging resistance 30, 43; see under individual cereals
Lolium multiflorum 12, T7, T16
Lolium temulentum 12, T7, T16
lupins 9, 103, 173, 178, T45
Lupinus albus 81, 82, T17
Lupinus angustifolius T43
Lupinus digitatus T43
Lupinus lutens 103, 104, F1, T43
Lupinus mutabilis 126, T17, T43
Lycopersicon esculentum 1, 28, 34, 52, 54, 65, 68, F1, T8
  disease resistance T23
  earliness 81, T18
  heterosis 90
  male sterility T13
Lycopersicon peruvianum T23
lysine 7, 131, 135, 143, 157, 158, T26, T29, T45; see also under Hordeum

M_1 generation 11, 12, 14ff., 86
macro-mutations 26
Macrophomina phaseoli T24
maize: see *Zea*
maize mosaic dwarf virus T22
maleic hydrazide T21
male sterility 6, 68ff., 90, T13
Malus sylvestris 52, 87, T9, T18, T24
malting quality 30
Medicago polymorpha 81, T17
Medicago sativa T3
Melilotus albus 13, 88, 129, F17, T43
Meloidogyne incognita T23

Mentha arvensis T44, T45
Mentha cardiaca T24
Mentha oil T44
Mentha piperita 5, 85, 88, 101, T24
methods for inducing mutations 9, 10ff.
methyl methane sulfonate T7, T16
micro-mutations 21, 26, 84, 125
microsporogenesis, effect of ms-genes 69
mildew resistance 5, 30, 31, 93, T19, T23
modifications 57, 80
modifiers 59
monstrosities 58
morphin T43
Morus 71, 74, T24
ms-genes 69ff., T13
multiple allelism 5, 55, 59, 75, 92
mutagens 1, 3, 4, 5, 10, 11, 31, T1; see also under EMS, EI, gamma rays, neutrons, X rays
mutant genes
    accumulation 21, F5
    inspecific action 21
    joint action 34ff., 69
    negative interactions 35ff.
    positive interactions 40ff.
    reaction to environmental factors 62; see also gene-ecology
mutants
    direct utilization 1
    indirect utilization 1, 13
    lethal 19, 65
    nutritional value 174ff.
    positive 20, 21, 26ff.; see also mutant varieties
    sterile 19
mutant sectors 14ff., 86, F2
mutant varieties 2, 3, 9, 11, 13, 27, 30–32, 45, 52–54, 61, 64, 67, 70, 73, 76, 81, 82, 85, 89, 91, 93, 97, 101, 103, F1, T1, T4, T5, T9–T11, T14–T19, T21–T24, T45
mutation spectra 4, 6
mutations
    desired 4, 5, 7
    dominant 12, 48, 53, 65, 69, 83, 93, 96, 101, T4, T7, T13, T16, T20
    frequency 4, 5, 16, 67, 86, 87
    homologous 62
    identical 12, 19, 62, 63
    in allogamous crops 2, 12, 13, 149
    of closely linked genes 12, 105, 108ff.
    several mutations in the same embryo 12, 20, 21, 35, 62, 63
    somatic 66, 68
    specific 4–7
    spontaneous 3, 68, 69, 72, 87, 89, 143, 155, 166, T10, T13
    systematic 54
mutator genes 63

neurotoxins 8
neutrons 3, 4, 10, 13, 18, 27, 31, 35, 40, 76, 82, 86, 96, T1–T24, T45
Nicotiana 11, F1
nitrate reductase 139
N-methyl-N-nitroso urethane 4, 10, 83, 126, 128, T2, T4, T5, T7, T13–T17, T20, T22, T23

oats: see *Avena sativa*
oil bodies 170
Olea europaea 52, T9
ornamentals 2, 30, 31, 43, 53, 66, 71, 85ff., T11
Oryza rufipogon 47, T5
Oryza sativa 1, 3, 5, 7, 8, 12, 23, 27, 31, 34, 68, 71, 116, 118, 129, 158, F1, F7, T2
    disease resistance 5, 91, 94ff., T15, T20
    dwarfs 46, 81, 82, T5
    earliness 27, 46, 47, 76, 81, 82, 102, 127, T15, T20
    erectoides types 46, 47, 118, T5, T20
    lateness 46, 53, 82, T10
    lodging resistance 27, 46, 53, 81, 82, 126, 127, T5, T20
    long internodes 53
    male sterility 69, T13
    milling quality 46
    photopriodic reaction 83
    pleiotropism 126
    protein bodies 157
    recombinants 40
    seed carbohydrates 169
    seed lipids 170
    seed proteins 46, 127, 131, 132, 134–136, 139, 156, 157, T26–T28, T45
    seed quality 46, 128
    seed shape 127, T15
    seed size 53, 126, 127
    semidwarfs 46, 47, T5, T15
    shattering resistance 104
    shedding resistance 104
    tillering 46, 47, 53, 54
    winterhardiness 103
outcrossing, natural 13
overdominance 90, T4

Panicum miliaceum T16
Papaver somniferum T43
Paspalum notatum 86, 87
pea: see *Pisum*

Subject Index

peanut: see *Arachis*
pearl millet: see *Pennisetum*
penetrance of mutant genes 20, 39, 40, 41, 55, 57, 58, 112ff., F22–F24
Pennisetum typhoides 7, 102, T2, T13, T16
  disease resistance 97, T22
Petunia T12
Phakipsora pachyrhizi T23
Phaseolus aureus 65, 71, T3, T17
Phaseolus mungo 129
Phaseolus vulgaris 4, 29, 81, 128, 136, T3, T17, T23, T44
  seed proteins 158–160, T45
phenols T43
Phleum nodosum 12
Phleum pratense 12
photoperiodic reaction 46, 52, 75, 83, 84; see also phytotron experiments
Physalis ixocarpa 65
Physalospora tucumanensis T22
phytic acid T44
phytoglycogen 165, 167
Phytophthora infestans 101, T23
phytotron experiments 38, 39, 83, 84, 123ff., 134, T14
Pisum arvense 34, 42
Pisum sativum 1, 4, 5, 12, 13, 16, 17, 18ff., 23, 26, 28, 29, 43, 52, 67, 71, 74, 102, 103, 177, F1, F4, F5, F8
  acacia 41, 42, 72, 163, F16
  afila 41, 42, 72, 73, F16
  branching 54, 80
  chlorophyll mutations 38, 62, 88
  cochleata 73, 125
  dim-segment 63, 110, F20
  dwarfs 51, F9, T8
  earliness 20, 22, 29, 34–40, 62, 75, 80–82, F9, F10, F23, T8, T17
    dependence on environment 37–39, 82, 118, 119, 124, 125, F10, T25
  epistasis 125
  gene-ecology 19, 37, 38, 39, 82, 107, 114, 117–122, 123ff. F10, F19, F24, F25
  heterosis 62, 88, 89, T10
  hypostasis 121, 122, 125
  interaction between mutant genes 35ff., 63, 124
  lateness 53, 57, 61, 82, 83, T10
  long internodes 37–39, 57, 58
  male sterility 69, T13
  multiple alleles 55, 59
  number of pods per plant 55, 60, 105, 112, F18, F19
  number of seeds per pod 34, 37, 58, 105, 122, 138, F18, F19, F23, F27, F32, T25
  penetrance 112ff., F22–F24, T25
  pleiofila 41, 42, 163, F16
  pleiotropism 105ff. 123, F18–F21
  polymery 112
  phytotron experiments 38, 39, 83, 84, 103, 123ff.
  proteins: see seed proteins
  recombinants 35ff., 52, 55, 57ff., 62ff., 72, 82, 114, 119, 120, 123ff., 139, 161, 163, F5, F9–F11, F16, F23, F31, F33, F37–F39, T10, T25
  resistance 120
  root nodules 140
  seed carbohydrates 165, 169, 170, T41, T42
  seed lipids 170
  seed proteins 20, 54, 134–136, 138, 139, 141, 160ff., 177, F28–F34, F36–F41, T27, T28, T45
    albumins 164, 165, F41
    globulins 164, 165
    influence of the climate 163, F38
    influence of other genes 161, 163, 165, 179, F37
  seed size 35, 58, 61–63, 110, 126, 127, 136ff., F9, F23, F30, F31
  semidwarfs 51, F9, T8
  short internodes 35–40, 43ff., 51, 57, 58, 61, 62, F8, F9, F11, T8
  stem bifurcation 20, 28, 35, 38–41, 54ff., 62, 102, 112ff., 121, F8, F12, F13, F22–F24, F26, T10, T25
  stem fasciation 12, 20–22, 28, 35–38, 40, 57, 58ff., 83, 84, 89, 90, 119, 124, 125, 127, 163, F8, F9, F14, F15, F26, F39, T10, T25
  tallness 53, 57, 61, T10
Pisum umbellatum 58
plant height 23, 43ff., 52, 53
plant hopper 102
plant weight 23
Plasmopara viticola T24
pleiotropism 12, 20, 32, 45, 48, 74, 78, 92, 93, 104, 105ff., 130, T4, F18–F21
Poa pratensis 8, 34, 87
Podosphaera leucotricha T24
pollen treatment 17
Polyanthes tuberosa 71
polygenic systems 5, 6, 23, 44, 46, 94
polymery 5, 44, 46, 55, 59, 96, 112
Portulaca grandiflora T11
prolamins 131, 132, 135, 143, 157, 174, T26, T29, T37
propane sultone 10
protein bodies 132, 150, 153, T29

protein mutants 6, 7, 130ff., 178ff.
  in cereals 142ff., F35, T29–T38
  in legumes 158ff., F36–F41
proteins: see seed proteins
protons 10
Prunus armeniaca T18
Prunus avium T9
Prunus cerasus T9, T18, T43
Prunus persica T18
Pseudomonas T20, T23
Psophocarpus tetragonolobus T17
Puccinia coronata T22
Puccinia glumarum T21
Puccinia graminis 96, T21
Puccinia recondita 96, T21
Puccinia striiformis 96, T21
pulses: see legumes
Pyricularia oryzae 94, T20
Pyrus communis 5, T9

quantitative characters 5, 7, 8, 23ff., 75ff., T2

radiosensitivity 4
recombinants 1, 32, 57, 62ff., 114, 154, 166, 167, 169, F5, F13, F23; see also under *Pisum*
recombination, intragenic 4
red clover 13, 65, 71
repressor genes 101
resistance
  general 5, 6, 8, 11, 21, 26, 30, 87, 88, 91ff.
  to animal pathogens 102
  to drought 103
  to fungi, bacteria, viruses 91ff.
  to herbicides 86
  to nematodes 86
  see also under the various crops
Rhododendron 10
Rhyzobium 134
Rhyzoctonia oryzae T20
rhyzomes 11, 86
Ribes nigrum 53, 87, T44
rice: see *Oryza*
Ricinus communis 52, 68, 76, 81, 102, 126, 171, T9, T18
ripening behaviour 6, 75ff.
root system 74
roses 53, 66, 67, 85, T11, T12
Rubia tinctoria T44
Rudbeckia T11
rye: see *Secale*
rye grass 103

Saccharum 8, 71, 76, 97, 173, T16, T22, T43, T45

Sclerospora graminicola 97, T22
Sclerotium hydrophilum 94, T20
Sclerotium oryzae 94, T20
Secale cereale 12, 50, 103, 158, T7
secalin T26
seed carbohydrates 6, 7, 23, 165ff., T39–T42
seed characters 126ff.
seed color 128
seed lipids 170, 171
seed oils 8
seed proteins 6, 7, 9, 23, 30, 130–165, F28–F34, T26–T38
  alteration through mutant genes 142ff.
  chemistry 131, 132, F41, T26, T30, T31, T36
  correlation with seed production 136ff., F30–F34
  influence of endogenous factors 135ff.
  influence of environmental factors 132ff.
seed quality 128, 130ff.
seed shape 127
seed size 7, 30, 46, 126, 127
seed storage substances 9, 130–171
seed weight 23, 126
segregations, unfavourable 14, 16, 17, F2
selection value 18ff., 56, 125, F4, F5
self-compatible mutants 70
self-incompatibility 12
semidwarfs 43ff., F4
Septoria nodorum 96, T21
Serradella 115
Sesamum indicum 52, 68, 104, T18
shattering resistance 27, 30, 81, 104
shedding resistance 104
shoot, alterations through mutant genes 43ff.
shoot fly 102
Sinapis 61, T45
snapdragon 68
sodium azide 5, 11, T5, T12, T19, T21
Solanum khasianum T45
Solanum tuberosum 83, 85, 86, T23, T43
Solanum viarum T43
solasodine T43, T45
Sorghum 7, 8, 12, 50, 54, 68, 82, 178
  dwarfs 50, 102, T7, T16
  lysine content 157, 176
  seed lipids 171
  seed proteins 131, 142, 157, 174, 176, F42, T26
  tannins 176
Sorghum bicolor 53, T7, T10, T13, T16
Sorghum subglabrescens 102, 128, T2
soybean: see *Glycine*
sprouting resistance 30

Subject Index

spur types 52, 87
starch T43; see seed carbohydrates
St. Augustinegrass 11
stem bifurcation 53, 54ff., T10; see *Pisum*
stem fasciation 53, 58ff., F14, F15, T10; see *Pisum*
stem rot T24
stolons 11, 86, 101
Streptocarpus 31, 67, 83, 85, 86, 87, T11, T12
sugar cane: see *Saccharum*
sugar content of seeds 7, 173, T41–T43; see also seed carbohydrates
sugars, free 166, 178, T29, T41
survival rates 16, 17, F3
suspension cultures 11
sweet clover: see *Melilotus*
sweet potato 9, 85, T43
Synchytrium endobioticum T23

tallness 53
tannins 8, 157, 159, 176
tea plant 87
threshability 30
Thuja gigantea 87
tillering 23, 54, 87
Tilletia triticoides 96, T21
tissue cultures 85
Tithonia rotundifolia T11, T12
Tobacco mosaic virus T23
tolerance 6, 11, 21, 103
tomato: see *Lycopersicon*
translocation-heterozygosity 4, 18
translocation-homozygosity 4, 8
Trifolium pratense 13, 65, 71
Trifolium subterraneum T44
triploid crops 11, 86
Triticum aestivum 1, 3, 4, 6–8, 12, 27, 28, 31, 34, 105, 106, 116, 118, 126, F1, T2
   disease resistance 91, 96, 126, T21
   dwarfs T6
   earliness 81, T16
   erectoides 48, T6
   lodging resistance 48, T6
   male sterility 69, T13
   pleiotropism 106, 107
   seed color 128
   seed lipids 171
   seed proteins 49, 131, 134–136, 139, 140, 142, 155, 156, 174, T6, T26–T28
   seed size 49
   semidwarfs 48, 49
   shattering resistance 27
   sprouting resistance 129
   winterhardiness 103

Triticum durum 1, 8, 28, 34, 116–119, F6, F7, T2
   disease resistance 96, T21
   dwarfs 49, T21
   earliness 34, 49, 50, 80–82, 119, T16, T21
   gene-ecology 116–119
   lodging resistance 34, 49, 50, 118, 119, F7, T6
   male sterility 69, T13
   recombinants 40
Tungro virus T20
turf grasses 86

Uromyces fabae T23
Ustilago nuda 94, T19
Ustilago scitaminea T22
Ustilago zeae T22

vegetatively propagated species 2, 9–11, 17, 30, 53, 66, 71, 83, 85ff., T1, T11, T18
Verticillium albo-atrum 101, T24
Verticillium dahliae T24
Vicia faba 13, T3, T23
   seed proteins 136, 159, T27
vicilin 132, 165
Vigna aconitifolia 29, 52
Vigna radiata 29, 62, 81, 103, T13, T17, T23
   seed proteins 159, T27
Vigna trilobata 29, 52, T8
Vigna unguiculata 12
Vitis vinifera T24

wheat: see *Triticum*
wilt-resistance 5, T24
winterhardiness 5, 11, 30, 45, 103, 104

Xanthomonas oryzae 94, T20
X rays 3, 4, 10, 16, 18, 31–33, 35, 53, 54, 80, 82, 83, 86, 93, 94, 96, 101, 129, 138, T1–T24, T43–T45

yam 85
yellow dwarf virus T19
yellow mosaic virus 94, T19, T23

Zea mays 12, 13, 31, 68, 158, 178, F1
   albumins 148, T31
   amino acids T30
   amylopectin 167, T29, T39
   amylose 167, 168, T39
   amylose extender 167, 168, T29, T39
   brittle 166–168, T29
   corn-grass mutant 54, 88
   De*30 T29
   disease resistance 97, 150, T22

Zea mays
    dull   167, T29, T39
    dwarfs   50, T7
    earliness   T26
    floury   6, 142, 143, 148–150, 171, 174, 175, T29–T31, T34
    free amino acids   146, T31
    free sugars   178, T29
    globulins   148, T31
    glutelins   148, T31
    heterosis   54, 88
    lodging resistance   T7
    lysine   143, 146, 148, 149, 175, T29, T30
    male sterility   T13
    methionine   T29, T30
    nutritional value of mutants   174, 175, F42
    opaque   6, 142, 143ff., 150, 166, 168, 171, 174, 175, 179, F42, T29–T33
    prolamins   146, 148, 150, T29
    protein bodies   150
    protein mutants   149, T29; see also floury and opaque
    seed carbohydrates   166ff., T39
    seed lipids   170
    seed proteins   131, 132, 134, 135, 139, 140, 143ff., 174, T27–T34
    shrunken   167, 168, T29, T39
    starch   165–168, T29, T39
    starch granules   168
    sugar content   166, T39
    sugary   7, 167, 168, 175, T29, T39
    tryptophane   143, 175, T29, T30
    waxy   167, 168, T29, T39
zein   143, T26
zertation   14, 16
Zinnia elegans   101, T24

Monographs on Theoretical and Applied Genetics

Editors: R. Frankel
(Coordinating Editor),
G. A. E. Gall, M. Grossmann,
H. F. Linskens, R. Riley

Springer-Verlag
Berlin
Heidelberg
New York

Volume 1
J. Sybenga

Meiotic Configurations

A Source of Information for Estimating Genetic Parameters
1975. 65 figures, 64 tables. X, 251 pages. ISBN 3-540-07347-7

"The present book treats meiotic chromosome configurations from a quantitative genetic point of view. Its purpose is not primarily to increase understanding of chromosome behaviour, but rather to construct generally applicable systems for estimating genetic parameters related to recombination.
There exist few previous books covering this important field of cytogenetics. Dr. Sybenga's book is therefore especially welcome ... Several tables and diagrams aid in the understanding of meiotic principles. Fortunately, the book contains simple figures usually explaining in a perfect manner the various phenomena ...
In conclusion the book is highly recommended for anyone interested in cytogenetics."
Norwegian Journal of Botany

Volume 2
R. Frankel, E. Galun

Pollination Mechanisms, Reproduction and Plant Breeding

1977. 77 figures, 39 tables. XI, 281 pages. ISBN 3-540-07934-3

"This book, which is based on an advanced course in plant breeding ... sets out to give a comprehensive account of the botanical, genetic and breeding aspects of the reproductive biology of spermatophytes, mainly with respect to angiosperms ... The line drawings are of a high quality, the photographs clear and the schematic representations of various phenomena described in the text and of breeding procedures well laid out. This ably written reference book will be of use to botanists, geneticists and plant breeders alike, bringing together as it does so many aspects of pollination and reproduction in plants."
Plant Breeding Abstracts

Volume 3
D. de Nettancourt

Incompatibility in Angiosperms

1977. 45 figures, 18 tables. XIII, 230 pages. ISBN 3-540-08112-7

"Rarely, in a new subject, is a book to be found that is both well balanced and up to date. *Incompatibility in Angiosperms* is just that. From its beginnings in 'classical' genetics the study of incompatibility mechanisms in plant breeding systems has emerged as a subject of considerable scientific and commercial interest, and rapid progress has been made from genetical, physiological and structural points of view. This book, written by an acknowledged expert in the field, manages to do justice to all these aspects of the work, and to preface them with first class introductory material ... The illustrations that accompany the text are in general of exceptional clarity..."
Phytochemistry

Monographs on Theoretical and Applied Genetics

Editors: R. Frankel (Coordinating Editor), G. A. E. Gall, M. Grossmann, H. F. Linskens, R. Riley

Volume 4
L. I. Korochkin

Gene Interactions in Development

Translated from the Russian and Edited by A. Grossman
1981. 109 figures. XIII, 318 pages. ISBN 3-540-10112-8

"This book deals with the genetic side if developmental biology. ... The logical framework of the book is most appealing. The themes are arranged in a more or less cyclic order. The first chapter is on differential gene activity as a basis for cell differentiation. The circle is closed by a final chapter on the organization of systems which control differential gene expression. In each individual chapter the author carefully explains in which way the fundamental concepts are rooted in classical genetics and developmental biology. At the same time, he emphasizes the contribution that molecular biology provides and will provide to the understanding of the basic mechanisms in development.
The result of this all is a comprehensive and coherent treatment which will be of interest for a large number of biologists especially those who look for some order in the tremendous amount of recent knowledge in this rapidly expanding field."

Theor. and Applied Genetics

Volume 5
K. H. Chadwick, H. P. Leenhouts

The Molecular Theory of Radiation Biology

1981. 236 figures. XVI, 377 pages. ISBN 3-540-10297-3

"... There are 12 chapters, each containing a wealth of information and provided with copious literature references.
There is no question that this is a thorough and erudite treatise, ... the authors have injected a great deal of order into a chaotic field. To quote from their preface: "Independent of whether the model be proved correct or not, if it has caused radiation biologists to think again about basic mechanisms; if it has given some a different insight in the field, and if it has stimulated renewed experimentation ... then it will have achieved its goal." This I think it has certainly done, and the book will retain its value for a long time to come."

British Journal of Radiology

"... The text is a careful and complete exposition of the implications of these assumptions. ... an important book for radiation and environmental biologists for, if nothing else, it is sure to stimulate new experiments and ideas."

Nature

"... the book gives clearly-written survey of a large area of cellular radiobiology, which would be a useful introduction for the non–specialist. From the point of view radiation protection the discussion of 'humped' dose-response curves is important..."

Radiation Protection Dosimetry

Volume 6

Heterosis

Editor: R. Frankel
1983. 32 figures. Approx. 320 pages. ISBN 3-540-12125-0

Contents: Biometrical Genetics of Heterosis. - Heterosis in Maize: Theory and Practice. - Heterosis and Hybrid Seed Production in Barley. - Hybrid Wheat. - Heterosis and Hybrid Seed Production in Fodder Grass. - Heterosis in Vegetable Crops.- Heterosis in the Tomato. - Heterosis and Hybrid Cultivars in Onions. - Heterosis in Ornamentals. - Heterosis and Intergenomic Complementation: Mitochondria, Chloroplast, and Nucleus. - Subject Index.

Springer-Verlag
Berlin
Heidelberg
New York